Polymer Blends and Composites in Multiphase Systems

ADVANCES IN CHEMISTRY SERIES **206**

Polymer Blends and Composites in Multiphase Systems

C. D. Han, EDITOR
Polytechnic Institute of New York

Based on a symposium cosponsored by
the Materials and Engineering
Sciences Division
of the American Institute
of Chemical Engineers
and the Polymer Division
of the American Chemical Society
at the 74th Annual Meeting
of the American Institute
of Chemical Engineers,
Los Angeles, California,
November 15–18, 1982

American Chemical Society, Washington, D.C. 1984

Library of Congress Cataloging in Publication Data

Polymer blends and composites in multiphase systems.

(Advances in chemistry series, ISSN 0065-2393; no. 206)

"Based on a symposium sponsored by AIChE, Materials and Engineering Sciences Division [and ACS] Polymer Division, at the 74th Annual Meeting of the American Institute of Chemical Engineers, Los Angeles, CA."

Includes bibliographical references and index.

1. Polymers and polymerization—Congresses. 2. Composite materials—Congresses.

I. Han, Chang Dae. II. American Institute of Chemical Engineers. Materials Engineering and Sciences Division. III. American Chemical Society. Division of Polymer Chemistry. IV. American Institute of Chemical Engineers. Meeting (74th: 1982: Los Angeles, Calif.) V. Series: Advances in chemistry series; 206.

QD1.A355 no. 206 [QD380] 540s [547.7] 83-24362
ISBN 0-8412-0783-6

PRINTED IN THE UNITED STATES OF AMERICA

Advances in Chemistry Series

M. Joan Comstock, *Series Editor*

FOREWORD

ADVANCES IN CHEMISTRY SERIES was founded in 1949 by the American Chemical Society as an outlet for symposia and collections of data in special areas of topical interest that could not be accommodated in the Society's journals. It provides a medium for symposia that would otherwise be fragmented, their papers distributed among several journals or not published at all. Papers are reviewed critically according to ACS editorial standards and receive the careful attention and processing characteristic of ACS publications. Volumes in the ADVANCES IN CHEMISTRY SERIES maintain the integrity of the symposia on which they are based; however, verbatim reproductions of previously published papers are not accepted. Papers may include reports of research as well as reviews since symposia may embrace both types of presentation.

ABOUT THE EDITOR

C. D. (PAUL) HAN is Professor of Chemical Engineering and Director of the Polymer Science and Engineering Program at Polytechnic Institute of New York. After receiving a B.Ch.E. degree in 1958 from Seoul National University in Korea, he came to M.I.T. in 1961 for graduate study, where he received his M.S. (1962) and Sc.D. (1964) degrees in Chemical Engineering. While working in industry as a process analyst and systems engineer, he continued graduate studies part time, first in Electrical Engineering (majoring in Automatic Control Theory) at the Newark College of Engineering and then in Mathematics at the Courant Institute of Mathematical Sciences of New York University. He earned two additional degrees: an M.S. in Electrical Engineering and an M.S. in Mathematics. He joined the faculty of the then Polytechnic Institute of Brooklyn in 1967 as Associate Professor of Chemical Engineering, and was promoted in 1972 to Full Professor. He was Head of the Chemical Engineering Department from 1974 to 1982 at Polytechnic Institute of New York.

His research interests include polymer rheology, polymer processing, process control, mathematical modeling of chemical processes, and polymerization reaction engineering. More specifically, his current research activities include rheological characterization of multiphase polymeric systems, processing of multiphase polymeric systems, and structure–property–processing relationships in polymeric systems. He has written over 170 technical papers and two research monographs, "Rheology in Polymer Processing" (Academic Press, 1976) and "Multiphase Flow in Polymer Processing" (Academic Press, 1981).

CONTENTS

PREFACE

MULTIPHASE POLYMER SYSTEMS have attracted the increasing attention of polymer researchers in both academic and industrial communities. This volume contains 19 technical papers presented at a symposium on "Multiphase Polymer Systems—Polymer Blends and Composites." This symposium was significant because it was the first one to be jointly sponsored by the American Chemical Society and the American Institute of Chemical Engineers.

This volume is divided into three parts: (1) compatibility and characterization of polymer blends; (2) rheology, processing and properties of heterogeneous polymer blends; and (3) polymer composites.

Broadly classified, the two types of polymer blends are homogeneous (i.e., miscible or compatible) blends and heterogeneous (i.e., immiscible or incompatible) blends. At a given temperature, homogeneous blends give rise to a single phase in which individual components are mutually soluble in one another. In most cases, compatible blends have mechanical properties superior to those of incompatible blends. Therefore, in the past, much effort has been spent on developing experimental techniques (e.g., electron microscopy, small-angle X-ray diffraction, light scattering, and dynamic mechanical analysis) to determine the compatibility of a pair of polymers. Some of the experimental techniques for determining whether or not polymer pairs are truly mixed on the molecular level and yield a single phase still present questions. Therefore, continuing development efforts are needed. The first 10 chapters in this volume discuss the compatibility or characterization of polymer pairs.

Heterogeneous polymer blends have many interrelated variables that affect their rheological behavior, processability, and the mechanical/physical properties of the finished product. For instance, the method of blend preparation (e.g., the method of mixing the polymers and the intensity of mixing) controls the morphology of the blend (e.g., the state of dispersion, the size of the dispersed phase, and the dispersed phase size distribution), which in turn controls the rheological properties of the blend. On the other hand, the rheological properties strongly dictate the choice of processing conditions, which in turn strongly influence the morphology and, therefore, the mechanical/physical properties of the finished product. In this volume, the next four chapters discuss one or more aspects of these problems.

The use of reinforced polymer composites in many commercial applications, with either thermoplastics or thermosets as the matrix, has increased tremendously in recent years and is expected to increase continuously. In spite of the great economic incentives for the use of composite materials, relatively little effort has been spent on understanding the admittedly very complicated relationships between the processing conditions and the mechanical properties, between the processing conditions and the microstructure, and between the mechanical properties and the microstructure. Continuing efforts are clearly needed for enhancing our understanding of the fundamental processing–morphology–property relationships of composite materials. The last five chapters address these problems.

Research activities on multiphase polymer systems, encompassing polymer blends and composites, have grown rapidly and will continue to do so. Although I fully recognize that the subjects covered in this volume are far from complete in covering current research activities on multiphase polymer systems, I believe that this volume represents some important research activities currently being undertaken by various research groups. I sincerely hope that the chapters collected here will stimulate further research on multiphase polymer systems.

I wish to express my sincere appreciation to the individual contributors whose strenuous efforts have made the publication of this volume possible. I wish also to thank Susan Robinson of the American Chemical Society, who has undertaken most of the editorial work needed for the publication of this volume.

C. D. HAN
Polytechnic Institute of New York
Brooklyn, NY
September, 1983

COMPATIBILITY AND CHARACTERIZATION OF POLYMER BLENDS

1

Gel Permeation Chromatography

Use in the Determination of Polymer–Polymer Interaction Parameters

VENKATARAMAN NARASIMHAN, CHARLES M. BURNS, and ROBERT Y. M. HUANG

Department of Chemical Engineering, University of Waterloo, Waterloo, Ontario, Canada N2L 3G1

DOUGLAS R. LLOYD

Department of Chemical Engineering, The University of Texas at Austin, Austin, TX 78712

The use of gel permeation chromatography for the quantitative analysis of composition in mixed polymer systems is presented. Equations are developed for the determination of polymer–polymer interaction parameters for monodisperse and polydisperse polymers. Typical interaction parameters determined from the experimental data for the polystyrene-polybutadiene systems are presented.

POLYMER INCOMPATIBILITY AND SUBSEQUENT PHASE SEPARATION have been the subjects of growing interest. The phenomenon of phase separation in mixtures of two polymers in a mutual solvent or two polymers in the solid state is known as incompatibility and is of considerable practical importance. Limited miscibility plays a role in the preparative and analytical fractionation of polymers; in the preparation of plastic films, including paint and varnish coatings; and in the determination of service properties of certain systems such as high impact styrene–butadiene products.

Many investigations dealing with polymer–polymer incompatibility in a common solvent have been conducted in the past 35 years (*1–31*). Phase separation between the two incompatible polymers polystyrene (PS) and polybutadiene (PBD) is of considerable industrial importance and has been studied in solution (*15–18, 24–26, 31*) and in the solid state (*32–36*). In solution, the binodal equilibrium curve on the triangular diagram has frequently been approximated by cloud point isotherms determined by

0065-2393/84/0206-0003$06.00/0

turbidimetric titration. Rigorous determination of the binodal curve, including tie lines and the critical point, requires lengthy equilibration and subsequent analysis of the conjugate phases. The method of analysis is severely restricted by the necessity of maintaining an antioxidant in the solution to inhibit the cross-linking of the PBD. The antioxidant normally used, 2,6-di-*tert*-butyl-4-methylphenol, masks the UV analysis of a solution of the two polymers. To overcome these problems of analysis, we used a gel permeation chromatograph equipped with both differential refractive index (RI) and UV absorbance detectors. The antioxidant was separated from the two polymers by the columns before the polymers entered the detectors. The use of gel permeation chromatography (GPC) for the determination of composition in the PS/PBD system is not new, although it has not been reported extensively (*20*, *37*, *38*). We successfully used GPC with sequential RI and UV detectors to give the compositional analysis of the conjugate phases in the incompatible system of PS and PBD with tetrahydrofuran (THF) as mutual solvent. Application of this method to equilibrated samples yields tie lines, binodal curves, and plait points. A detailed discussion of the experimental procedure and of the determination of the tie lines, the plait point, and the binodal curves was given earlier (*27*, *28*, *39*).

The object of the present work is to evaluate the polymer–polymer interaction parameter χ_{23} from the quantitative analysis of mixtures of PS and PBD in toluene by using GPC. This parameter is valuable as a means of characterizing the incompatibility of the two polymers.

In the past, several methods have been used for the determination of polymer–polymer interaction parameters. These methods were mostly based on the study of ternary systems consisting of the two polymers in question and a common solvent. Stockmayer and Stanley (*40*) calculated χ_{23} from light scattering measurements; Sakurada et al. (*41*) calculated χ_{23} by measuring the extent of swelling of polymers by a swelling agent; and Allen et al. (*15*) and Berek et al. (*30*) calculated χ_{23} from the parameters of phase equilibrium by using polymer–solvent interaction parameters from two component systems.

We now discuss the theoretical basis for the determination of χ_{23} from the phase equilibrium data.

Theoretical Discussions

Scott (*6*) and Tompa (*7*) were the first to investigate different mathematical treatments of the Flory–Huggins theory to derive expressions that would help in studying the behavior of polymer–polymer–solvent systems. Scott (*6*) discussed mixtures of two polymers in the presence of a solvent; that is, a three-component mixture. He obtained equations that lead to Gibbs free energy of mixing. Tompa (*7*) developed equations to express spinodals for such ternary systems.

Hsu and Prausnitz (*42*) described a numerical procedure for predicting the compositions of the coexisting phases, for establishing tie lines, and for tracing the binodal and determining the critical point. These calculations were also based on the Flory–Huggins theory.

A simple mathematical scheme can be developed to calculate the interaction parameters starting from the Flory–Huggins expression for the Gibbs free energy of mixing (ΔG_m) for a system consisting of two polymers and one solvent:

$$\frac{\Delta G_m}{RT} = n_1 \ln \phi_1 + n_2 \ln \phi_2 + n_3 \ln \phi_3 + (\chi_{12}\phi_1\phi_2 + \chi_{13}\phi_1\phi_3 + \chi_{23}\phi_2\phi_3) \times (m_1 n_1 + m_2 n_2 + m_3 n_3) \tag{1}$$

where n_i is the number of moles of ith component in the mixture, ϕ_i is the volume fraction of ith component, χ_{ij} is the Flory–Huggins interaction parameter, and m_i is the ratio of the molar volume of i to that of the reference component. Subscripts 2 and 3 denote polymers 2 and 3, and 1 denotes the solvent. The chemical potentials of each component (*8*) can be obtained by differentiation of the Gibbs free energy of mixing with respect to n:

$$\frac{\Delta\mu_1}{RT} = \ln \phi_1 + \left[1 - \left(\frac{1}{m_2}\right)\right]\phi_2 + \left[1 - \left(\frac{1}{m_3}\right)\right]\phi_3 + \chi_1(\phi_2 + \phi_3)^2 + \chi_2\phi_2^2 + \chi_3\phi_3^2 \tag{2}$$

$$\frac{\Delta\mu_2}{RT} = \ln \phi_2 + \left[1 - \left(\frac{m_2}{m_3}\right)\right]\phi_3 + (1 - m_2)\phi_1 + m_2[\chi_2(\phi_3 + \phi_1)^2 + \chi_3\phi_3^2 + \chi_1\phi_1^2] \tag{3}$$

$$\frac{\Delta\mu_3}{RT} = \ln \phi_3 + \left[1 - \left(\frac{m_3}{m_2}\right)\right]\phi_2 + (1 - m_3)\phi_1 + m_3[\chi_3(\phi_2 + \phi_1)^2 + \chi_2\phi_2^2 + \chi_1\phi_1^2] \tag{4}$$

where

$$\chi_1 = \tfrac{1}{2}(\chi_{12} + \chi_{13} - \chi_{23}) \tag{5a}$$

$$\chi_2 = \tfrac{1}{2}(\chi_{12} + \chi_{23} - \chi_{13}) \tag{5b}$$

$$\chi_3 = \tfrac{1}{2}(\chi_{13} + \chi_{23} - \chi_{12}) \tag{5c}$$

and m_2 and m_3 are the molar volume ratios of the polymers to the reference volume V_0. The reference volume V_0 is the molar volume of solvent V_1; m_1 is, therefore, equal to unity, and

$$m_2 = V_2/V_1 \tag{6a}$$

$$m_3 = V_3/V_1 \tag{6b}$$

Equations 2, 3, and 4, mathematically independent, are equivalent to Equations 5a, 5b, and 5c of Ref. 6, respectively.

At equilibrium the chemical potential of each component must be the same in both phases. Denoting the two conjugate phases by single and double primes, $\Delta\mu_1' = \Delta\mu_1''$, $\Delta\mu_2' = \Delta\mu_2''$, and $\Delta\mu_3' = \Delta\mu_3''$. Thus, Equation 2 will give

$$\chi_1[(\phi_2' + \phi_3')^2 - (\phi_2'' + \phi_3'')^2] + \chi_2(\phi_2'^2 - \phi_2''^2) + \chi_3(\phi_3'^2 - \phi_3''^2)$$
$$= \ln\left(\frac{\phi_{1''}}{\phi_{1'}}\right) + \left(1 - \frac{1}{m_2}\right)(\phi_2'' - \phi_2') + \left(1 - \frac{1}{m_3}\right)(\phi_3'' - \phi_3') \tag{7}$$

Similarly, Equation 3 yields

$$\chi_2[(\phi_3' + \phi_1')^2 - (\phi_3'' + \phi_1'')^2] + \chi_3(\phi_3'^2 - \phi_3''^2) + \chi_1(\phi_1'^2 - \phi_1''^2)$$
$$= \frac{1}{m_2}\ln\left(\frac{\phi_2''}{\phi_2'}\right) + \left(\frac{1}{m_2} - \frac{1}{m_3}\right)(\phi_3'' - \phi_3') + \left(\frac{1}{m_2} - 1\right)(\phi_1'' - \phi_1') \tag{8}$$

and Equation 4 yields

$$\chi_3[(\phi_2' + \phi_1')^2 - (\phi_2'' + \phi_1'')^2] + \chi_2(\phi_2'^2 - \phi_2''^2) + \chi_1(\phi_1'^2 - \phi_1''^2)$$
$$= \frac{1}{m_3}\ln\left(\frac{\phi_3''}{\phi_3'}\right) + \left(\frac{1}{m_3} - \frac{1}{m_2}\right)(\phi_2'' - \phi_2') + \left(\frac{1}{m_3} - 1\right)(\phi_1'' - \phi_1') \tag{9}$$

Subtracting Equation 7 from Equation 8 and simplifying yield

$$2\chi_2(\phi_2' - \phi_2'') + 2\chi_1(\phi_1'' - \phi_1') = \ln\left(\frac{\phi_1''}{\phi_1'}\right) - \frac{1}{m_2}\ln\left(\frac{\phi_2''}{\phi_2'}\right) \tag{10}$$

Subtracting Equation 7 from Equation 9 and simplifying yield

$$2\chi_3(\phi_3'' - \phi_3') + 2\chi_1(\phi_1' - \phi_1'') = \frac{1}{m_3}\ln\left(\frac{\phi_3''}{\phi_3'}\right) - \ln\left(\frac{\phi_1''}{\phi_1'}\right) \tag{11}$$

Substituting the values for χ_1, χ_2, and χ_3 from Equations 5a, 5b, and 5c yields for Equation 10

$$\chi_{23}[(\phi_2' - \phi_2'') - (\phi_1'' - \phi_1')] - \chi_{13}[(\phi_2' - \phi_2'')] + \chi_{12}[(\phi_2' - \phi_2'')$$
$$+ (\phi_1'' - \phi_1')] = \ln(\phi_1''/\phi_1') - \frac{1}{m_2}\ln(\phi_2''/\phi_2') \tag{12}$$

and for Equation 11

$$\chi_{23}[(\phi_3'' - \phi_3') + (\phi_1'' - \phi_1')] + \chi_{13}[(\phi_3'' - \phi_3') - (\phi_1'' - \phi_1')] - \chi_{12}[(\phi_3'' - \phi_3') + (\phi_1'' - \phi_1')] = \frac{1}{m_3}\ln(\phi_3''/\phi_3') - \ln(\phi_1''/\phi_1') \quad (13)$$

Thus we have two equations in terms of the concentrations of the conjugate solutions and the three interaction parameters. Adding Equations 12 and 13 and simplifying yield

$$\chi_{23} = \frac{\frac{1}{m_3}\ln(\phi_3''/\phi_3') - \frac{1}{m_2}\ln(\phi_2''/\phi_2') - (\chi_{13} - \chi_{12})(\phi_1' - \phi_1'')}{(\phi_3'' - \phi_3' + \phi_2 - \phi_2'')} \quad (14)$$

Calculating the interaction parameters from measured equilibrium concentrations by using the above equations will be discussed in the section entitled "Results and Discussion."

Equations can also be developed for the interaction parameters when dealing with polydisperse polymers. The free enthalpy of mixing function Z employed in this development is given by Koningsveld et al. (*43*).

$$Z = \frac{\Delta G_m}{NRT} = \phi_1 \ln \phi_1 + \sum_{i=1}^{k} \phi_{2,i} m_{2,i}^{-1} \ln \phi_{2,i} + \sum_{j=1}^{l} \phi_{3,j} m_{3,j}^{-1} \ln \phi_{3,j} + \psi(\phi_2, \phi_3, T) \quad (15)$$

where ψ is the interaction function given by

$$\psi = \psi_{12} + \psi_{13} + \psi_{23} = \phi_1\phi_2\chi_{12}(\phi_2, \phi_3) + \phi_1\phi_3\chi_{13}(\phi_2, \phi_3) + \phi_2\phi_3\chi_{23}(\phi_2, \phi_3)$$

and where N is the total number of moles; ΔG_m is the free enthalpy (Gibbs free energy) of mixing; R is the gas constant; T is the absolute temperature; ϕ_1 is the volume fraction of the low molecular weight solvent; $\phi_{2,i}$ is the volume fraction of species i in polymer 2; $m_{2,i}$ is the relative chain length of species i in polymer 2; $\phi_{3,j}$ is the volume fraction of species j in polymer 3; $m_{3,j}$ is the relative chain length of species j in polymer 3; $\phi_2 = \sum_i \phi_{2,i}$ is the volume fraction of the whole polymer 2; $\phi_3 = \sum_j \phi_{3,j}$ is the volume fraction of the whole polymer 3; and χ_{12}, χ_{13}, and χ_{23} are the interaction parameters. There are k and l components, respectively, in polymer 2 and polymer 3

$$N = n_1 + \sum_{i=1}^{k} n_{2,i}m_{2,i} + \sum_{j=1}^{l} n_{3,j}m_{3,j} \quad (16)$$

where n is the number of moles of component 1, 2, or 3.

As in the case of monodisperse polymers, the chemical potentials of each component can be obtained by differentiation of Equation 15 with respect to n. Partial differentiation yields the following equations for cases in which polymer 2 alone is polydisperse and for cases in which polymers 2 and 3 are both polydisperse.

When polymer 2 alone is polydisperse, we get

$$\frac{\Delta\mu_1}{RT} = \ln\phi_1 + \left[1 - \left(\frac{1}{m_2}\right)\right]\phi_2 + \left[1 - \left(\frac{1}{m_3}\right)\right]\phi_3 + \chi_1(\phi_2 + \phi_3)^2 + \chi_2\phi_2^2 + \chi_3\phi_3^2 - \left\{\phi_1\phi_2\left(\phi_2\frac{\partial\chi_{12}}{\partial\phi_2} + \phi_3\frac{\partial\chi_{12}}{\partial\phi_3}\right) + \phi_1\phi_3\left(\phi_2\frac{\partial\chi_{13}}{\partial\phi_3} + \phi_3\frac{\partial\chi_{13}}{\partial\phi_3}\right) + \phi_2\phi_3\left(\phi_2\frac{\partial\chi_{23}}{\partial\phi_2} + \phi_3\frac{\partial\chi_{23}}{\partial\phi_3}\right)\right\} \quad (17)$$

$$\frac{\Delta\mu_2}{RT} = \frac{N}{n_2}\left[\frac{1}{N}\left\{\sum_{i=1}^{k}\ln\phi_{2i}n_{2i} + \phi_1\left(n_2 - \Sigma\frac{m_{2i}n_{2i}}{m_1}\right) + \phi_3\left(n_2 - \Sigma\frac{m_{2i}n_{2i}}{m_3}\right)\right\} + \phi_2\{\chi_2(\phi_3 + \phi_1)^2 + \chi_3\phi_3^2 + \chi_1\phi_1^2\} + \phi_2\left\{\phi_1\phi_2\left(\frac{\partial\chi_{12}}{\partial\phi_2}(1 - \phi_2) - \phi_3\frac{\partial\chi_{12}}{\partial\phi_3}\right) + \phi_2\phi_3\left(\frac{\partial\chi_{23}}{\partial\phi_2}(1 - \phi_2) - \phi_3\frac{\partial\chi_{23}}{\partial\phi_3}\right) + \phi_3\phi_1\left(\frac{\partial\chi_{13}}{\partial\phi_2}(1 - \phi_2) - \phi_3\frac{\partial\chi_{13}}{\partial\phi_3}\right)\right\}\right] \quad (18)$$

$$\frac{\Delta\mu_3}{RT} = \ln\phi_3 + \left[1 - \left(\frac{m_3}{m_2}\right)\right]\phi_2 + (1 - m_3)\phi_1 + m_3[\chi_3(\phi_2 + \phi_1)^2 + \chi_2\phi_2^2 + \chi_1\phi_1^2] + m_3\left\{\phi_1\phi_2\left(\frac{\partial\chi_{12}}{\partial\phi_2}(1 - \phi_2) - \phi_3\frac{\partial\chi_{12}}{\partial\phi_3}\right) + \phi_2\phi_3\left(\frac{\partial\chi_{23}}{\partial\phi_2}(1 - \phi_2) - \phi_3\frac{\partial\chi_{23}}{\partial\phi_3}\right) + \phi_3\phi_1\left(\frac{\partial\chi_{13}}{\partial\phi_2}(1 - \phi_2) - \phi_3\frac{\partial\chi_{13}}{\partial\phi_3}\right)\right\} \quad (19)$$

When polymers 2 and 3 are both polydisperse,

$$\frac{\Delta\mu_1}{RT} = \ln\phi_1 + \left[1 - \left(\frac{1}{m_2}\right)\right]\phi_2 + \left[1 - \left(\frac{1}{m_3}\right)\phi_3 + \chi_1(\phi_2 + \phi_3)^2 + \chi_2\phi_2^2 + \chi_3\phi_3^2 - \left\{\phi_1\phi_2\left(\phi_2\frac{\partial\chi_{12}}{\partial\phi_2} + \phi_3\frac{\partial\chi_{12}}{\partial\phi_3}\right)\right.$$

$$+ \phi_1\phi_3\left(\phi_2 \frac{\partial\chi_{13}}{\partial\phi_2}\right) + \phi_3\left(\frac{\partial\chi_{13}}{\partial\phi_3}\right) + \phi_2\phi_3\left(\phi_2 \frac{\partial\chi_{23}}{\partial\phi_2} + \phi_3 \frac{\partial\chi_{23}}{\partial\phi_3}\right)\Bigg\} \quad (20)$$

$$\frac{\Delta\mu_p}{RT} = \frac{N}{n_p}\left[\frac{1}{N}\left\{\sum_{i=l}^{k} \ln \phi_{pi}n_{pi} + \phi_1\left(n_p - \Sigma \frac{m_{pi}n_{pi}}{m_1}\right)\right.\right.$$

$$+ \phi_q\left(n_p - \Sigma \frac{m_{pi}n_{pi}}{m_q}\right)\Bigg\} + \phi_p\{\chi_p(\phi_q + \phi_1)^2 + \chi_q\phi_q^2 + \chi_1\phi_1^2\}$$

$$+ \phi_p\Bigg\{\phi_1\phi_2\left(\frac{\partial\chi_{12}}{\partial\phi_2}(1 - \phi_2) - \phi_3 \frac{\partial\chi_{12}}{\partial\phi_3}\right) + \phi_2\phi_3\left(\frac{\partial\chi_{23}}{\partial\phi_2}(1 - \phi_2)\right.$$

$$\left.\left.\left. - \phi_3 \frac{\partial\chi_{23}}{\partial\phi_3}\right) + \phi_3\phi_1\left(\frac{\partial\chi_{13}}{\partial\phi_2}(1 - \phi_2) - \phi_3 \frac{\partial\chi_{13}}{\partial\phi_3}\right)\right\}\right] \quad (21)$$

for $p \neq q$; $p = 2, 3$; and $q = 2, 3$ and where χ_1, χ_2, and χ_3 are given by Equation 5a, 5b, and 5c, respectively.

From these equations, suitable expressions can be derived to calculate interaction parameters by using phase equilibrium data, as discussed earlier for the monodisperse polymers, and by using information about the molecular weight distributions obtained by GPC.

Experimental

The system of PS and PBD was selected to demonstrate the case of narrow molecular weight distributions. The importance of the two polymers in the polymer industry and the commercial availability of these two polymers in narrow molecular weight distribution samples made this system a logical choice. The characteristics of the particular polymer samples employed are given in Table I. Toluene was selected because it is a good mutual solvent and because the χ_{12} values for PS–toluene are published (*44*).

The procedure adopted for the sample preparation is given earlier (*27*). Equilibrium was attained at 23 °C and 1 atm.

A detailed discussion of GPC with sequential RI and UV detectors for the quantitative analysis of the conjugate phases of the incompatible system of PS and PBD with THF as solvent is also given earlier (*27, 28*).

Results and Discussion

The binodal curves, tie lines, and plait points for the two PS–PBD systems studied are given in Figures 1 and 2.

The polymer–solvent interaction parameter χ_{12} for PS–toluene was obtained from the work of Scholte (*44*) and used in Equations 12 and 13 to solve simultaneously for the interaction parameters χ_{13} (PBD–toluene) and χ_{23} (PS-PBD). The results are presented in Tables II and III.

Table I. Characteristics of Polymer Samples

Sample	$\bar{M}_w \times 10^{-3}$	$\bar{M}_n \times 10^{-3}$	$\bar{M}_w/\bar{M}_n$
PS 37,000	36.0	33.0	<1.06
PS 100,000	100.0	100.0	<1.06
PBD 170,000[a]	170.0 ± 17	135.0 ± 13	1.26

NOTE: Manufacturers' data were supplied by Pressure Chemical Co. (PS samples) and Phillips Petroleum Co. (PBD samples).

[a] 47.1% *cis*, 44.5% *trans*, 8.4% vinyl, 0.04% antioxidant.

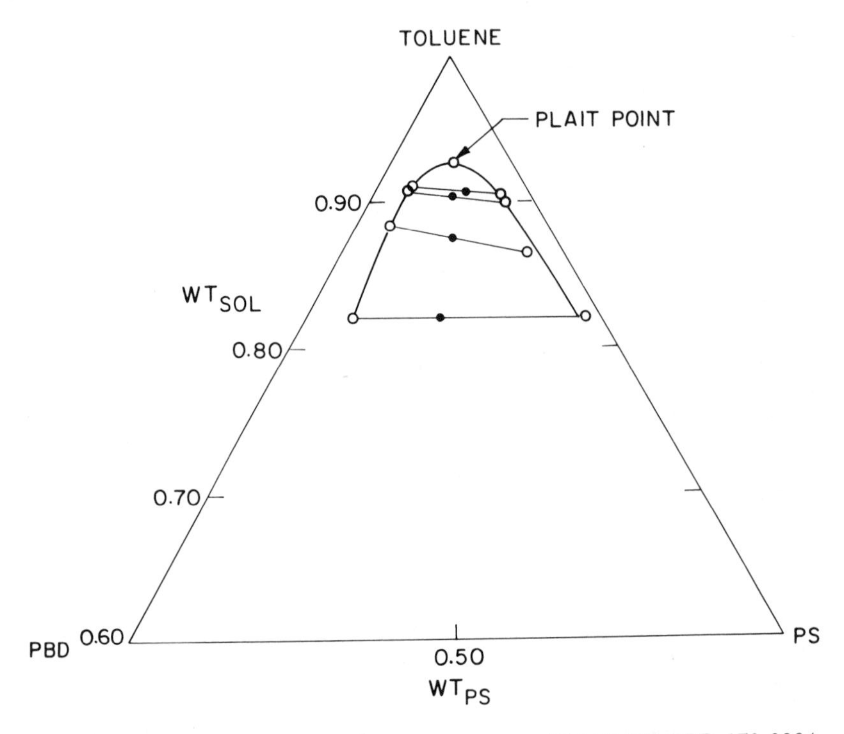

Figure 1. Phase diagram for the ternary system PS 100,000/PBD 170,000/toluene at 23 °C and 1 atm. Solid symbols (●) represent mix point compositions; open symbols (○) represent equilibrium phase compositions; tie lines connect conjugate points via mix points; WT_{PS} is the weight fraction of PS in the total polymer; and WT_{SOL} is the weight fraction of solvent toluene in the total polymer–solvent mixture.

The results for various tie lines are tabulated in an orderly fashion. Each table corresponds to a distinct polymer molecular weight; each table starts with the bottom tie line of the binodal curve, progresses toward the pure solvent apex, and ends with the tie line at the top of the binodal curve nearest to the plait point.

For a given system, the value of χ_{23} is generally found to have its lowest value for the tie lines belonging to the lower portions of the binodal curve. The value progressively increases as the plait point composition is approached. That is, χ_{23} increases with decreasing total polymer concentration in any particular system. Berek et al. (*30*) observed χ_{23} values of the same order of magnitude with a similar trend in the equilibria of fractionated polypropylene (PP) and PS in toluene. The experimental error in the

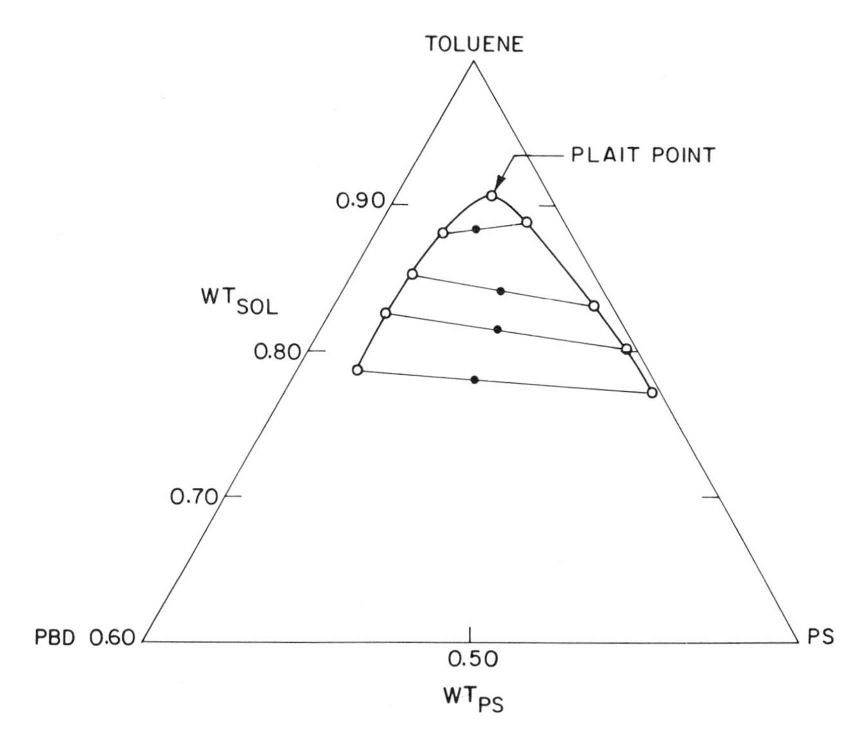

Figure 2. Phase diagram for the ternary system PS 37,000/PBD 170,000/toluene at 23 °C and 1 atm. Solid symbols (●) represent mix point compositions; open symbols (○) represent equilibrium phase compositions; tie lines connect conjugate points via mix points; WT_{PS} is the weight fraction of PS in the total polymer; and WT_{SOL} is the weight fraction of solvent toluene in the total polymer–solvent mixture.

Table II. Interaction Parameters for PS 100,000/PBD 170,000/Toluene

Wt % Solvent (mean)	χ_{12}[a]	χ_{13}	χ_{23}
82.00	0.402	0.459	0.010
87.43	0.411	0.385	0.019
90.36	0.414	0.427	0.023
90.68	0.414	0.442	0.021

[a] Ref. 44.

Table III. Interaction Parameters for PS 37,000/PBD 170,000/Toluene

Wt % Solvent (mean)	χ_{12}[a]	χ_{13}	χ_{23}
77.91	0.385	0.417	0.026
81.39	0.390	0.382	0.029
84.10	0.392	0.376	0.030
88.41	0.398	0.458	0.026

[a] Ref. 44.

values of χ_{23} was estimated to be less than 0.006 by "propagation of error" methods.

The plait points of the binodal curves were obtained by extrapolation of the midpoints of the tie lines to intersect the interpolated binodal curve. Hence the composition of the plait point could be determined graphically. The value χ_{23} at the plait point was calculated from the measured composition by using the following equation (*6*) (Table IV).

$$\chi_{23,\mathrm{crit}} = \frac{1}{2}\left\{\left(\frac{1}{m_2}\right)^{1/2} + \left(\frac{1}{m_3}\right)^{1/2}\right\}^2 \left\{\frac{1}{(1 - \phi_1)}\right\} \tag{22}$$

The value of $\chi_{23,\mathrm{crit}}$ calculated by using Equation 22 is higher than the value of χ_{23} obtained for the tie line closest to the plait point. This difference is to be expected because χ_{23} decreases with total polymer concentration. A correlation exists between the location of the plait point and $\chi_{23,\mathrm{crit}}$ for the two systems studied. The system of higher molecular weight polymers, which has its plait point at a higher solvent concentration, has a

Table IV. Plait Point Composition

Variable	*Figure 1*	*Figure 2*
	weight percent	
Toluene	92.5	90.5
PS	3.75	5.75
PBD	3.75	3.75
	volume fraction	
ϕ_1 (toluene)	0.934	0.920
ϕ_2 (PS)	0.030	0.045
ϕ_3 (PBD)	0.037	0.037
	volume ratios	
m_2	847.62	313.62
m_3	1787.1	1787.1
	interaction parameter	
$\chi_{23,\mathrm{crit}}$	0.026	0.039

lower value of $\chi_{23,crit}$. In other words, $\chi_{23,crit}$ increases as we move from plait points at a higher solvent concentration toward the plait point at a lower solvent concentration. It would be possible to predict, with reasonable accuracy, the relative positions of the plait points of two systems of PS/PBD with different molecular weights of each polymer from their $\chi_{23,crit}$ values.

Comparing the binodal curves for the two PS/PBD systems shows that the system with the combination of the higher molecular weight of these polymers (PS 100,000 and PBD 170,000) has the plait point at a higher solvent concentration.

Conclusions

GPC provides methods for using data from polydisperse polymers in conjunction with Scott's equations for interaction parameters and methods for eliminating the effects of low molecular weight additives on the analysis of phases.

The polymer–polymer interaction parameter χ_{23} was evaluated from the parameters of phase equilibria of the system PS–PBD–toluene in which the polymers have narrow molecular weight distributions. We found that, for any pair of polymers, χ_{23} increases as the total polymer concentration decreases and reaches a maximum at the plait point. The values of χ_{23} at the plait points ($\chi_{23,crit}$) were calculated and were found to decrease as the plait points move toward the solvent apex of the triangular diagram.

Nomenclature

χ_{ij} Flory–Huggins interaction parameter for components i and j
ΔG_m Gibbs free energy of mixing
n_i Number of moles of the ith component
ϕ_i Volume fraction of the ith component
m_i Ratio of the molar volume of component i to that of the reference component
$\Delta \mu_i$ Chemical potential of component i
Z Free enthalpy of mixing function
N Total number of moles
ψ Interaction function
$\phi_{2,i}$ Volume fraction of species i in polymer 2
$m_{2,i}$ Relative chain length of species i in polymer 2
k Number of components in polymer 2
l Number of components in polymer 3
p, q Polymers 2 and 3 in the polydisperse case
$\chi_{23,crit}$ Polymer–polymer interaction parameter at the plait point

Acknowledgments

The authors thank N. Sri Namachchivaya and V. Ramakrishnan for the valuable help and discussions. Support from the Natural Sciences and Engineering Research Council of Canada and the National Science Foundation is gratefully appreciated. V. Narasimhan was supported in part by a Dean of Engineering Scholarship, University of Waterloo.

Literature Cited

1. Koningsveld, R. *Adv. Colloid Interface Sci.* **1968**, *2*, 151.
2. Flory, P. J. *J. Chem. Phys.* **1941**, *9*, 660.
3. Flory, P. J. *J. Chem. Phys.* **1942**, *10*, 51.
4. Huggins, M. L. *J. Chem. Phys.* **1941**, *9*, 440.
5. Huggins, M. L. *Ann. N. Y. Acad. Sci.* **1942**, *43*, 1.
6. Scott, R. L. *J. Chem. Phys.* **1949**, *17*, 279.
7. Tompa, H. *Trans. Faraday Soc.* **1949**, *45*, 1142.
8. Tompa, H. "Polymer Solutions"; Butterworths: London, 1956; Chapter 7.
9. Prigogine, I.; Trappeniers, N.; Mathot, V. *Discuss. Faraday Soc.* **1953**, *15*, 93.
10. Prigogine, I.; Trappeniers, N.; Mathot, V. *J. Chem. Phys.* **1953**, *21*, 559.
11. Patterson, D. *Macromolecules* **1969**, *2*, 672.
12. Lacombe, R. H.; Sanchez, I. C. *J. Phys. Chem.* **1976**, *80*, 2568.
13. Dobry, A.; Boyer-Kawnoki, F. *J. Polym. Sci.* **1947**, *2*, 90.
14. Kern, R. J. *J. Polym. Sci.* **1956**, *21*, 19.
15. Allen, G.; Gee, G.; Nicholsen, J. P. *Polymer* **1960**, *1*, 56.
16. Turley, S. G. *J. Polym. Sci., Part C* **1963**, *1*, 101.
17. Paxton, T. R. *J. Appl. Polym. Sci.* **1963**, *7*, 1499.
18. Angelo, R. J.; Ikeda, R. M.; Wallack, M. L. *Polymer* **1965**, *6*, 141.
19. Koningsveld, R. *Discuss. Faraday Soc.* **1970**, *49*, 144.
20. White, J. L.; Sallady, D. G.; Quisenberry, D. O.; Maclean, D. L. *J. Appl. Polym. Sci.* **1972**, *16*, 2811.
21. Parent, R. P.; Thompson, E. V. *Polymer Prepr. Am. Chem. Soc., Div. Polym. Chem.* **1977**, *18*(*2*), 507.
22. Parent, R. P.; Thompson, E. V. *Polymer Prepr. Am. Chem. Soc., Div. Polym. Chem.* **1978**, *19*(*1*), 180.
23. Thompson, E. V. *Org. Coat. Plast. Chem.* **1979**, *40*, 751.
24. Welygan, D. G.; Burns, C. M. *J. Polym. Sci., Polym. Lett. Ed.* **1973**, *11*, 339.
25. Welygan, D. G.; Burns, C. M. *J. Appl. Polym. Sci.* **1974**, *18*, 521.
26. Lloyd, D. R.; Burns, C. M. *J. Appl. Polym. Sci.* **1978**, *22*, 593.
27. Lloyd, D. R.; Narasimhan, V.; Burns, C. M. *Polymer Prepr. Am. Chem. Soc., Div. Polym. Chem.* **1979**, *20*(*1*), 897.
28. Narasimhan, V.; Lloyd D. R.; Burns, C. M. *J. Appl. Polym. Sci.* **1979**, *23*, 749.
29. Bristow, G. M. *J. Appl. Polym. Sci.* **1959**, *2*, 120.
30. Berek, D.; Lath, D.; Durdovic, V. *J. Polym. Sci., Part C* **1967**, *16*, 659.
31. Welygan, D. G., M.A. Sc. Thesis, Univ. of Waterloo, Waterloo, Ontario, 1973.
32. Starita, J. M. *Trans. Soc. Rheol.* **1972**, *16*, 339.
33. Bucknal, C. B. *Br. Plast.* **1967**, *40*(*11*), 118.
34. Bucknal, C. B. *Br. Plast.* **1967**, *40*(*12*), 84.
35. Curtis, A. J.; Covitch, M. J.; Thomas, P. A.; Sperling, L. H. *Polymer* **1972**, *12*, 101.
36. Slonimskii, G. L. *J. Polym. Sci.* **1958**, *30*, 625.
37. Runyon, J. R.; Barnes, D. E.; Rudd, J. F.; Tung, L. M. *J. Appl. Polym. Sci.* **1969**, *13*, 2359.

38. Cantow, H. J.; Probst, J.; Stojanow, C. *Kautsch. Gummi, Kunstst.* **1968,** *21,* 609.
39. Lloyd, D. R.; Narasimhan, V.; Burns, C. M. *J. Liq. Chromatrogr.* **1980,** *3,* 1111.
40. Stockmayer, W. H.; Stanley, H. E. *J. Chem. Phys.* **1950,** *18,* 153.
41. Sakurada, I.; Nakajima, H.; Aoki, H. *J. Polym. Sci.* **1959,** *35,* 507.
42. Hsu, C. C.; Prausnitz, J. M. *Macromolecules* **1974,** *7,* 3250.
43. Koningsveld, R.; Chermin, H. A. G.; Gordon, M. *Proc. R. Soc. London* **1970,** *A319,* 331.
44. Scholte, Th. G. *Eur. Polym. J.* **1970,** *6,* 1063.

RECEIVED for review January 20, 1983. ACCEPTED July 26, 1983.

2

A Polymer Blend Exhibiting Both Upper and Lower Critical Solution Temperature Behavior: Polystyrene/Poly(*o*-chlorostyrene)

S. L. ZACHARIUS[1], W. J. MACKNIGHT, and F. E. KARASZ

Department of Polymer Science and Engineering, University of Massachusetts, Amherst, MA 01003

Polymer–polymer interactions of polystyrene (PS) and poly(o-chlorostyrene) (PoClS) blends have been investigated by using a vapor sorption technique. These results, combined with heat of mixing experiments, indicate that both upper and lower critical solution temperature behavior are exhibited by this blend. Previous data showed that the miscibility of the PS/PoClS system is extremely molecular weight sensitive. This sensitivity is shown to be related to the existence of a critical double point in the system. Model calculations using the Flory equation-of-state theory illustrate these points.

POLYMER–POLYMER INTERACTIONS AND THE GENERAL PHASE BEHAVIOR of the polystyrene/poly(*o*-chlorostyrene) (PS/PoClS) system were investigated with experimental ternary solution methods. Ryan (*1*) studied PS/PoClS blends as a function of the molecular weight of PS. The phase behavior of the polymer blends was investigated by using differential scanning calorimetry (DSC) to measure the glass transition temperature (T_g) of the blends after the blends were annealed at a series of temperatures between 150 and 400 °C. The enthalpy of mixing was also measured at 35 and 68 °C. Because direct measurement of the enthalpy of mixing is not possible, the required parameters were determined by using differential heats of solution measurements and Hess's law. (*1*)

[1]Present address: The Aerospace Corporation, Los Angeles, CA 90009

0065-2393/84/0206-0017$06.00/0

The results of the annealing studied showed that, when PoClS of molecular weight 100,000 was blended with PS of molecular weight equal to or greater than 32,400, the blend could be phase separated. When the same PoClS was blended with PS with a molecular weight equal to or less than 26,700, the blend exhibited one T_g and could not be phase separated.

The results of the enthalpy of mixing experiments showed small but nearly always positive values. The molecular weight of the PS did not significantly affect these values.

To explain these experimental results, we postulate that blends with the higher molecular weight PS have an hourglass-type phase diagram; that is, the upper critical solution temperature (UCST) and the lower critical solution temperature (LCST) have merged. The temperature of this merger was determined to be about 550 K. The blends with the lower molecular weight PS are single phase systems. As the molecular weight of the PS decreases, the UCST decreases and the LCST increases to the extent that both critical points are beyond experimental detection. Although a change in the critical points is theoretically predicted, this large a change as a result of such a small change in molecular weight has not been observed before. A change in the molecular weight of less than 4000 results in separation of the UCST and LCST by at least 250 °C.

We combined the experimental results of Ryan with experimental vapor sorption results to conclude that the PS/PoClS blend does exhibit both UCST and LCST behavior. In addition, we present theoretical evidence that the extreme molecular weight sensitivity of the UCST and the LCST is the result of the existence of a critical double point in this system. Sample calculations illustrate this theory.

The familiar Flory–Huggins theory (*2–4*), which is based on a simple lattice representation for polymer solutions, can only predict the UCST phenomenon. To predict the LCST phenomenon, the differences in the equation-of-state properties of both pure components must be taken into account. Several theories (*5*, *6*), including Flory's equation-of-state theory (*7*, *8*), have been developed to describe both the UCST and LCST behavior.

The equation-of-state theory considers the role of free volume in polymer solution thermodynamics. During the mixing process, the free volume of each component is changed. An intermediate value for the free volume is approached that is characteristic of the mixture. The difference in free volume is particularly significant when the mixture consists of a polymer and a solvent. Polymers typically have very low degrees of thermal expansion and free volume compared to solvents. This difference is reflected in different equation-of-state properties. The difference in free volume between polymers and solvents is independent of any chemical difference between the two components.

The change in free volume results in an overall change in the total volume of mixing (ΔV_M), the enthalpy of mixing (ΔH_M), the entropy of mixing (ΔS_M), and, therefore, also the free energy of mixing (ΔG_M). In general, the overall volume decreases and results in a negative contribution to both the enthalpy and entropy of mixing. That is,

$$\Delta H_{M, \text{f.v.}} < 0 \qquad (1)$$

and

$$\Delta S_{M, \text{f.v.}} < 0 \qquad (2)$$

The total free volume (f.v.) contribution is predicted to be positive (*9*)

$$\Delta H_{M, \text{f.v.}} - T\Delta S_{M, \text{f.v.}} > 0 \qquad (3)$$

This free volume contribution has an unfavorable effect on mixing. It becomes more significant as the temperature increases. The equation-of-state theory is thus able to predict phase separation as the temperature increases, LCST. The UCST is a result of a positive ΔH_M arising from the breaking of like contacts and the forming of unlike contacts. This positive ΔH_M effect decreases as the temperature increases.

For a mixture of two polymers, the combinatorial entropy of mixing will be very small. Therefore, usually a negative enthalpy of mixing resulting from specific interactions is necessary to attain miscibility. Only if the free volume contribution is very small can a small positive enthalpy of mixing be tolerated. McMaster (*10*) has shown that in this case the polymer–polymer system can exhibit both UCST and LCST behavior. We believe that the PS/PoClS system is such a system.

Experimental

Polystyrene. Two different molecular weight samples of PS with narrow distributions were used. The lower molecular weight sample and characterization data were obtained from Pressure Chemical Company. The PS is atactic batch 41220, prepared by anionic polymerization. The number average molecular weight ($\bar{M}_n$) as determined by membrane osmometry is 15,000 ± 6%. The polymer was used as received without further purification.

The higher molecular weight sample and characterization data were obtained from Goodyear Chemicals. It is anionically polymerized atactic PS sample CDS-S-6. The $\bar{M}_n$ as determined by membrane osmometry is 80,800, and 75,300 as determined by GPC. The weight average molecular weight ($\bar{M}_w$) as determined by light scattering is 82,900, and 82,100 as determined by GPC. The polymer was dissolved in toluene, filtered through a very fine sintered glass funnel, precipitated into methanol, and then dried under vacuum at 80 °C for several days before being used.

Poly(*o*-chlorostyrene). The polymer was synthesized by free radical polymerization (*1*). The $\bar{M}_n$ as determined by GPC was 77,700; the $\bar{M}_w$ as determined by GPC was 169,000. The polymer was then fractionated with a preparative GPC. The fraction used in the vapor sorption measurements had $\bar{M}_n = 100{,}500$ and $\bar{M}_w = 147{,}100$, as determined by GPC.

Decahydronaphthalene (Decalin). The decalin, obtained from Fisher Scientific Company, was refluxed over LiAlH for 3.5 h at atmospheric pressure and 180 °C. It was then distilled under partial vacuum at 83 °C.

Film Preparation. Films containing PS, PoClS, and their mixtures were cast from a methylene chloride solution onto aluminum pans. Initial polymer concentration in methylene chloride was approximately 4%. The ratio of PS to PoClS is on weight basis. This technique provided a sample that was thin, on the order of 0.1 mm. The solvent was allowed to evaporate at room temperature for several hours. The films were then dried under high vacuum (10^{-4} mm Hg) at 80 °C for several days in an abderhalden drying tube. The absence of residual solvent was confirmed by a constant T_g.

Results and Discussion

Vapor Sorption Studies. Vapor sorption technique is used to study the thermodynamic properties of polymer–solvent systems and polymer blends (*11*). A polymer solvent solution is allowed to come to equilibrium with pure solvent of known partial pressure. As a first approximation, the activity of the solvent is equal to the relative vapor pressure $P/P°$, where $P°$ is the vapor pressure of the pure solvent. For this study, the approximate activities were corrected for deviation from the perfect gas law by using Equation 4:

$$\ln a_1 - \ln(P/P°) = (P - P°)/\{RT(\alpha/RT - \beta)\} \tag{4}$$

where α and β are the two van der Waals constants. The activity coefficient a_1 of the solvent in a binary or ternary solution is related to the chemical potential $\Delta\mu_1$ by

$$\ln a_1 = \Delta\mu_1/RT \tag{5}$$

For a binary solution

$$\Delta\mu_1/RT = \ln\phi_1 + [1 - 1/r_2]\phi_2 + \chi_{12}\phi_2^2 \tag{6}$$

where ϕ_1 and ϕ_2 are the volume fractions of the solvent and the polymer, with a degree of polymerization r_2. The Flory–Huggins parameter is χ_{12}. For a ternary solution (*12, 13*)

$$\Delta\mu_1/RT = \ln\phi_1 + (1 - 1/r_2)\phi_2 + (1 - 1/r_3)\phi_3 + (\chi_{12}\phi_2 + \chi_{13}\phi_3)(1 - \phi_1) - \chi_{23}\phi_2\phi_3 \tag{7}$$

where subscript 1 refers to the solvent and subscripts 2 and 3 to the two polymers. The value of χ_{23} of the two polymers can be calculated from three separate vapor sorption experiments, one with each of the polymer components separately and a third with the blend.

Polymer–Solvent Systems. Two different molecular weight PS-decalin systems were investigated, PS 15,000 and PS 80,000. The vapor sorption measurements were made at 80 °C. The T_g is 100 °C for both samples of PS. Once a certain amount of solvent has been absorbed, the PS is sufficiently plasticized so that the experimental temperature is now above T_g and the system is no longer glassy. Equation 6 is used to evaluate the χ_{12} parameter from the equilibrium sorption data of the plasticized systems. Results are shown in Figures 1 and 2 for PS 15,000 and PS 80,000, respectively. The solid lines represent a linear least squares analysis of the data. The results are $\chi_{12} = -1.00\ \phi_1 + 0.797$ for PS 15,000, and $\chi_{12} = -0.565\ \phi_1 + 0.688$ for PS 80,000. The molecular weight of PS has a considerable influence on the composition dependence of χ_{12} in the investigated composition range.

Vapor sorption measurements were also made on the decalin/PoClS system. The T_g for PoClS is 133 °C. Therefore, the polymer must absorb more solvent before the T_g is depressed below the experimental temperature of 80 °C. The χ_{13} parameter for the plasticized system is shown in Figure 3. The solid line is a linear least squares fit by $\chi_{13} = -1.42,\ \phi_1 +$ 1.144.

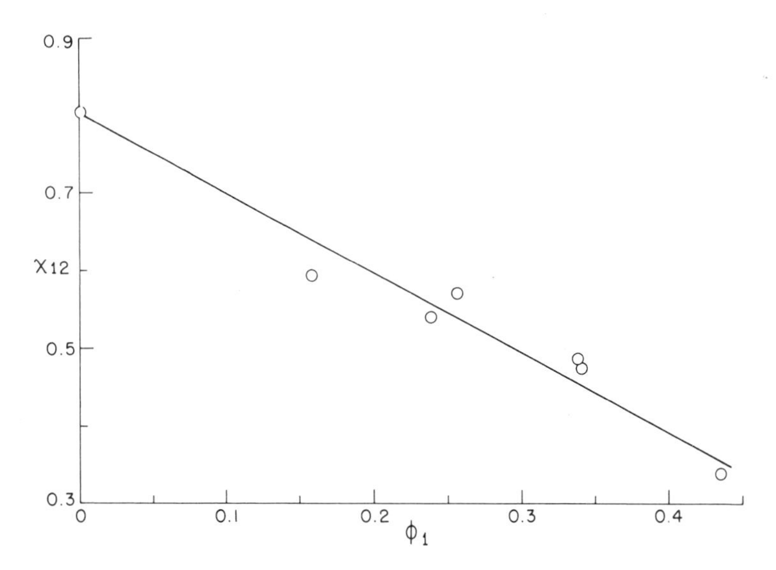

Figure 1. The concentration dependence of the χ parameter for the system $C_{10}H_{18}$/PS 15,000 as measured by vapor sorption experiments.

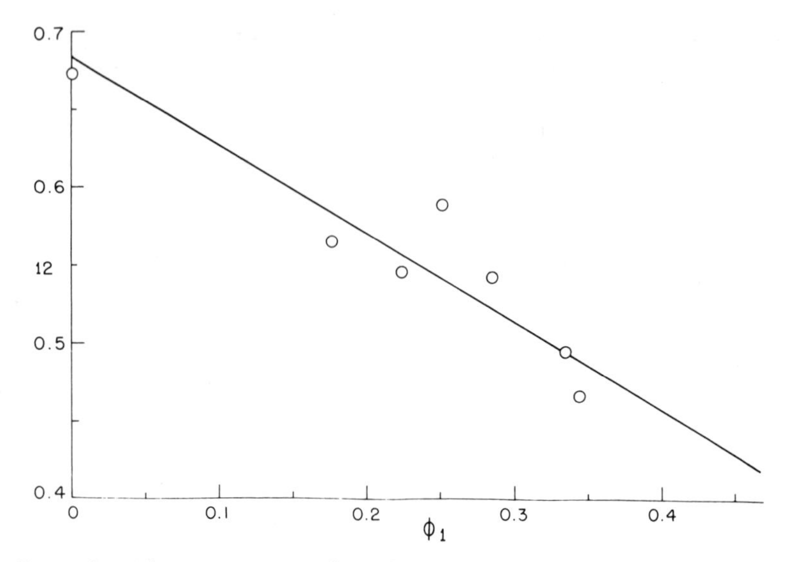

Figure 2. The concentration dependence of the χ parameter for the system $C_{10}H_{18}$/PS 80,000 as measured by vapor sorption experiments.

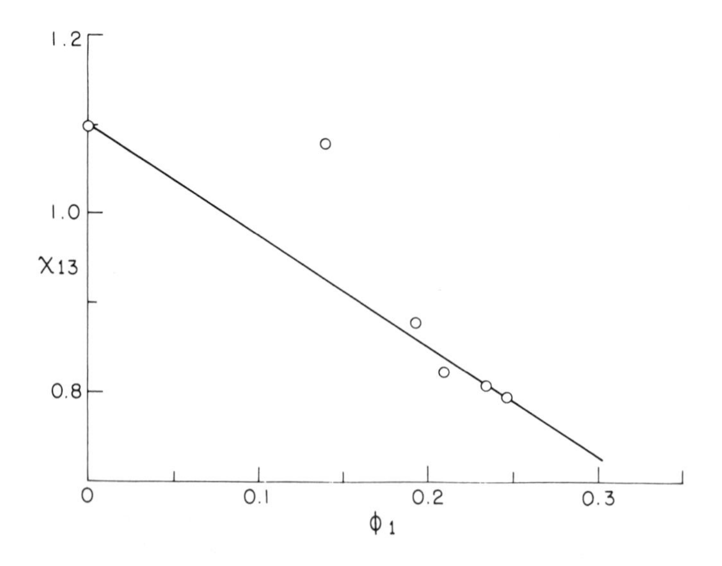

Figure 3. The concentration dependence of the χ parameter for the system $C_{10}H_{18}$/PoClS as measured by vapor sorption experiments.

Polymer–Polymer–Solvent Systems. Three different ternary systems were investigated: PS 15,000/PoClS/decalin with polymer–polymer ratios of 25:75, 50:50, and 75:25. The blend of the high molecular weight PS is an immiscible system, and Equation 7 does not apply. The values of χ_{23}, calculated by using Equation 7 and the previously determined values of χ_{12} and χ_{13}, are shown in Table I. The χ_{23} values show a considerable scatter. Except for the 25:75 systems the values are small and positive, an indication for a possible absence of specific interactions. Because the χ parameter for the latter systems is negative, it is difficult to draw a definite conclusion regarding the presence or absence of specific interactions. The results are, however, clearly different from those of various other miscible polymer systems where unequivocally negative values are found for the χ parameter (*14–17*). This difference indicates a much stronger interaction between both components of those blends than between PS and PoClS.

The variation with composition of the χ parameter is generally attributed to a difference in surface-to-volume ratio of the two components (*18*). This difference might be quite large for polymer–solvent systems. Our results on the polymer–solvent systems are in agreement with this viewpoint. Accordingly, for two polymers that are very much alike, one does not expect a significant composition dependence of the χ parameter as found for the PS 15,000/PoClS system.

A common criticism of mutual solvent techniques in general is that the solvent effect must be "subtracted out." The validity of this approach is questionable. The absolute quantities determined may not be accurate, but these experiments provide a great deal of qualitative information. The only alternative for determining polymer–polymer interactions is to investigate low molecular weight analogs.

Enthalpy of Mixing. According to the equation-of-state theory (*7, 8*) or the lattice fluid theory (*5*), the χ parameter consists of two contributions,

Table I. Flory–Huggins Polymer–Polymer Parameter at 80 °C

Percent PS 15,000 in Film	χ_{23}
75	0.31
	0.07
	0.35
	0.14
50	0.04
	0.23
	0.23
	0.08
25	−0.43
	−0.08
	−0.63
	−0.94

an exchange interaction term (X_{23}) and a free volume term. The X_{23} contribution dominates the low temperature behavior. It is positive for dispersive forces and negative when specific interactions like weak hydrogen bonds or charge transfer complexes are present. The free volume contribution is positive and becomes increasingly important at higher temperatures. A well-known schematic representation for χ_{12} as a function of temperature is presented in Figure 4 (*19, 20*). Curve 1 shows the contribution to χ resulting from contact energy dissimilarities between components 1 and 2. This contribution is considered in both the Flory–Huggins theory and the equation-of-state theory. In the absence of specific interactions, curve 1 decreases with temperature. Curve 2 shows the contribution to χ resulting from equation-of-state parameters for the two components. Curve 3 is the total of these two contributions. Figure 4 illustrates why the Flory–Huggins theory predicts only the UCST and why the equation-of-state theory predicts two critical points: one as the temperature decreases, the UCST; and one as temperature increases, the LCST. Figure 4 also illustrates that two poly-

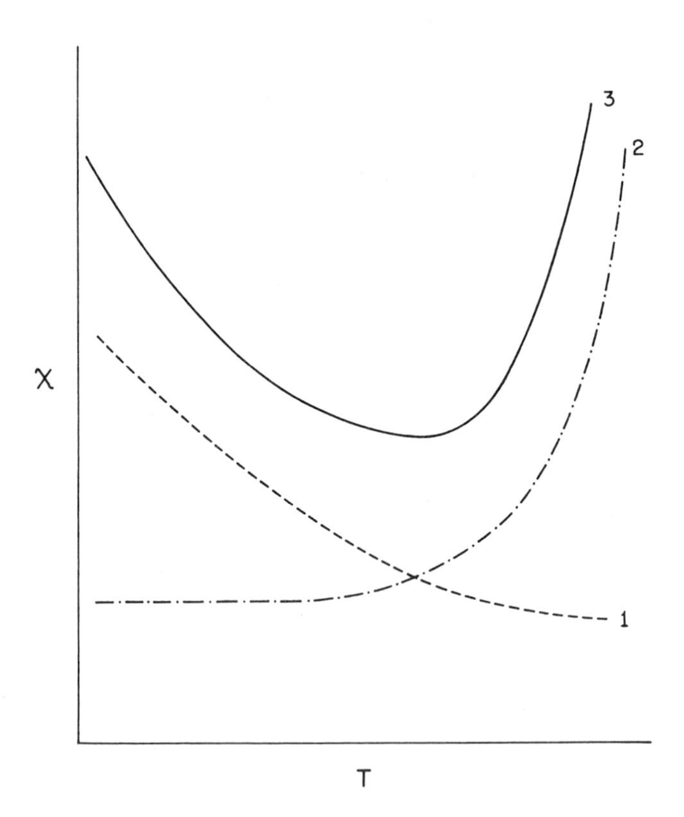

Figure 4. Schematic representation of the temperature dependence of the χ parameter. See *text for explanation of curves.*

mers with a specific interaction, that is a negative value for χ_{23}, will exhibit only an LCST. Curve 1 will not be applicable in this situation.

The phase behavior of the system will be determined by the value of the χ parameter. The following is the expression for the critical value of χ_{23} (*12, 13*)

$$\chi_{cr} = \frac{1}{2}(r_2^{-1/2} + r_3^{-1/2})^2 \tag{8}$$

where r_2 and r_3 are the degree of polymerization for components 2 and 3, respectively. The system is miscible for all temperatures for which $\chi_{23}(T) < \chi_{cr}$ and partially miscible for all temperatures for which $\chi_{23}(T) > \chi_{cr}$.

The equation-of-state expression for the free energy of mixing can be written by using the Flory–Huggins expression where the χ parameter consists of two terms, an interactional term and a free volume term.

$$\Delta G_M/RT = (\phi_2/r_2) \ln \phi_2 + (\phi_3/r_3) \ln \phi_3 + \chi_{23}(T)\, \phi_2 \phi_3 \tag{9}$$

where ϕ_2 and ϕ_3 are volume fractions of components 2 and 3, respectively.

The enthalpy of mixing is related to the free energy of mixing by

$$\Delta H_m = \Delta G_m - T(\partial \Delta G_m/\partial T)_P \tag{10}$$

Combining Equations 9 and 10 results in

$$\Delta H_m = -RT^2(\partial \chi_{23}/\partial T)\phi_2 \phi_3 \tag{11}$$

The sign of the enthalpy of mixing is determined by the first derivative of the χ parameter. Prigogine and Defay (*21*) have shown that ΔH_m is negative near an LCST and positive near a UCST, unless an inflection point exists in the composition dependence of ΔH_m.

With this in mind, the implications of Ryan's results become clear. Ryan (*1*) measured the enthalpy of mixing of PS/PoClS at 35 and 68 °C for various molecular weight samples of PS while keeping the molecular weight of PoClS constant. In nearly all cases a small but positive heat of mixing was found, with no clear dependence on the molecular weight of PS. These results show that 35 and 68 °C are in the temperature range in which a UCST is possible. Consequently, blends of PS and PoClS constitute an example of miscibility without specific interactions. McMaster (*10*) has already shown that this exceptional situation can occur if the exchange interaction contribution is sufficiently small and if the *PVT* properties of both components do not differ too much.

Model Calculations. The analysis so far showed that the experimental results for PS/PoClS blends clearly point to the possibility of critical double points in these systems. Although we argued that the enhanced sensitivity of the phase behavior to a change in molar mass is probably a result of this fact, such a small change may not indeed create a miscibility region of over 250 °C. An obvious model to investigate this is given by Flory's equation-of-state theory (*7, 8*).

The theory is based on the assumption that the degrees of freedom of a molecule in a liquid can be separated into internal and external degrees of freedom. This assumption was first made by Prigogine (*22*). The external degrees of freedom per polymer segment are less than those for a similar small molecule. This condition is generally denoted as $3c\,(c \leq 1)$ external degrees of freedom. The internal degrees of freedom depend on intramolecular chemical bond forces. The external degrees of freedom depend only on the intermolecular forces. Each pure component is characterized by three equations-of-state parameters: the characteristic temperature (T^*), pressure (p^*), and specific volume (v^*_{sp}). On the basis of these assumptions, the partition function for a system of N r-mers is:

$$Z(T, V) = Z_{int}(T)\, Z_{ext} \tag{12}$$

The partition function associated with the internal degrees of freedom Z_{int}, is assumed to be density independent. It is unaffected by neighboring segments in the liquid and therefore does not contribute to the PVT equation of state. The external contribution is equal to

$$Z_{ext} = (2\pi mkT/h^2)^{3Ncr/2} Q \tag{13}$$

where

$$Q = Q_{(comb)}\,(4\pi\gamma/3\,(v^{1/3} - v^{*1/3})^{3Ncr} \exp(-E_0/kT) \tag{14}$$

In these equations, m is the mass of one segment, k is Boltzmann's constant, h is Planck's constant, V is the total volume, $v = V/rN$ is the volume per mer, v^* is the hard core volume of the mer, Q is the configurational integral, γ is a geometric factor, and E_0 is the mean intermolecular energy.

The first term on the right of Equation 13 is a kinetic contribution that is omitted in Flory's original derivation. It was added by McMaster (*10*).

The resulting equation of state is

$$\tilde{p}\tilde{v}/\tilde{T} = \tilde{v}^{1/3}/(\tilde{v}^{1/3} - 1) - 1/\tilde{v}\tilde{T} \tag{15}$$

where the tilde (~) represents reduced parameters: $\tilde{p} = p/p^*$; $\tilde{v} = v/v^*$; and $\tilde{T} = T/T^*$. The three characteristic parameters, V^*, p^*, and T^*, can be determined from the experimental values of the thermal expansion coefficient α, the thermal pressure coefficient γ, and the specific volume v_{sp}. With the aid of several combining rules, this theory is easily adapted to mixtures. The mer volumes v^*_1 and v^*_2 are chosen in such a way that they have equal hard core volumes. The total number of volume dependent degrees of freedom is $3\bar{r}Nc$, where

$$c = \sum_{i=1}^{n} \psi_i c_i \tag{16}$$

where ψ_i is the segment fraction of component i. The additional parameter c_{12} introduced by Lin (*23*), which characterizes the deviation from additivity of the number of external degrees of freedom per segment in the mixture, has been neglected here.

The characteristic pressure of the mixture is defined as

$$p^* = \psi_1 p_1^* + \psi_2 p_2^* - \psi_1 \theta_2 X_{12} \tag{17}$$

where θ_2 is the surface fraction of component 2 and X_{12} is the exchange energy parameter for unlike interactions. The characteristic temperature is obtained from

$$1/T^* = (1/p^*)(\psi_1 \tilde{T}_1 p_1^* + \psi_2 \tilde{T}_2 p_2^*) \tag{18}$$

Eichinger and Flory (*24*) have also added an entropic correction parameter Q_{12} to the theory to obtain better agreement between the theory and experimental results. It accounts for the entropy of interaction between unlike segments. This approach is consistent with Guggenheim's contention that intermolecular interactions are not entirely energetic in nature.

By using the above combining rules, the equation of state for the two-component mixture is found to be identical to Equation 15.

The equation for the free energy of mixing for a multicomponent system is

$$\Delta G_M = \Delta H_M - T\Delta S_M \tag{19}$$

$$\begin{aligned}\Delta G_M/kT = {} & \sum_{i=1}^{n} N_i \ln \psi_i + \sum_{i=1}^{n} 3r_i N_i (c_i - c) \ln[(2\pi m_i kT)^{1/2}/h] \\ & + 3 \sum_{i=1}^{n} r_i N_i c_i \ln[(\tilde{v}_i^{1/3} - 1)/(\tilde{v}^{1/3} - 1)] \\ & + \bar{r}Nv^*/kT [\sum_{i=1}^{n} \psi_i \theta_i^* (\tilde{v}_i^{-1} - \tilde{v}^{-1}) \\ & + \sum_{j=2}^{n} \sum_{i=1}^{j-1} \psi_i \theta_j (X_{ij}/\tilde{v} - T\tilde{v}_i Q_{ij})]\end{aligned} \tag{20}$$

The enthalpy of mixing is defined in Equation 10. For a binary system this is

$$\Delta H_M = \bar{r}Nv^*[\psi_1 p_1^* (\tilde{v}_1^{-1} - \tilde{v}^{-1}) + \psi_2 p_2^* (\tilde{v}_2^{-1} - \tilde{v}^{-1}) + \psi_1 \theta_2 X_{12}/\tilde{v}] \tag{21}$$

If the *PVT* data are independent of molar mass then the enthalpy of mixing is also independent of the molar mass of the polymer. Equation 22 is the same for Flory's original formulation without McMaster's modification. The additional term added by McMaster is therefore an entropic contribution.

The chemical potential of each component in a multicomponent system is

$$\Delta\mu_k/kT = (\partial\Delta G_M/\partial N_K) + (\partial\Delta G_M/\partial\tilde{v})(\partial\tilde{v}/\partial N_K) \qquad (22)$$

At low pressures, the last term of Equation 23 is very small and can be neglected. Substitution of Equation 20 into Equation 22 yields

$$\begin{aligned}\Delta\mu_1/kT = {} & \ln\psi_1 + (1 - r_1/r_2)\psi_2 + 3r_1\psi_2(c_1 - c_2)\ln(m_1/m_2)^{1/2} \\ & + 3r_1c_1\ln[(\tilde{v}_1^{1/3} - 1)/(\tilde{v}^{1/3} - 1)] \\ & + r_1v^*/kT[p_1^*(\tilde{v}_1^{-1} - \tilde{v}^{-1}) + \theta_2^2(X_{12}/\tilde{v} - T\tilde{v}_1Q_{12})] \qquad (23)\end{aligned}$$

for a binary system.

The relation between the Flory–Huggins interaction parameter χ_{12} and the exchange energy parameter X_{12} is

$$\begin{aligned}\chi_{12} = {} & 3r_1(c_1 - c_2)\ln(m_1/m_2)^{1/2} + (3r_1c_1/\psi_2^2) \\ & \ln[(\tilde{v}_1^{1/3} - 1)/(\tilde{v}^{1/3} - 1)] + (v^*/kT\,\psi_2^2) \\ & [p_1^*(\tilde{v}_1^{-1} - \tilde{v}^{-1}) + \theta_2^2(X_{12}/\tilde{v} - T\tilde{v}_1Q_{12})] \qquad (24)\end{aligned}$$

The application of the Flory equation-of-state theory requires the knowledge of the specific volume v_{sp}, the coefficient of thermal expansion α, and the thermal pressure coefficient γ. With these parameters, the characteristic parameters v^*, p^*, and T^* can be evaluated by using the following equations:

$$v_{sp}^* = v_{sp}[(1 + T\alpha)/(1 + 4\alpha T/3)]^3 \qquad (25)$$

$$\tilde{v} = v_{sp}/v_{sp}^* \qquad (26)$$

$$\tilde{T} = (\tilde{v}^{1/3} - 1)/\tilde{v}^{4/3} \qquad (27)$$

$$\tilde{T} - T/T^* \qquad (28)$$

$$p^* = \tilde{v}^2T\gamma \qquad (29)$$

$$\tilde{p} = p/p^* \qquad (30)$$

The *PVT* properties of PS have been investigated (*26–33*), but no *PVT* data are available for PoClS. Only the specific volume is available for PoClS (*34*). The equation-of-state parameters used for PS were those determined by Flory (*33*). The coefficient of thermal expansion and the thermal pressure coefficient for PoClS were estimated to be 5–6% different from the values for PS. The coefficient of thermal expansion for PoClS was taken to be 6% larger than that of PS, because a survey of literature values for other polymers showed that dense polymers (those with small specific volumes) tended to have larger coefficients of thermal expansion than less dense polymers (*35–37*). The experimentally determined specific volumes show that PoClS is more dense than PS. The thermal pressure coefficient was taken to be 5% less than the value for PS for the same reason. The *PVT* data and the characteristic parameters for PS and PoClS are shown in Table II.

The last quantity necessary for the full characterization of a mixture of the equation-of-state theory is X_{ij}, the exchange energy parameter. The exchange energy parameter X_{23} is expected to be positive for a nonpolar system (*38*). On the basis of the chemical similarities of the two components, one would expect that the value of X_{23} would also be very small. The small positive value for the enthalpy of mixing determined by Ryan indicates that a very small but positive value for X_{23} is appropriate for this system. This suggestion is further supported by the small positive values calculated for χ_{23}. With these facts in mind, we calculated spinodal curves for various values of X_{23}.

The simulated spinodals for PS of various molecular weights blended with PoClS (100,000) with $X_{23} = 0.0225$ cal/cm^3, $Q_{23} = 0$, and $s_2/s_3 = 1$ are shown in Figure 5. The spinodals are located in the PS-rich region of the phase diagram. This finding is to be expected because the PS is the lower molecular weight component.

As Figure 5 illustrates, the blends with PS of molecular weight 32,000 or greater have an hourglass-shaped spinodal. When the molecular weight of the PS is dropped by only 2000, the blend is now miscible for all compositions between approximately 465 and 530 K. A further reduction of 1000 to

Table II. Pure Component Properties at 200 °C

Property	*PS*	*PoClS*
V_{sp}, cm^3/g	1.033	0.877
$\alpha \times 10^{-4}$, K^{-1}	5.80	6.30
γ, cal/cm^{3}K	0.167	0.157
P^*, cal/cm^3	120	116
T^*, K	8697	8301
V^*_{sp}, cm^3/g	0.839	0.706
M_w	variable	100,000

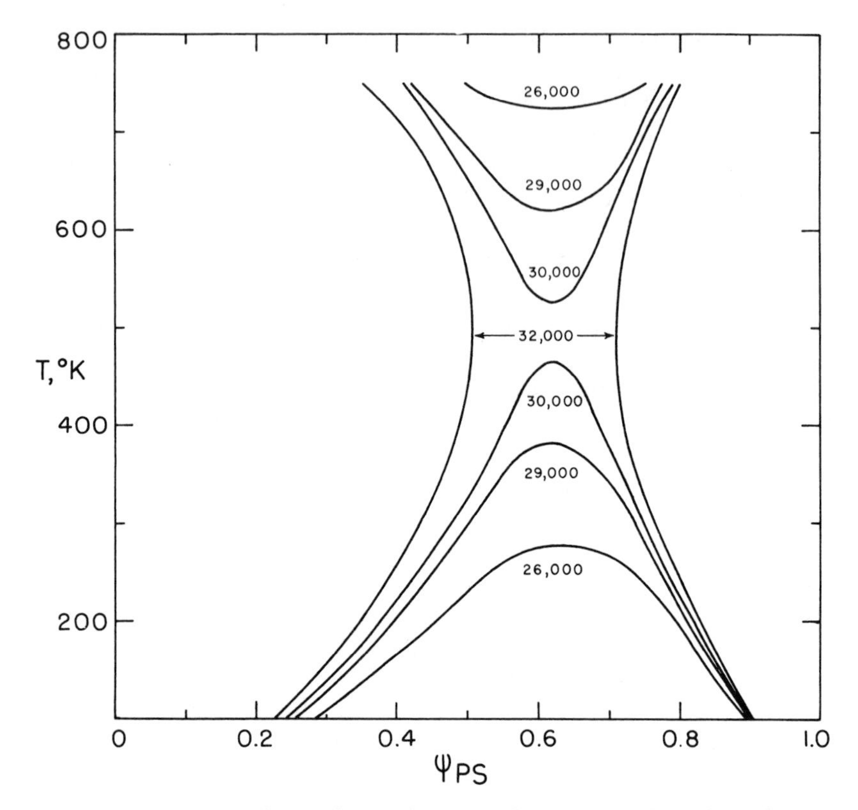

Figure 5. Simulated spinodal for the PS/PoClS system for the indicated molecular weights of PS.

PS 29,000 results in a UCST at 380 K and an LCST at 620 K. The LCST is now above the degradation temperature. Figure 5 demonstrates that under the right circumstances the equation-of-state theory is capable of predicting the type of mixing behavior that has been proposed for PS–PoClS system.

The situation is more clearly illustrated in Figure 6. This figure shows the temperature dependence of χ for the PS–PoClS blend with segment fraction equal to 0.5 and $X_{23} = 0.0225$ cal/cm^3. This curve is equivalent to curve 3 in Figure 4. The χ is independent of molecular weight, if we assume that the equation-of-state parameters are independent of molecular weight. This assumption is considered reasonable for the molecular weight range under consideration here. Therefore, Figure 6 represents the χ vs T curve for all PS–PoClS blends when $X_{23} = 0.0225$. The critical value of χ is very dependent on molecular weight as shown by Equation 8. The horizontal lines in Figure 6 represent the critical value of χ for PS with the indicated molecular weight. The first intersection of a horizontal line with the curve represents the UCST for that molecular weight system. The second intersection represents the LCST. When the molecular weight of PS is

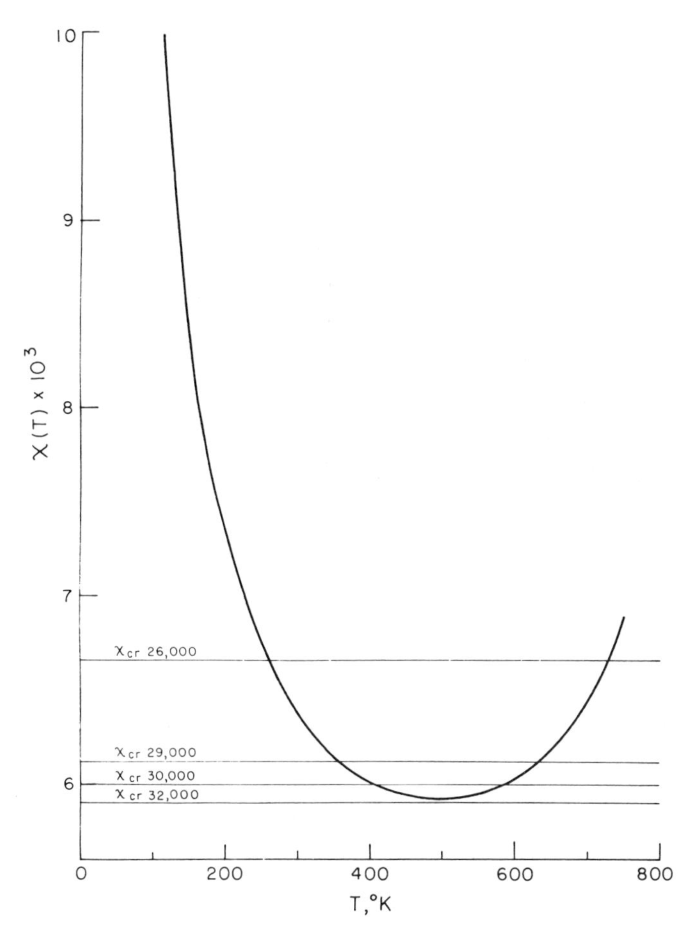

Figure 6. Calculated temperature dependence of χ parameter for the PS/PoClS system. The horizontal lines represent the critical values of the χ parameter for the indicated molecular weights of PS.

32,000, the LCST and UCST merge at about 490 K. The point at which the LCST and UCST merge is called the critical double point. Any system with a PS molecular weight greater than 32,000 has no UCST or LCST; the blend is immiscible. Clearly, the flatness of the χ vs. T curve in the vicinity of the critical double point is responsible for the large temperature change in the critical temperatures with very small changes in molecular weight.

All predictions of the equation-of-state theory presented here for the PS/PoClS blends have been based on estimated equation-of-state parameters for PoClS. These predictions are intended to give a qualitative picture of the mixing behavior of the polymer blend. The numbers are significant only in terms of their relative magnitudes and signs.

In summary, the experimental observations indicate the existence of both a UCST and an LCST in the PS/PoClS system. The equation-of-state theory has been used to show the connection between the occurrence of a critical double point and the enhanced sensitivity of the phase behavior to small changes in molecular weight.

Acknowledgments

The preparative GPC was constructed at the University of Connecticut, Storrs, CT, in the laboratories of Julian Johnson.

Literature Cited

1. Ryan, C. L., Ph.D. Thesis, Univ. of Massachusetts, Amherst, Mass., 1979.
2. Flory, P. J. *J. Chem. Phys.* **1942,** *10*, 51.
3. Huggins, M. L. *Ann N. Y. Acad. Sci.* **1942,** *43*, 1.
4. Shultz, A. R.; Flory, P. J. *J. Am. Chem. Soc.* **1952,** *74*, 4760.
5. Lacombe, R. H.; Sanchez, I. C. *J. Phys. Chem.* **1976,** *80*, 2568.
6. Scott, R. L.; van Konynenburg, P. H. *Discuss. Faraday Soc.* **1970,** *49*, 87.
7. Flory, P. J.; Orwoll, R. A.; Vrij, A. *J. Am. Chem. Soc.* **1964,** *86*, 3507, 3515.
8. Flory, P. J. *J. Am. Chem. Soc.* **1965,** *87*, 1833.
9. Patterson, D.; Robard, A. *Macromolecules* **1978,** *11*, 690.
10. McMaster, L. P. *Macromolecules* **1973,** *6*, 760.
11. Olabisi, O.; Robeson, L. M.; Shaw, M. T. "Polymer–Polymer Miscibility"; Academic Press: New York; 1979.
12. Scott, R. L. *J. Chem. Phys.* **1949,** *17*, 179.
13. Tompa, H. *Trans. Faraday Soc.* **1949,** *45*, 1142.
14. Paul, D. R.; Barlow, J. W. *J. Macromol. Sci.* **1980,** *C18*, (1), 109.
15. Kwei, T. K.; Nishi, T.; Roberts, R. F. *Macromolecules* **1974,** *7*, 667.
16. Olabisi, O. *Macromolecules* **1975,** *9*, 316.
17. Ziska, J. J.; Barlow, J. W.; Paul, D. R. *Polymer* **1981,** *22*, 918.
18. Flory, P. J.; Hocker, H. *Trans. Faraday Soc.* **1971,** *67*, 2258.
19. Patterson, D. *Macromolecules* **1969,** *2*, 672.
20. Patterson, D. *Polym. Eng. Sci.* **1982,** *22*, 64.
21. Prigogine, I.; Defay, R. "Chemical Thermodynamics"; Longmans, Green: London; 1952.
22. Prigogine, I. "The Molecular Theory of Solutions"; North–Holland Pub.: Amsterdam; 1957.
23. Lin, P. H., Ph.D. thesis, Washington University, St. Louis, Mo., 1970.
24. Eichinger, B. E.; Flory, P. J. *Trans. Faraday Soc.* **1968,** *64*, 2053, 2061, 2066.
25. Guggenheim, E. A. *Proc. R. Soc. London, Ser. A* **1932,** *135*, 181.
26. Spencer, R. S.; Gilmore, G. D. *J. Appl. Phys.* **1949,** *20*, 502.
27. Toor, H. L.; Egleton, S. D. *J. Appl. Chem.* **1953,** *3*, 351.
28. Matsuoka, S.; Maxwell, B. *J. Polym. Sci.* **1958,** *32*, 131.
29. Hellwege, K. H.; Knappe, W.; Lehmann, P. *Kolloid-Z.* **1962,** *183* 110.
30. Breuer, H.; Rehage, G. *Kolloid-Z.* **1967,** *216*, 159.
31. Warfield, R. W. *Polym. Eng. Sci.* **1966,** *6*, 176.
32. Quach, A.; Simha, R. *J. Appl. Phys.* **1971,** *42*, 5492.
33. Hocker, H.; Blake, G. J.; Flory, P. J. *Trans. Faraday Soc.* **1971,** *67*, 2251.
34. Kubo, K.; Ogino, K. *Bull. Chem. Soc. Jpn.* **1971,** *44*, 997.
35. Ten Brinke, G.; Eshuis, A.; Roerdink, E.; Challa, G. *Macromolecules* **1981,** *14*, 867.
36. Chang, Y. H.; Bonner, D. C. *J. Appl. Polym. Sci.* **1975,** *19*, 2457.
37. "Polymer Handbook," Second edition; Brandup, J., Immergut, E. H., Eds.; J. Wiley & Sons: New York; 1975.
38. Hocker, H.; Flory, P. J. *Trans. Faraday Soc.* **1971,** *67*, 2270.

RECEIVED for review January 20, 1983. ACCEPTED July 30, 1983.

3

Phase Equilibria in Polymer Melts by Melt Titration

M. T. SHAW and R. H. SOMANI

Department of Chemical Engineering and Institute of Materials Science, University of Connecticut, Storrs, CT 06268

The miscibilities of poly(methyl methacrylate) (PMMA) fractions with poly(styrene) (PS) were measured by melt titration and found to depend on molecular weight, but not as strongly as anticipated from the Flory equation-of-state theory. The observed miscibility ranged from 3.4 to 7.5 ppm as the molecular weight of the PMMA fraction changed from 160,000 to 75,000. Fits of these data by the theory resulted in segmental interaction parameters (X_{12}) for PS and PMMA of 0.11 and 0.20 cal/cm^3, respectively. Efforts to improve the agreement—obtain interaction parameters independent of molecular weight—centered on the correction of density and thermal expansion coefficient for molecular weight, and the introduction of a segmental entropy parameter (Q_{12}). None of these refinements, however, was capable of removing the molecular weight dependence of X_{12}.

THE SCIENCE AND TECHNOLOGY OF POLYMER BLENDS cover a wide variety of topics that may often be usefully classified according to the physical nature of subject blend. Two convenient categories are miscible blends (those with a single glass transition) and immiscible blends (those with two or more phases under normal conditions).

Although the miscible blends often prove to be of great commercial interest, they are relatively rare. For this reason, growing attention has been focused on blends exhibiting limited miscibility. The phase behavior and properties of these blends are often sharply dependent on composition, temperature, and pressure, as well as on the details of the structure of each component. To unify these observations and to provide a sound basis for the design of new blends, scientists have developed a number of theories describing the phase behavior of binary polymer systems. This chapter re-

0065-2393/84/0206-0033$06.00/0

ports on a direct experimental assessment of the ability of one of these theories to handle changes in molecular weight in a mixture of poly(styrene) (PS) and poly(methyl methacrylate) (PMMA), a system of low miscibility under most conditions.

Flory, et al. (*1*, *2*) first described the usefulness of the equation-of-state theory in the analysis of liquid–liquid phase equilibria and the calculation of the thermodynamic functions of polymer mixtures. The Flory equation-of-state theory (*3–8*) has been shown to predict correctly the qualitative nature of the phase diagram of polymer–polymer systems. McMaster (*5*) showed, by using the theory, that the lower critical solution temperature (LCST) should be a common phenomenon in high molecular weight polymer blends. Also, the dependencies of the phase diagram on factors such as molecular weight, thermal expansion coefficients, thermal pressure coefficients, the interaction energy parameter, and pressure were discussed.

The experimental study of phase equilibria of polymer blends is more difficult than that of polymer solutions or solutions of low molecular weight compounds. Consequently it has not been possible to catalog for many polymer–polymer systems the adjustable parameters present in the theory: the segmental interaction energy parameter X_{12} and the interaction entropy parameter Q_{12}. In principle, if one has these parameters at hand, the entire phase diagram and many physical properties of the polymer mixture can be calculated by using Flory's equation-of-state theory.

In Flory's theory, the interaction parameters (X_{12} and Q_{12}) are believed to be characteristic of the polymer pair. These parameters are dependent only on the chemical structure of the components and are independent of the composition, temperature, and molecular weight. Zhikuan et al. (*9*) suggested that both X_{12} and Q_{12} may have some molecular weight dependence. We used the melt titration technique (*10–12*) to determine the miscibility, at constant temperature and pressure, of two monodisperse PMMA samples in PS. The results from the melt titration technique were used to evaluate the interaction energy parameter for these blends. A significant difference was found in the calculated interaction energy parameters for high and low molecular weight polymers. This difference could not be explained by incorporating the effect of molecular weight on physical properties (i.e., thermal expansion coefficient and density) of the polymer (PMMA). By using the average value of the interaction energy parameter (X_{12}), the effect of the segmental interaction entropy parameter (Q_{12}) on the computed phase diagrams was evaluated. These results lead to the conclusion that either the interaction energy parameter or the interaction entropy parameter may have molecular weight dependence.

Theory

The Flory equation-of-state theory is based on an assumption first made by Prigogine (*13*) that the motions of a molecule in a liquid can be separated into fully independent internal and external degrees of freedom. External degrees of freedom attributable to a monomer in a polymer chain are less than those for a similar small molecule. Thus, a monomer will only have $3C$ ($C < 1$) instead of three degrees of freedom. External degrees of freedom are assumed to depend only on intermolecular forces, whereas internal degrees of freedom are associated with intramolecular chemical bond forces. Flory assumed that the intersegmental energy arises from interactions between the surfaces of adjoining segments. On the basis of these assumptions, a configurational partition function for the mixture is obtained and the free energy of mixing is evaluated by using the standard equations of statistical thermodynamics. The expression for chemical potential is then found by differentiating the free energy equation. For component 1 in a binary mixture this expression is given by Equation 1 (*14*).

$$\frac{\Delta\mu_1}{kT} = \ln\phi_1 + (1 - r_1/r_2)\phi_2 + \tfrac{3}{2}r_1c_1\ln\left(\frac{m_1}{m}\right) - \tfrac{3}{2}r_1c\phi_2\left(\frac{m_1 - m_2}{m}\right) + 3r_1c_1\ln\left[\frac{\tilde{V}_1^{1/3} - 1}{\tilde{V}^{1/3} - 1}\right] + \frac{r_1V^*}{kT}\left[P_1^*\left(\frac{1}{\tilde{V}_1} - \frac{1}{\tilde{V}}\right) + \frac{\theta_2^2X_{12}}{\tilde{V}} - \theta_2^2T\tilde{V}_1Q_{12}\right] - \frac{r_1c_1E_1(\phi,\tilde{V})}{\tilde{V}^{2/3}(\tilde{V}^{1/3} - 1)} + \frac{P^*r_1V^*E_1(\phi,\tilde{V})}{kT\tilde{V}^2} \tag{1}$$

where

$$E_k(\phi,\tilde{V}) = \left|\frac{\left(\dfrac{rN_T}{r_k}\right)\dfrac{\partial\tilde{P}}{\partial N_k} - \dfrac{1}{\tilde{T}}\left(\tilde{P} + \dfrac{1}{\tilde{V}^2}\right)\left(\dfrac{rN_T}{r_k}\dfrac{\partial\tilde{T}}{\partial N_k}\right)}{\dfrac{n+1}{\tilde{V}^2} - \dfrac{\tilde{T}(\tilde{V}^{1/3} - \frac{2}{3})}{\tilde{V}^{5/3}(\tilde{V}^{1/3} - 1)}}\right| \tag{2}$$

$$N_T = N_1 + N_2 \tag{3}$$

$$\phi_i = \frac{r_iN_i}{\sum_{i=1}^{2} r_iN_i} \tag{4}$$

$$\phi_i = \frac{S_ir_iN_i}{\sum_{i=1}^{2} S_ir_iN_i} \tag{5}$$

$$m = \phi_1m_1 + \phi_2m_2 \tag{6}$$

$$c = \phi_1c_1 + \phi_2c_2 \tag{7}$$

$$P^* = \phi_1 P_1^* + \phi_2 P_2^* - \phi_1 \theta_2 X_{12} \quad (8)$$

$$V^* = V_1^* = V_2^* \quad (9)$$

In these equations, X_{12} is the interaction energy parameter; Q_{12} is the interaction entropy parameter; ϕ_i is the segmental volume fraction of the component i; S_i is the surface per segment of component i; V^*, T^*, and P^* are the characteristic volume, temperature, and pressure, respectively; and $\tilde{V}$, $\tilde{T}$, and $\tilde{P}$ are the reduced volume, the reduced temperature, and the reduced pressure, respectively.

To evaluate the chemical potential, the characteristic parameters V^*, T^*, and P^* are needed. These parameters can be obtained from the physical properties (density, thermal expansion coefficient, and thermal pressure coefficient) of the pure components by using the relationships

$$V^* = \frac{1}{\rho}\left[\frac{1 + T\alpha}{1 + 4T\alpha/3}\right]^3 \quad (10)$$

$$P^* = \tilde{V}^2 T\gamma \quad (11)$$

$$T^* = \frac{T\tilde{V}^{4/3}}{(\tilde{V}^{1/3} - 1)} \quad (12)$$

where ρ is the density, α is the thermal expansion coefficient, and γ is the thermal pressure coefficient.

Experimental

The melt titration technique, described in detail elsewhere (*10–12*), was used to determine the composition of PMMA in PS at equilibrium. A schematic diagram of the experimental apparatus is shown in Figure 1. The first component, PS, was fed to an extruder used as a mixer. At constant temperature (150 °C) and pressure, the second component, PMMA dissolved in methylene chloride, was added slowly at a constant flow rate. The presence of a second phase (at the cloud point composition or equilibrium composition) was detected by light scattering in a special die and optical system. The equilibrium composition was calculated from the location of the cloud point and the flow rate and concentration of the second component.

One high molecular weight sample of PMMA (M_w 160,000) and one low molecular weight sample (M_w 75,000) were used as second components. Both samples were obtained from Polysciences, Inc. The PS (M_w 520,000) was manufactured by Dow Chemical Co. (trade name Styron).

Results and Discussion

The results from the melt titration technique are listed in Table I. The miscibility of low molecular weight polymer is higher than the miscibility of the high molecular weight polymer, as expected.

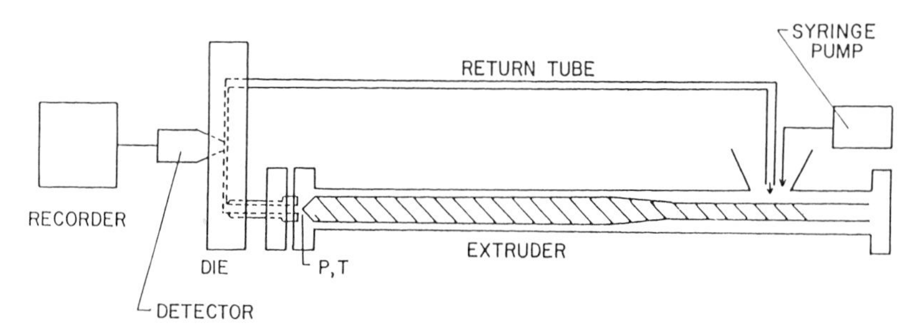

Figure 1. Apparatus for melt titration technique. (Reproduced from Ref. 10. Copyright 1981, American Chemical Society.)

Table I. Melt Titration Results for PS/PMMA System, and Physical Properties of the Two Components

	Physical Properties						Melt Titration Results[a]
Polymer	M_w	M_{mer}	$\alpha \times 10^4$, K^{-1}	γ, atm/K	ρ, g/cm³	Temp., °C	ϕPMMA/PS $\times 10^6$
PS[b]	520,000	104	6.0800	7.23	1.01000	150	—
PMMA[c]	160,000	100	5.0045	7.13	1.13415	150	3.4
PMMA[c]	75,000	100	5.0095	7.13	1.13384	150	7.5

Note: S_1/S_2 = 1.0 (Ratio of surface per segment of the two components.)
[a]Ref. 12.
[b]Ref. 10.
[c]$\alpha \times 10^4 = 5.0 + 714/M_w$. $Vsp = 0.670 + 5.0 \times 10^{-4}\, T + (714 \times 10^{-4}\, T + 4.3)/M_w$ (Ref. 15 and 16)

The Flory equation-of-state theory was used to evaluate the interaction energy parameter from the melt titration results. The physical properties of the two components (the density, the thermal expansion coefficient, and the thermal pressure coefficient) used to calculate the chemical potential are given in Table I. At equilibrium the chemical potential of each component in the two phases is the same. We have two nonlinear algebraic equations that can be solved for the volume fractions of the two components on the two sides of the phase diagram by using a standard nonlinear optimization routine. A set of interaction energy parameter values was assumed and the equations for chemical potential were solved to obtain the equilibrium composition. The values assumed for X_{12} were such that the resultant equilibrium compositions on one side of the phase diagram (that is, ϕ_{2A}) were close to the result from the melt titration. The interaction parameters from several trials were then plotted against logarithms of the calculated volume fractions.

Figure 2 shows the results of these calculations for the PS/PMMA

(160,000) and PS/PMMA (75,000) systems. The relationship between X_{12} and log (ϕ_{2A}) is very linear. This linearity allows the interaction energy parameter for the polymer pair to be easily evaluated by straight line interpolation corresponding to the equilibrium composition (ϕ_{2A}) obtained from the melt titration technique. From such an analysis, X_{12} is equal to 0.20 cal/cm^3 for the PS/PMMA (75,000) system and 0.11 cal/cm^3 for the PS/PMMA (160,000) system.

McMaster has shown (5) that small differences in physical properties of the polymer components can have significant effects on the equilibrium compositions. To understand the difference in the X_{12} for the two polymer–polymer systems, the effects of molecular weight on the density and the thermal expansion coefficient for PMMA were incorporated in the calculation of equilibrium compositions. The interaction parameters were then reevaluated by using the same procedure.

The results of this exercise were that the incorporation of molecular weight effects on density and expansivity can account for only a 5% change in X_{12}, whereas the experimental difference is about 90%. The reason for this small change is that both PMMA samples have fairly large molecular weight and the difference in their physical properties is very small (Table I). The correction of the physical properties for molecular weight does pro-

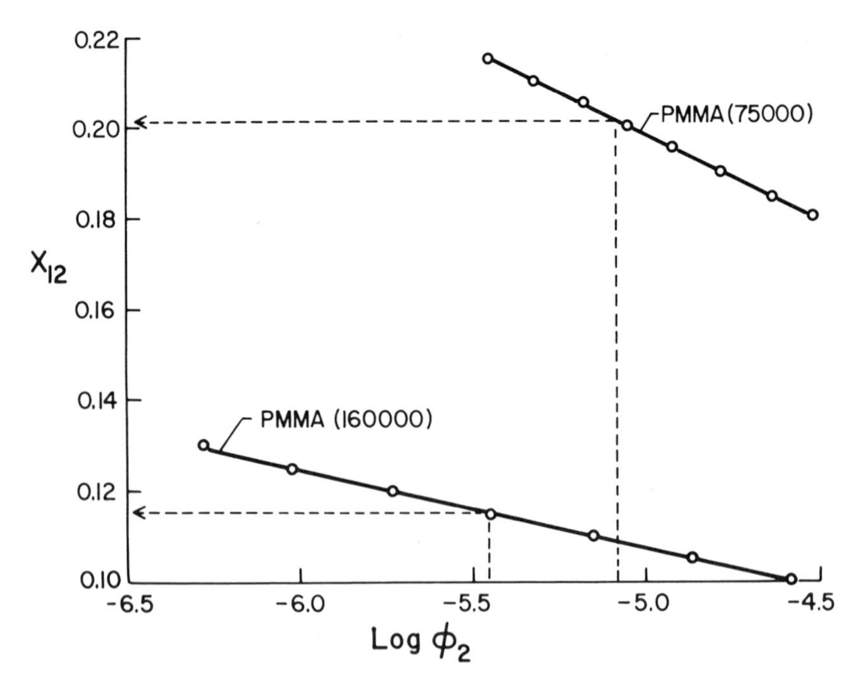

Figure 2. Interaction energy parameter for PS/PMMA (160,000) and PS/PMMA (75,000) systems at 423 K.

duce a quantitatively correct change; i.e., the molecular weight dependence of the interaction parameter was reduced.

The equilibrium composition of each component on the two sides of the phase diagram is also very sensitive to the interaction entropy parameter, Q_{12}. The interaction entropy parameter takes into account the entropic contribution to the chemical potential other than the combinatorial term. To evaluate the effect, a value for Q_{12} was assumed, and the corresponding equilibrium compositions were determined at various temperatures. These calculations resulted in a phase diagram (binodals) corresponding to each Q_{12}.

Figures 3 and 4 show the phase diagrams for PS/PMMA (160,000) and PS/PMMA (75,000), respectively. Also shown is the equilibrium composition obtained from the melt titration technique. If the experimental data are fit to the predicted binodal curves, a positive value of Q_{12} (between 0 and 0.0001 cal/cm^3 K) must be assumed for PS/PMMA (160,000) system. On the other hand a negative value of Q_{12} (between 0 and -0.0001 cal/cm^3 K) must be assigned to the PS/PMMA (75,000) system. Thus, at least at the low equilibrium concentrations characteristic of the PS/PMMA system, the interaction parameters must depend on molecular weight.

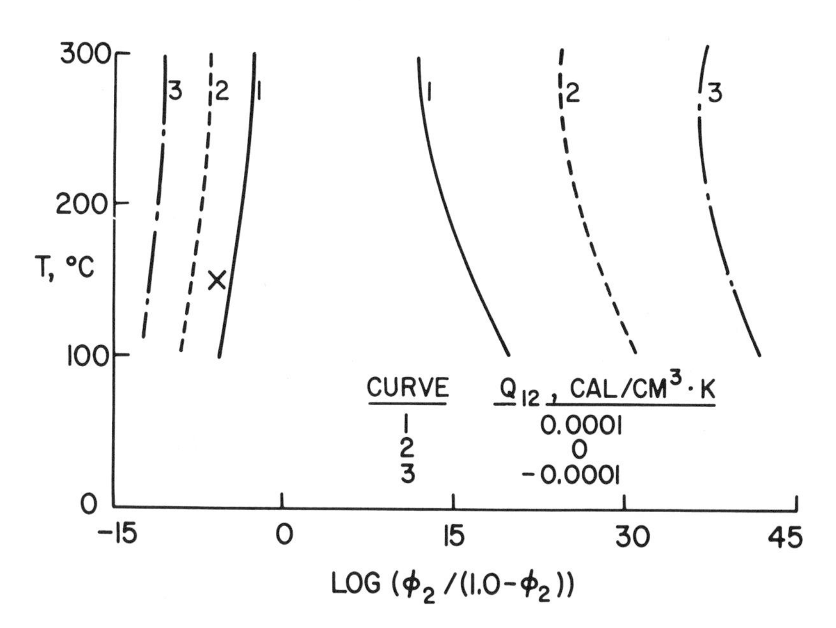

Figure 3. Calculated phase diagrams for PS/PMMA (160,000) system and experimentally observed cloud point (×) X_{12} = 0.1591. Log scale is natural log.

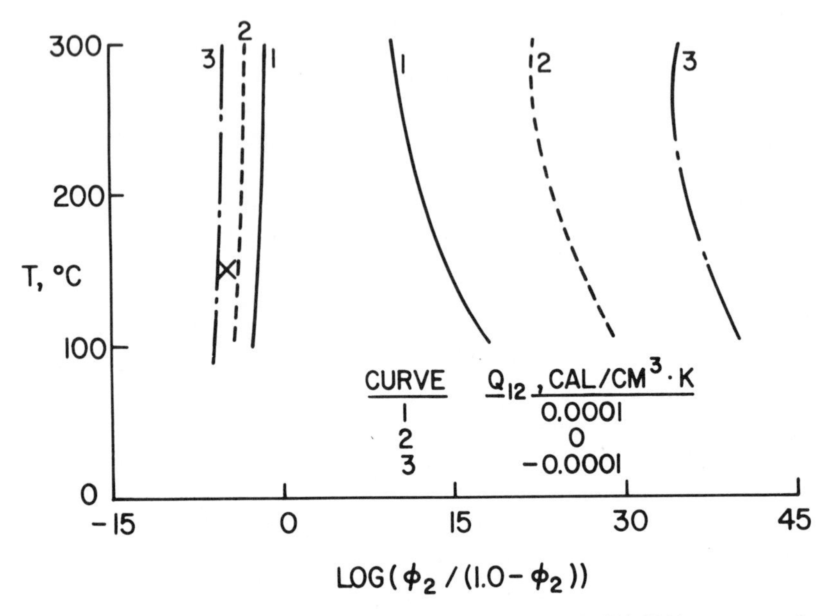

Figure 4. Calculated phase diagrams for PS/PMMA (75,000) system and experimentally observed cloud point (×); X_{12} = 0.1591. Log scale is natural log.

Two points concerning the experiment should be added as caveats: Although the melt titration experiment can handle systems of low miscibility, the linear mixing rules required for any equation-of-state theory may not be appropriate at the very low concentration found in this study. Second, concerning the solvent, a third component that may influence the phase behavior of the system (*17*), several tests have shown that the amount is very low and will not concentrate in the second phase; however, exact solvent effects are not known.

Conclusion

The analysis of the results obtained from the melt titration technique by using Flory's equation-of-state theory gives two significantly different interaction energy parameters (X_{12}) and interaction entropy parameters (Q_{12}) for the PS/PMMA (160,000) and PS/PMMA (75,000) polymer–polymer systems. The calculated difference in X_{12} could not be explained by incorporating the effect of molecular weight on the physical properties of the second component (PMMA). Therefore, we conclude that either or both X_{12} and Q_{12} have some molecular weight dependence. A similar analysis for the system PE/PS is in progress.

Nomenclature

T_i^* Characteristic temperature for component i
P_i^* Characteristic pressure for component i
V_i^* Characteristic volume for component i
$\tilde{P}_i$ Reduced pressure for component i
$\tilde{V}_i$ Reduced volume for component i
$\tilde{P}$ Reduced pressure for mixture
$\tilde{V}$ Reduced volume for mixture
V^* Characteristic volume for mixture
P^* Characteristic pressure for mixture
$\tilde{T}$ Reduced temperature for mixture
N_i Number of molecules of component i
N_T Total number of molecules in mixture
S_i Surface area per segment for component i
θ_i Segment surface area fraction for component i
r_i Number of segments of component i
ϕ_i Volume fraction of component i
C_i Degrees of freedom for component i
C Degrees of freedom for mixture
m_i Mass-per-segment for component i
m Mass-per-segment for mixture
X_{12} Interaction energy parameter
Q_{12} Interaction entropy parameter
$\Delta\mu_i$ Chemical potential for component i
K Boltzman constant
T Absolute temperature
ρ Density
α Thermal expansion coefficient
γ Thermal pressure coefficient

Acknowledgments

This chapter is based upon work supported by the National Science Foundation under Grant DMR 78–18078, Polymers Program.

Literature Cited

1. Abe, A.; Flory, P. J. *J. Am. Chem. Soc.* **1966**, *88*, 2887.
2. Flory, P. J.; Eichlinger, B. E.; Orwoll, R. A. *Macromolecules* **1968**, *1*, 287.
3. Shiomi, T.; Izumi, Z.; Hamada, F.; Nakajima, A. *Macromolecules* **1980**, *13*, 1149.
4. ten Brinke, G.; Eshuis, A.; Roerdink, E.; Challa, G. *Macromolecules* **1981**, *14*, 867.
5. McMaster, L. P. *Macromolecules* **1973**, *6*, 760.

6. Hamada, F.; Chiomi, T.; Fujisawa, K.; Nakajima, A. *Macromolecules* **1980**, *13*, 729.
7. Sanchez, I. C. In "Polymer Blends"; Paul, D. R.; Newman, S., Eds.; Academic Press: New York, 1978; Vol. 1, Chapter 3.
8. Olabisi, O.; Robeson, L. M.; Shaw, M. T. "Polymer–Polymer Miscibility"; Academic Press: New York, 1979.
9. Zhikuan, C.; Ruona, S.; Walsh, D. J.; Higgins, J. S. *Polymer*, in press.
10. Somani, R. H.; Shaw, M. T. *Macromolecules* **1981**, *14*, 886.
11. Anavi, S.; Shaw, M. T.; Johnson, J. F. *Macromolecules* **1979**, *12*, 1227.
12. Somani, R. H. M. S. Thesis, University of Connecticut, Storrs, Conn., 1981.
13. Prigogine, I. "The Molecular Theory of Solutions"; North–Holland Publishing Co.: Amsterdam, 1957.
14. Somani, R. H.; Shaw, M. T., presented at the 1982 NATEC, Society of Plastics Engineers, Bal Harbour, Fla., October 25–27, 1982.
15. Pezzin, G.; Zillo-Grandi, F.; Sanmartin, P. *Eur. Polym. J.* **1970**, *6*, 1053.
16. Martin, J. R.; Johnson, J. F.; Cooper, A. R. *J. Macromol. Sci., Rev. Macromol. Chem.* **1972**, *C-8 (1)*, 175.
17. Bank, M.; Leffingwell, J.; and Thies, C. *Macromolecules* **1971**, *4*, 43.

RECEIVED for review December 28, 1982. ACCEPTED July 15, 1983.

4

A Comparison of Miscible Blend Binary Interaction Parameters Measured by Different Methods

J. E. HARRIS, D. R. PAUL, and J. W. BARLOW

Department of Chemical Engineering and Center for Polymer Research, The University of Texas, Austin, TX 78712

Miscible blends of the polyhydroxy ether of bisphenol-A (Phenoxy) with a series of aliphatic polyesters were studied by melting point depression analysis and sorption to obtain the Flory–Huggins polymer–polymer interaction parameter, B. The B values obtained from these measurements agreed well in sign, magnitude, and variation with ester repeat structure. These values also agreed with B values measured calorimetrically for mixtures of low molecular weight compounds with structures that are analogous to those of the polymers. These comparisons suggest that the same mechanisms are responsible for the exothermic heats of mixing measured directly for the analog compounds and indirectly for the miscible polymer blends. For this general system, we suggest that hydrogen bond formation between the hydroxyl group on Phenoxy and the ester moiety is probably responsible for the exothermic interactions and polymer blend miscibility observed.

A PRIORI PREDICTIONS OF WHICH POLYMER PAIRS will form miscible binary mixtures must necessarily involve consideration of component molecular structure as it influences the thermodynamics of mixing and phase behavior, and of means to quantify the relationship between structure and intermolecular interactions, so that calculations are possible. These efforts are hampered by a lack of good bases for mathematically describing specific interactions even in mixtures of low molecular weight materials. Another difficulty is a similar lack of quantitative measurements, primarily result-

0065-2393/84/0206-0043$06.00/0

ing from the poor theoretical bases and the inherent difficulties associated with direct measurements of excess thermodynamic functions (such as the heat of mixing), for high polymer mixtures.

Despite these difficulties, progress can be made, by a combination of experimentation and thermodynamic reasoning, toward the initial goal of characterizing the intermolecular interactions in a thermodynamic context. These interactions can then be related in a quantitative way to the structural features of the polymer molecules. For example, Cruz et al. (*1*) demonstrated a strong correlation between the ability of two polymers to form a miscible binary and the ability of low molecular weight analogs of the polymers to show exothermic heats of mixing. Similarly, the presence of exothermic heat of mixing has been demonstrated for several systems within the context of the Flory–Huggins equations for mixing as applied to the observed depression in the melting point of a crystallizable component in a miscible blend with increasing amorphous component content (*2–5*). These results have led to the conclusions that polymer pairs are miscible if they show exothermic heats of mixing and that entropic contributions to the free energy of mixing can be safely ignored (*6*, *7*).

Despite these conceptual successes, no quantitative comparison between the magnitudes of heats of mixing observed for analog compounds and those observed from melting point depression of the crystallizable component in the blend has been obtained in these prior works for a variety of reasons. Nor has the assumption that the crystals in the blend are thick enough to avoid melting point depression for morphological reasons (*8*) been thoroughly tested by obtaining the excess heat by some other experiment.

This chapter presents a thorough study of interaction parameters obtained for several miscible blends of polyesters with the polyhydroxy ether of bisphenol-A (Phenoxy) (*5*) by melting point depression analysis, by a newly developed sorption technique (*9*), and by calorimetric measurements of analog compounds. This comparison will demonstrate that, within experimental error, all three methods give consistent magnitudes for the interaction parameters. Also, the parameter varies with magnitude in a smooth consistent manner with the polyester chemical structure. This variation suggests an optimum density of carbonyl species on the polyester that maximizes miscibility with Phenoxy.

Materials

Six different aliphatic polyesters were found to be miscible with Phenoxy by Harris et al. (*5*). The properties of these materials, their abbreviated nomenclature, and their sources of supply are given in Table I. Of these six, only poly(ϵ-caprolactone) (PCL), poly(1,4-butylene adipate) (PBA), and poly(ethylene adipate) (PEA) possessed sufficiently rapid crystallization rates to allow melting point depression analyses to be carried out. Similarly, only four materials, PEA, PBA, PCL, and poly(1,4-cyclohexane-dimethanol succinate) (PCDS), showed sufficient thermal stability to allow sorption measurements to be carried out above their melting points.

The structures and sources of the low molecular weight materials used as analog compounds of the polymers in the calorimetric determination of heats of mixing are shown in Table II.

Analog Calorimetry

An adiabatic calorimeter, Figure 1, was employed to measure the heats of mixing for low molecular weight compounds with structures similar to those of the polymeric materials. This device was operated at 100 °C for all work. Its description and operation are thoroughly discussed elsewhere (*1*). Some insight into the mechanism for Phenoxy/polyester miscibility is suggested by the comparison of the magnitudes of the heats of mixing of the analog materials shown in Figure 2. Diphenoxypropanol (DPP) has a chemical structure that closely simulates the Phenoxy repeat unit. The heat of mixing of DPP with the diethyl adipate analog of a polyester is significantly more exothermic than that of diphenoxyethane (DPE). This result suggests that the formation of a hydrogen bond between the hydroxyl group on the Phenoxy repeat unit and the ester moiety on the polyester is responsible for the observed miscibility of Phenoxy with a variety of aliphatic polyesters.

Phenoxy is not miscible with all aliphatic polyesters. Therefore one expects that the sign and magnitude of the excess heat of mixing must vary in a logical way with the composition of the polyester. To the extent that miscibility of the polymers is dependent only on enthalpic interactions and to the extent that DPP is a reasonable analog for Phenoxy, the excess heat of mixing of DPP with various esters should vary with the carbonyl content of the ester. Some results are shown in Table III. Comparison of these excess heats with the ester structures in Table II leads to the conclusion that an optimum concentration of carbonyl content in the ester exists for maximum interaction with DPP.

A convenient, although empirical, way to express the ester composition dependence on the heat of mixing with DPP is to first characterize the heat of mixing per unit volume through a binary interaction parameter, B, via

$$\Delta H_m = \phi_1(1 - \phi_1)B \tag{1}$$

where ϕ_1 is the volume fraction of DPP, and then to watch the variation of B with the content of carbonyl or aliphatic groups in the ester. This method is illustrated by the solid line in Figure 6 where a smooth trend exists with a maximum exothermic interaction occurring for linear esters containing a volume fraction of about 0.20 carbonyl ester moieties. This method of

Table I. Properties of Polymers Used in This Study

Polymer	*Structure*	*Abbreviation*	*Amorphous Density at 25 °C (g/cc)*	*Intrinsic Viscosity (dL/g)*	*Molecular Wt.*
Polyhydroxyether of bisphenol-A[a]	$-C_6H_5C(CH_3)_2-C_6H_5OCH_2CH(OH)CH_2O-$	Phenoxy	1.20	0.29[b]	$\bar{M}_n = 23{,}000$ $\bar{M}_w = 80{,}000$
Poly(1,4-butylene adipate)[c]	$-(CH_2)_4OC(=O)(CH_2)_4C(=O)O-$	PBA	1.13	0.157[d]	—
Poly(ethylene adipate)[c]	$-(CH_2)_2OC(=O)(CH_2)_4C(=O)O-$	PEA	1.21	0.110[d]	—
Poly(2,2-dimethyl-1,3-propylene succinate)[c]	$-CH_2C(CH_3)_2CH_2OC(=O)(CH_2)_2C(=O)O-$	PDPS	1.17	0.080[d]	—
Poly(2,2-dimethyl-1,3-propylene adipate)[e]	$-CH_2C(CH_3)_2CH_2OC(=O)(CH_2)_4C(=O)O-$	PDPA	—	0.084[d]	—

Poly(1,4-cyclohexanedimethanol succinate)[c]	$-CH_2-\langle S \rangle-CH_2OC(=O)(CH_2)_2C(=O)O-$	PCDS	1.16	0.342[d]	—
Poly(hexamethylene sebacate)[c]	$-(CH_2)_6OC(=O)(CH_2)_8C(=O)O-$	PHS	1.03	0.510[d]	$\bar{M}_w = 16{,}500$
Poly(ethylene succinate)[c]	$-(CH_2)_2OC(=O)(CH_2)_2C(=O)O-$	PES	1.32	0.111[f]	—
Poly(ε-caprolactone)[g]	$-(CH_2)_5OC(=O)-$	PCL	1.095	0.67[h]	$\bar{M}_n = 15{,}500$ $\bar{M}_w = 40{,}500$ $\bar{M}_v = 46{,}700$
Poly(pivalolactone)[i]	$-CH_2C(CH_3)_2OC(=O)-$	PPL	—	—	—

[a] Source: Union Carbide PKHH.
[b] In 2-butanone at 25 °C.
[c] Source: Scientific Polymer Products.
[d] In benzene at 25 °C.
[e] Source: Hooker Chemicals, Rucoflex Polyester S–1016–55.
[f] In chloroform at 25 °C.
[g] Source: Union Carbide, PCL 700.
[h] In benzene at 30 °C.
[i] Source: Polysciences.

Table II. Model Compound Information

Designation	*Structure*
PP[a]	$CH_3(CH_2)_2C(=O)OCH_2CH_3$
PB[a]	$CH_3(CH_2)_2C(=O)O(CH_2)_2CH_3$
BB[a]	$CH_3(CH_2)_3C(=O)O(CH_2)_2CH_3$
DMS[a]	$CH_3C(=O)O(CH_2)_2C(=O)OCH_3$
DES[a]	$CH_3CH_2C(=O)O(CH_2)_2C(=O)OCH_2CH_3$
DBS[a]	$CH_3(CH_2)_3C(=O)O(CH_2)_2C(=O)O(CH_2)_3CH_3$
DEA[a]	$CH_3CH_2C(=O)O(CH_2)_4C(=O)OCH_2CH_3$
DESb[b]	$CH_3CH_2C(=O)O(CH_2)_8C(=O)OCH_2CH_3$
EPV[a]	$C(CH_3)_3C(=O)OCH_2CH_3$
DPP[c] (MP, 81–2 °C)	$C_6H_5OCH_2CH(OH)—CH_2OC_6H_5$
DPE[c] (MP, 95–6 °C)	$C_6H_5OCH_2CH_2OC_6H_5$

[a]Source: Pfaltz and Bauer.
[b]Source: Eastman Organic Chemicals.
[c]Source: Aldrich Chemical Company.

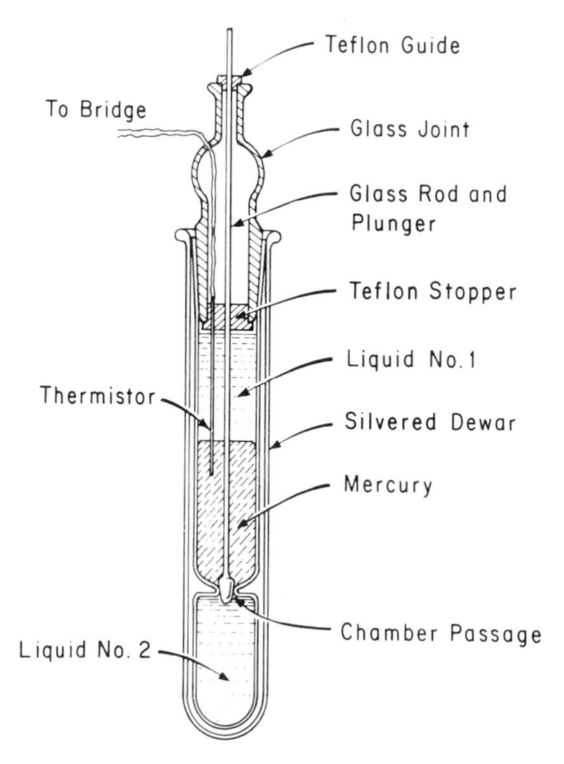

Figure 1. Schematic of solution calorimeter. (Reproduced from Ref. 1. *Copyright 1979, American Chemical Society.)*

characterizing the carbonyl ester content in the ester differs somewhat from previous work (*1*), where the content of the ester was simply expressed by the number fraction of aliphatic carbons to ester groups. It has the advantage, however, of more properly accounting for the differences in carbonyl moiety density between carbonyls on polymer molecules (no chain ends) and carbonyls on the low molecular weight esters.

Melting Point Depression Analysis

The melting point depression of the crystallizable component in a miscible blend can be used to estimate the interaction parameter B between the blend components (*5*, *10*, *11*) via

$$\frac{-B}{RT_{m2}}\phi_1^2 = \left(\frac{1}{T_{m2}} - \frac{1}{T_{m2}^{\circ}}\right)\frac{\Delta H_{2u}}{RV_{2u}} + \frac{\ln \phi_2}{V_2} + \phi_1\left(\frac{1}{V_2} - \frac{1}{V_1}\right) \quad (2)$$

where component 2 is the crystallizable component, the polyester; T_{m2}° is its equilibrium melting temperature; $\Delta H_{2u}/V_{2u}$ is its heat of fusion per unit

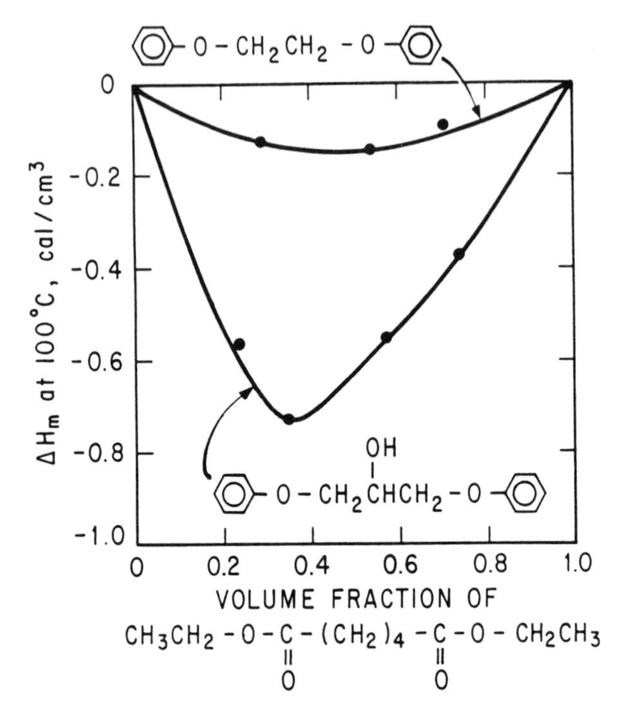

Figure 2. Heats of mixing DEA with DPE and with DPP at 100 °C.

volume of repeat unit for 100% crystalline material; V_2 is its molar volume; and ϕ_2 is its volume fraction in the blend. For miscible blends of high molecular weight materials, V_1 and V_2 are large and Equation 2 can be rearranged to

$$T^{\circ}_{m2} - T_{m2} = \frac{-BV_{2u}}{H_{2u}} T^{\circ}_{m2} \phi_1^2 \tag{3}$$

from which B can be directly evaluated by plotting $T^{\circ}_{m2} - T_{m2}$ vs ϕ_1^2. Blends of low molecular weight materials can be similarly analyzed, if the molar volumes are known, by plotting the right-hand side of Equation 2 versus ϕ_1^2. Figure 3 shows typical constructions that seem to follow Equation 3 insofar as a linear dependence is observed. Unfortunately, the molecular weights of the PBA and PEA used in this study are unknown. However, a rough estimate of these values made from intrinsic viscosities (Table I) suggests that they lie in the 2000–10,000 range. Table IV shows the B values computed from the observed depression in melting point, the estimates of molecular weight, and literature values of $\Delta H_{2u}/V_{2u}$ via Equation 2. As shown here, the B values computed by using the lower estimate of molecular weight are roughly 10% lower than those calculated by assuming the higher molecular weight. This result is a natural consequence of the greater weighting given to

Table III. Heat of Mixing of DPP With Esters at 100 °C

Compound	*Volume Fraction DPP*	*Heat of Mixing (cal/mL)*
DMS	0.30	−0.31
	0.45	−0.36
	0.65	−0.34
DES	0.21	−0.30
	0.40	−0.36
	0.46	−0.40
	0.75	−0.39
DEA	0.26	−0.37
	0.43	−0.55
	0.57	−0.76
	0.65	−0.73
	0.76	−0.56
DBS	0.44	−0.25
	0.56	−0.28
	0.70	−0.19
DESb	0.34	−0.44
	0.47	−0.52
	0.59	−0.46
PP	0.35	−0.22
	0.49	−0.30
	0.60	−0.32
PB	0.31	−0.04
	0.49	−0.05
	0.66	−0.04
BB	0.31	+0.08
	0.44	+0.13
	0.69	+0.08
EPV	0.45	+0.32
	0.48	+0.25

entropic contributions in Equation 2 because of the lower molecular weight. At any rate, the B values do not change substantially with molecular weights greater than 2000. The primary contribution to the melting point depression is a result of enthalpic interactions that lead to negative B values.

These B values are also shown in Figure 6 to vary with the carbonyl content of the polyester repeat structure in a manner that is similar in both trend and magnitude to the B values obtained from heats of mixing data of the analog compounds. This comparison suggests that DPP is a reasonable analog compound for Phenoxy and that the mechanisms responsible for exothermic heats of mixing of analog compounds are also responsible for the melting point depression observed in miscible polymer blends that have comparable structures. The additional observation that the B values from melting point depression studies are somewhat more negative than those from analog calorimetry may be indicative of the additional melting point depression that results from a reduction in crystalline lamella thickness as suggested by Hoffman and Weeks (8). This may also simply be a result of the inaccuracies in the measurements of melting point depressions and in the $\Delta H_{2u}/V_{2u}$ values used.

Sorption Measurements

In principle, many of the difficulties inherent in the melting point depression method for obtaining polymer–polymer interaction parameters could be eliminated by following, instead, the deviations from tie line additivity of the equilibrium sorption of small concentrations of solvent probe mole-

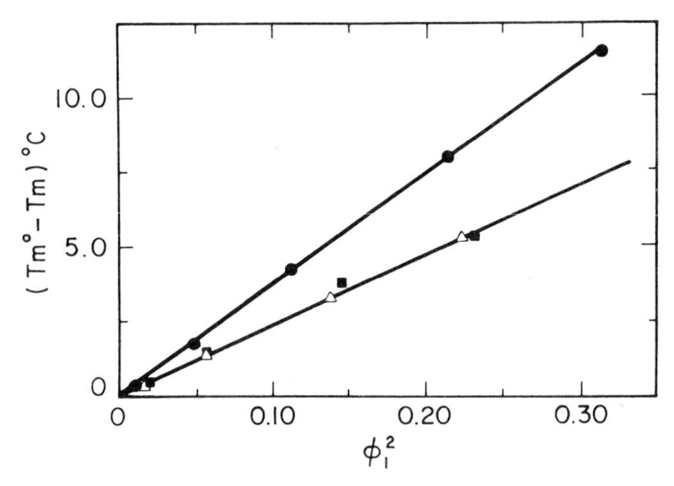

Figure 3. Melting point depressions of PBA (●), PEA (Δ), and PCL (■) in Phenoxy. (Reproduced with permission from Ref. 5. Copyright 1982, J. Appl. Polym. Sci.)

Table IV. Estimated *B* Values from Melting Point Depression Analysis

System	T°_{m2} (°C)	$\Delta H_{2u}/V_{2u}$ (cal/mL)	$\frac{BT^{\circ}_{m2}V_{2u}}{\Delta H_{2u}}$ (°C)	B at T°_{m2} (cal/mL)
PBA/Phenoxy	61	34.5	– 37.5[a]	– 3.87[a]
			– 35.0[b]	– 3.6[b]
PEA/Phenoxy	49	31.0	– 24.0[a]	– 2.31[a]
			– 21.2[b]	– 2.04[b]
PCL/Phenoxy	56	35.0	– 22.7	– 2.41

[a]Molecular weight of polyester is assumed to be infinite.
[b]Molecular weight of polyester is assumed to be 2000.

cules with blend composition. The governing equation for this method, via the Flory–Huggins analysis (*12*), is

$$\ln a_1 = \ln \phi_1 + (1 - \phi_1) + (1 - \phi_1)^2 \chi_{1b} \tag{4}$$

$$\chi_{1b} = \chi_{12}\phi_2' + \chi_{13}\phi_3' - \chi_{23}\phi_2'\phi_3' \tag{5}$$

where a_1 is the activity of the solvent vapor in equilibrium with the polymer (a quantity that is simply the ratio of the vapor partial pressure to that at saturation if the vapor is ideal); ϕ_1 is the volume fraction of solvent sorbed in the polymer blend, ϕ_2' and ϕ_3' are the volume fractions of polymer 2 and polymer 3 on a solvent free basis in the blend, respectively; and the parameters χ_{12}, χ_{13}, and χ_{23} are the interaction parameters associated with solvent 1 and polymer 2, solvent 1 and polymer 3, and polymer 2 and polymer 3 interactions, respectively. Of particular interest, the polymer–polymer interaction parameter, χ_{23}, is related to B via

$$B = \chi_{23} RT / V_1 \tag{6}$$

where V_1 is the molar volume of the sorbed solvent and T is the absolute temperature.

Equation 4 suggests that the interaction parameter contributions will be most important when the volume fraction of solvent in the polymer, ϕ_1, is very small or when sorption is carried out in the Henry's law region to yield

$$P_1/P_1^\circ = a_1 = \phi_1 \exp(1 + \chi_{1b}) \tag{7}$$

At these operating conditions, the basic problem with conventional sorption measurements using gravimetric techniques becomes clear. Relatively large quantities of polymer must be used to accurately measure the small increases in mass by sorption. The large polymer quantities lead to extraordinarily long equilibrium times because of the poor diffusivity of solvents in most polymers (*13–15*).

The solution to this problem is to use a vibrating piezoelectric crystal as a microbalance (*16–18*). A very thin, typically 1-μm polymer film is deposited from solution on the surfaces of a 7-MHz piezoelectric crystal and carefully dried. The presence of the film increases the mass of the crystal by about 0.1 mg and decreases its resonant frequency by about 12 kHz. Because the frequency can be easily measured to within $\pm$ 1 Hz, the crystal microbalance is theoretically capable of detecting a weight fraction pickup of solvent vapor by the film as small as 8×10^{-5}. More typically however, uncertainties in frequency change caused by variations in temperature, pressure, coating uniformity and characteristics, and electronic drift (*9*) limit the absolute reproducibility of solvent weight fraction measurements to about $\pm 3 \times 10^{-4}$ for weight fractions of solvent in the 6×10^{-4}–$2 \times$

10^{-2} range. More detail concerning the entire experimental system will be published in the near future. Extreme care in all aspects of the measurement technique is required to achieve good reproducibility. If that care is taken, construction of an entire sorption isotherm in one afternoon is possible because of the rapidity with which equilibrium is reached.

Very little good quality data exists with which to compare the results of this method, especially in the region corresponding to less than 5% solvent uptake. Some idea of the quality of results can be obtained by examining Figure 4. The interaction parameters for polyisobutylene with two solvents, calculated by using Equation 4 on a point by point basis, are examined and compared with the work of Bonner and Prausnitz (*14*), who used a quartz spring balance. Compared with their work, the piezoelectric microbalance method gives much more consistent values of the polymer–solvent interaction parameter, χ_{12}. Multiple measurements for χ_{12} generally replicate within less than ± 0.05 units as compared to severe scatter when the quartz spring method is employed. Interestingly, at very low values of solvent uptake the χ_{12} values appear to become less positive. With reference to either Equation 4 or Equation 7, this trend suggests that more solvent is being sorbed at very low levels of sorption than can be adequately predicted by the Flory–Huggins equation. Hill and Rowen (*19, 20*) and Bonner (*14*) developed theories based on combined localized Langmuir sorption and bulk absorption to account for this behavior, which is present

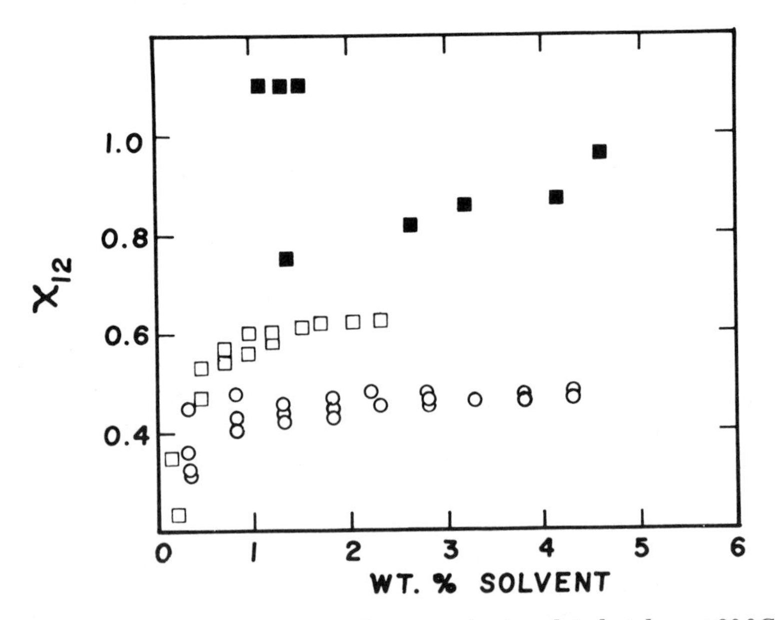

Figure 4. Interaction parameters from sorption in polyisobutylene at 80 °C of benzene (■, □) and of carbon tetrachloride (○). Closed symbols are results of Bonner and Prausnitz (14).

in many polymeric systems. Of particular interest to this work is the observation that χ_{12} is nearly independent of solvent concentration for solvent uptakes greater than 1%. This is presumably the correct χ_{12} and reflects the dominance of bulk sorption (*12*).

Figure 5 summarizes values of χ_{1b} measured for Phenoxy–PBA blends and pure components with two different solvents at 115 °C, a temperature above the melting temperature of the PBA. Maximum solvent uptakes at all compositions were less than 7% at this temperature. This uptake corresponds to maximum solvent partial pressures of less than 750 mm Hg, and the χ_{1b} values were found to be independent of solvent uptake for uptakes greater than 1%. Of particular interest, here, is the positive deviation in χ_{1b} from the tie line constructed through the pure component end points. This deviation suggests, by reference to Equation 5, that the Phenoxy/PBA interaction parameter, χ_{23}, is negative, and its magnitude can be obtained as suggested in Figure 5 by constructing the best fit parabola through the data according to Equation 5. This procedure was used to obtain all of the χ_{23} parameters for the miscible polyester/Phenoxy blends studied, and these results using the alternate form, Equation 6, are presented in Figure 6. Interestingly, the maximum deviations from the tie lines in Figure 5 appear to be independent of the solvent probe used. This result is expected,

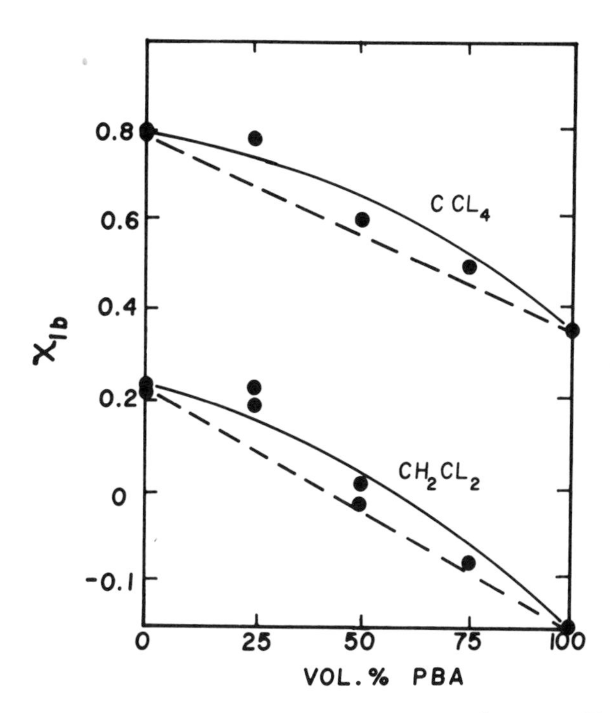

Figure 5. Interaction parameters from sorption of carbon tetrachloride and dichloromethane in Phenoxy–PBA blends at 115 °C.

intuitively and on the basis of Equation 4. It does contrast with previous work using gas chromatography (*21*), however, and additional work will be necessary to adequately resolve this issue.

Summary

Figure 6 summarizes the comparison of polymer–polymer interaction parameters, *B*, and their variation with aliphatic carbon content in the ester structural unit. Within the limits of error, all three methods—analog calorimetry, melting point depression, and sorption—give substantially the same values for *B* despite the fact that these measurements were made at different temperatures ranging from 100 °C, for calorimetry, to 115 °C, for sorption, and at the melting temperatures of the various polyesters, 49–61 °C. This observation suggests that the interaction parameter, *B*, is not strongly temperature dependent; a situation that is consistent with the idea that *B* is negative as the result of hydrogen bond formation and with the idea that *B* is primarily an enthalpic parameter with little or no entropic contribution. The consistency of observations also suggests that a Van Laar type heat of mixing expression, Equation 1, used in the calorimetry experi-

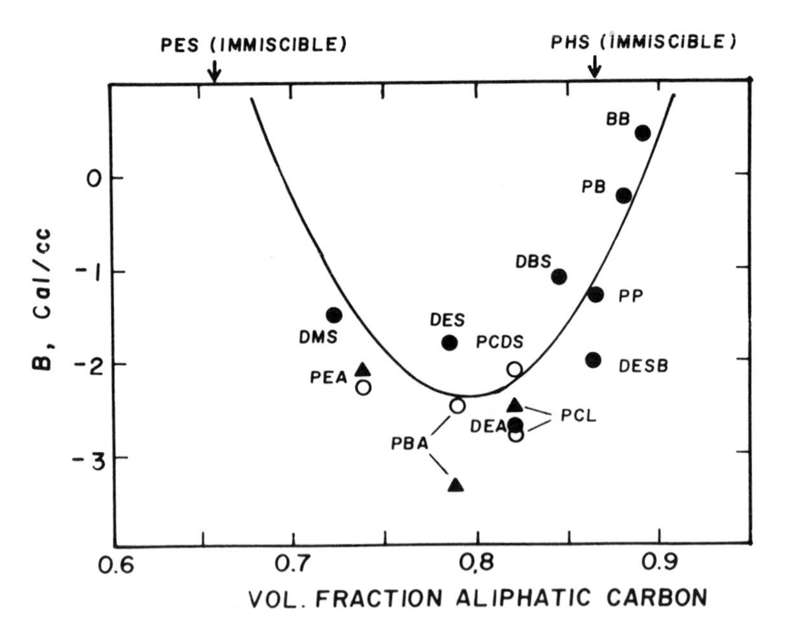

Figure 6. A comparison of B *parameters obtained by sorption (○), calorimetry (●), and polymer melting point depression (▲) vs the aliphatic content of the esters and polyesters examined.*

ments as well as in the Flory–Huggins analysis of both melting point depression and sorption in blends, adequately represents the composition dependence of the heat of mixing to a first approximation. Finally, these results reaffirm earlier conclusions (*1*) concerning the direct correspondence between analog calorimetry and polymer–polymer miscibility in a fairly quantitative manner and reinforce the idea that polymer–polymer miscibility depends on the presence of an exothermic heat of mixing.

The reasons for the "miscibility window" of ester structure in Figure 6 are still not totally clear, although the shape of the curve does suggest some competition between exothermic and endothermic contributions to the heat of mixing. Attempts to model this behavior are currently being made and the results of this work will be reported.

Nomenclature

Symbol	*Units*	*Description*
a_1	—	Activity of solvent vapor in equilibrium with polymer
B	cal/mL	Interaction energy density parameter
ΔH_{2u}	cal/gmole	Heat of fusion per mole of crystallizable units in semicrystalline polymer 2
P	mm Hg	Pressure of solvent vapor
P°	mm Hg	Saturation pressure of solvent vapor at temperature T
R	cal/gmole K	Ideal gas constant
T	K	Absolute temperature
T_{m2}	K	Equilibrium melting temperature of polymer 2 in the blend
T_{m2}°	K	Equilibrium melting temperature of pure polymer 2
V_i	mL/gmole	Molar volume of component i
V_{2u}	mL/gmole	Volume per mole of crystallizable units in semicrystalline polymer 2
ϕ_i	—	Volume fraction of component i in mixture
ϕ_i'	—	Volume fraction of component i in mixture on a solvent free basis
χ_{ij}	—	Flory–Huggins interaction parameter describing interaction between components i and j in the mixture
χ_{1b}	—	Flory–Huggins interaction parameter describing interaction between solvent 1 and the polymer blend

Acknowledgments

Acknowledgment is made to the donors of the Petroleum Research Fund, administered by the American Chemical Society, for their support of this research.

Literature Cited

1. Cruz, C. A.; Barlow, J. W.; Paul, D. R. *Macromolecules* **1979**, *12*, 726.
2. Bernstein, R. E.; Wahrmund, D. C.; Barlow, J. W.; Paul, D. R. *Polym. Eng. Sci.* **1978**, *18*, 1220.
3. Ziska, J. J.; Paul, D. R.; Barlow, J. W. *Polymer* **1981**, *22*, 918.
4. Barnum, R. S.; Goh, S. H.; Paul, D. R.; Barlow, J. W. *J. Appl. Polym. Sci.* **1981**, *26*, 3917.
5. Harris, J. E.; Goh, S. H.; Paul, D. R.; Barlow, J. W. *J. Appl. Polym. Sci.* **1982**, *27*, 839.
6. Barlow, J. W.; Paul, D. R. *Ann. Rev. Mater. Sci.* **1981**, *11*, 299.
7. Paul, D. R.; Barlow, J. W. *J. Macromol. Sci., Rev. Macromol. Chem.* **1980**, C *18(1)*, 109.
8. Hoffman, J. D.; Weeks, J. J. *J. Chem. Phys.* **1962**, *37*, 1723.
9. Harris, J. E. Ph.D. Dissertation, University of Texas, TX, 1981.
10. Nishi, T.; Wang, T. T. *Macromolecules* **1975**, *8*, 909.
11. Imken, R. L.; Paul, D. R.; Barlow, J. W. *Polym. Eng. Sci.* **1976**, *16*, 593.
12. Flory, P. J. "Principles of Polymer Chemistry"; Cornell University Press: Ithaca, New York, 1953; Chap. 13.
13. Bawn, C. E. H.; Patel, R. D. *Trans. Faraday Soc.* **1956**, *52*, 1664.
14. Bonner, D. C.; Prausnitz, J. M. *J. Polym. Sci., Polym. Phys. Ed.* **1974**, *12*, 51.
15. Eichinger, B. E.; Flory, P. J. *Trans. Faraday Soc.* **1968**, *64*, 2061.
16. Mason, W. P. "Piezoelectric Crystals and Their Applications to Ultrasonics"; D. Van Nostrand Co.: New York, 1950.
17. King, W. H. *Anal. Chem.* **1964**, *36*, 1735.
18. Saeki, S.; Bonner, D. C. *Polymer* **1978**, *19*, 319.
19. Hill, T. L.; Rowen, J. W. *J. Polym. Sci.* **1952**, *9*, 93.
20. Rowen, J. W. *J. Polym. Sci.* **1958**, *31*, 199.
21. Olabisi, O. *Macromolecules* **1975**, *8*, 316.

RECEIVED for review March 14, 1983. ACCEPTED June 22, 1983.

5

Compatibility Studies of Poly(vinylidene fluoride) Blends Using Carbon-13 NMR

THOMAS C. WARD and T. S. LIN

Department of Chemistry and Polymer Materials and Interfaces Laboratory, Virginia Polytechnic Institute and State University, Blacksburg, VA 24061

Measurements of the degree of mixing at the molecular level between poly(vinylidene fluoride) and three vinyl polymers were obtained by using solid state ^{13}C NMR. These measurements showed that the specific interactions were different for different carbons. High resolution solid state ^{13}C NMR with magic angle spinning, proton dipolar decoupling, and cross polarization was applied to poly(vinylidene fluoride) blends with poly(vinyl acetate), poly(methyl methacrylate), and poly(vinyl methyl ether). Nondecoupled fluorine-19 attenuation of the carbon-13 peak intensities was observed and indicated the extent of the mixing process. Varying sample preparation resulted in different degrees of intermolecular association.

INTEREST IN POLYMER–POLYMER INTERACTIONS in the solid state and the continued growth in commercial applications of polymer blends have resulted in the use of a number of experimental techniques to investigate polymer–polymer miscibility (*1*, *2*). However, different methods reveal different facets of the nature of the mixing process; consequently, no single method has answered all the interesting questions at the molecular level (*3*, *4*). Thus, we wanted to extend the range of the potential tools available for analysis in this important area and to compare the results to the previous efforts.

Pulsed proton NMR experiments have been extensively used to study polymer–polymer miscibility through the measurement of the spin–lattice relaxation times (T_1), spin–spin relaxation times (T_2), and rotating frame relaxation times ($T_{1\rho}$) directly from the time domain free induction decay (FID) (*5–7*). However, high resolution frequency

0065-2393/84/0206-0059$06.00/0

domain ^{1}H spectra in the solid state are difficult to obtain because of direct nuclear dipolar couplings and strong chemical shift anisotropy that exists for samples in an immobilized matrix. With the development of magic angle spinning (MAS), heteronuclear dipolar–dipolar decoupling (DD), and cross-polarization (CP) techniques, high resolution solid state NMR instruments for examining some relatively rare nuclear spins, that is, those on carbon-13 and nitrogen-15, have become commercially available (*8–11*).

We used the new NMR instrumental advances to examine polymer blends of poly(vinylidene fluoride) (PVF_2) from a molecular, and unique, point of view. High resolution solid state ^{13}C NMR previously was used to study polymer–polymer miscibility through the measurement of T_1, T_2, $T_{1\rho}$ and other relaxation parameters, in particular the cross-polarization transfer rates and nuclear Overhauser factors (*12*, *13*).

For most solid organic polymers, the use of MAS in combination with high power ^{13}C–^{1}H dipolar decoupling and cross polarization will result in high resolution carbon-13 NMR spectra. With all three techniques applied, however, polymers that contain an abundance of other types of nuclei with strong nuclear magnetic moments have spectra that usually will be quite broad, or in some cases completely featureless, as a consequence of dipolar coupling between carbon-13 nuclei and the other nuclei. For instance, when PVF_2 is analyzed without simultaneous decoupling of the fluorine-19, the ^{13}C frequency domain spectrum is completely devoid of peaks. Perhaps, therefore, a powerful method for investigating the miscibility of polymer blends containing PVF_2 might be the study of the intermolecular dipolar interactions between the ^{19}F nuclei in the PVF_2 and the ^{13}C nuclei in the second polymer. Strong internuclear distance dependence would be expected for such forces. The major advantage of this experiment over the previous ^{1}H–^{19}F work (*7*) would be the quantitative and specific identification of all of the sites of interaction. If a polymer was indeed immiscible with PVF_2, then its ^{13}C-NMR signal intensities in the "blend" simply would be directly proportional to those shown by the pure polymer; the intensity reduction would be accounted for by sample dilution. On the other hand, if a polymer was extensively and intimately intermixing with PVF_2, then its ^{13}C-NMR signal intensities in the blend would be additionally attenuated by the ^{19}F coupling, if we assume an absence of ^{19}F decoupling.

Magic angle spinning is primarily a technique for overcoming the effect of ^{13}C chemical shift anisotropies that result from multiple orientations of the nuclear moment with respect to the applied field. However, MAS can also decouple weak dipolar interactions, those interactions with strengths (in Hertz) less than that of the frequency of the spinning (*11*, *12*). The spinning rates in our experiments were about 2.0 kHz, and a limit of the resolution of estimations of the intimacy of mixing was in the range of 3–4 Å. Beyond this limit MAS decoupling will reduce any signal attenuation tendencies of ^{19}F on ^{13}C.

MAS prevents the ^{13}C–^{19}F dipolar coupling effect from being a simple broadening of the observed ^{13}C peaks. Once MAS is established, varying the spinning rate does *not* affect the lineshape of the central band; however, the location and intensity of spinning sidebands and the intensity of the central peak are modified. This central band lineshape is independent of the ^{13}C–^{19}F internuclear distance. The essential concept is that the central band intensity will vary with the internuclear distance. An account of this analysis may be helpful (*27*).

Our purpose was to demonstrate a new approach to the study of polymer–polymer miscibility at the molecular level. In this regard, a series of polymer blends with PVF_2 that had been extensively explored by other techniques (*14–19*) was desirable. Future work will be directed toward other, less investigated blends, especially as functions of composition and temperature, and to further evaluations of the instrumental factors affecting this work.

Experimental

Sources and Characterization of Homopolymers. The PVF_2 (M_w 100,000), poly(methyl methacrylate) (PMMA) (M_w 800,000), and poly(vinyl methyl ether) (PVME) (M_w 18,000) were obtained from Polysciences, Inc. Our solution ^{13}C NMR spectrum of the PMMA showed it to be 75% syndiotactic. The pure PMMA was compression molded for the solid state NMR studies. The PVME solution ^{13}C NMR spectrum showed it to be 50% syndiotactic. The PVME was packed neat into the NMR rotor.

Poly(vinyl acetate) (PVAc) (M_w 70,000) was obtained from Aldrich. It was about 65% syndiotactic according to its ^{13}C NMR spectrum in solution. It was also compression molded for solid state NMR studies.

Preparation of Blends. The PVF_2/PMMA blends were prepared in three ways: with a melt extruder and with two mutual solvents, methyl ethyl ketone (MEK) and dimethyl formamide (DMF). The PVF_2/PVAc and the PVF_2/PVME blends were both produced by using DMF as the common solvent.

Extruded blends were prepared on a CSI–MAX mixing extruder at 190 °C; residence time at 190 °C was relatively short (minutes). Blends using MEK as the common solvent were made by first dissolving the individual polymers at about a 2-g/100 mL concentration, then appropriately mixing these solutions. When mixed, the solutions were clear in all proportions studied. These mixtures were poured into aluminum pans and subsequently air-dried for 2 days, followed by further drying in a vacuum oven for 24 h at room temperature. Blends using DMF as a common solvent were made by the same procedures as those using MEK, except that the mixtures were first dried in air for at least 1 week, then further dried in a vacuum oven for 48 h at room temperature.

Melted-and-quenched PVF_2/PMMA samples were prepared from solution blends by first heating the dried blends in a vacuum oven at 190 °C until the PVF_2 crystals totally melted, then immediately immersing the materials into liquid nitrogen for 5 min.

Films from all the above preparations were punched to form circular discs that were weighed and stacked into the MAS rotor.

NMR. The DD/MAS solid state ^{13}C experiments were carried out at 15.0 MHz on a JEOL–60Q spectrometer with a Kel–F "bullet" type rotor and approximately 2.0-kHz spinning speeds at room temperature. For pure PVME and the PVF_2/PVME blends, a single 90° pulse sequence was used (90° pulse = 7.6 μs). For pure PVAc and the PVF_2/PVAc blends, and pure PMMA and the PVF_2/PMMA blends, matched spin-lock cross-polarization transfer of protons of 33 kHz with 0.75- and 3-ms single contact times were employed, respectively. Contact times, pulse delays, and the number of scans per spectrum will be discussed later.

All the spectra were obtained from 1000 FT, with zero filling to a total of 4000 points. A spectral width of 8000 Hz and an acquisition time of 64 ms were selected for all the spectra. Appropriate pulse delay times were used to ensure full relaxation between scans.

NMR Signal Intensities and Number of Accumulations. The NMR signal intensities are assumed to be directly proportional to the number of total accumulations for each sample. However, when a large number of scans are stored by using only time domain accumulations, considerable deterioration of the accumulation efficiency may exist because of the limited computer bit number. The JEOL–60Q instrument employed a technique of progressive decrement of the A to D converter bit number to avoid data overflow. The dynamic range of the spectra was thereby reduced, which would make the assumption of signal-to-noise ratio (*S*/*N*) less reliable, especially for the weak peaks. Thus, to avoid this problem, all the spectra were carried out at conditions such that the final A to D converter bit number remained at least five bits. Table I gives ^{13}C NMR signal intensities of PMMA obtained at various number of accumulations. The table shows that with our experimental conditions, the assumptions concerning *S*/*N* remained effective.

The spectra were externally referenced to liquid tetramethylsiloxane (TMS), on the basis of substitution of hexamethylbenzene (HMB) as the secondary reference and assigning 132.3 and 16.9 ppm to the shifts of the aromatic and aliphatic carbons, respectively, of HMB relative to liquid TMS.

DSC. Thermal analysis of PVF_2 and its blends was carried out on a Perkin–Elmer DSC–2 differential scanning calorimeter at a heating rate of 20 °C/min. Melting points were taken as the temperature at which the last detectable trace of crystallinity disappeared (*14*).

Results

Figure 1 shows the high resolution solid state ^{13}C NMR spectra of pure PMMA, PVAc, and PVME. Peak assignments were made according to the respective solution spectra in the literature (*20–22*). Except for the methylene carbon of PMMA, all of the carbon resonances were separable from one another at half peak height.

Figure 2 gives one example of how pure PMMA signals were attenuated when placed in a blend with PVF_2. Spectra a, b, and c were obtained

Table I. Carbon-13 NMR Signal Intensities of PMMA Obtained at Various Numbers of Accumulations

Number of Accumulations	*Word Length (bits)*	*Final AD Converter (bits)*	*^{13}C Peak Intensities (arbitrary unit)*			
			Carbonyl	*Methoxyl*	*Quaternary*	*α-Methyl*
800	16	8	3565×2^8	4761×2^8	5844×2^8	1882×2^8
1600	16	7	3488×2^9	5095×2^9	5323×2^9	1730×2^9
3200	16	6	3468×2^{10}	4805×2^{10}	5592×2^{10}	1731×2^{10}
6400	16	5	3446×2^{11}	4732×2^{11}	5787×2^{11}	1852×2^{11}

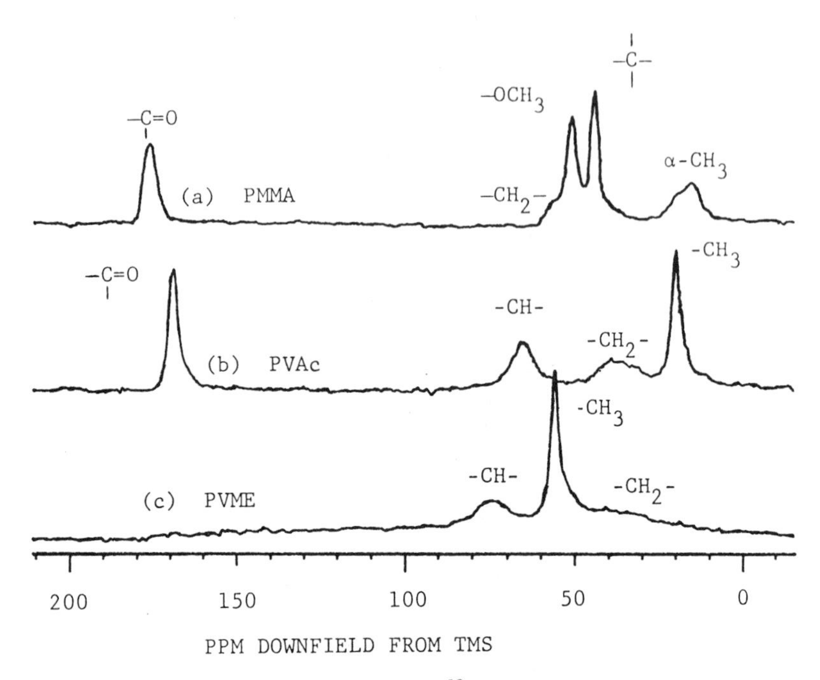

Figure 1. The 15.0 DD-MAS solid state ^{13}C NMR spectra of (a) PMMA—obtained from 1000 FID accumulations with a CP contact time of 3 ms and a pulse delay time of 3 s, (b) PVAC—obtained from 1492 FID accumulations with a CP contact time of 0.75 ms and a pulse delay time of 3 s, and (c) PVME—obtained from 4800 FID accumulations with a single ^{13}C 90° pulse of 7.6 μs and a pulse delay time of 2 s.

for PMMA, a 50:50 MEK blended PVF_2/PMMA, and a 50:50 MEK, melted-and-quenched PVF_2/PMMA blend, respectively. The number of accumulations (scans) of b and c were selected to ensure that the three spectra would have equal sensitivity (signal-to-noise ratio) if there was no intermixing between PVF_2 and PMMA.

Sample calculations involving the chosen number of accumulations and the residual PMMA signal intensities in the blends are found in Table II. All calculations are based on the following assumptions: the absolute signal intensity was directly proportional to the number of nuclei (weight of PMMA) and to the number of accumulations; the sensitivity (S/N) was proportional to the number of nuclei and to the square root of the number of accumulations (*23*); and, finally, the data handling capabilities of the instrument computer were not limiting as the number of scans increased. Thus, for any particular carbon in a blended polymer the ratio of expected signal intensity to observed signal intensity will provide the percentage of attenuation that was ascribed to ^{19}F in the immediate vicinity of that carbon, if all other factors of data reduction are assumed constant.

Tables III, IV, and V present an analysis of the data in terms of the

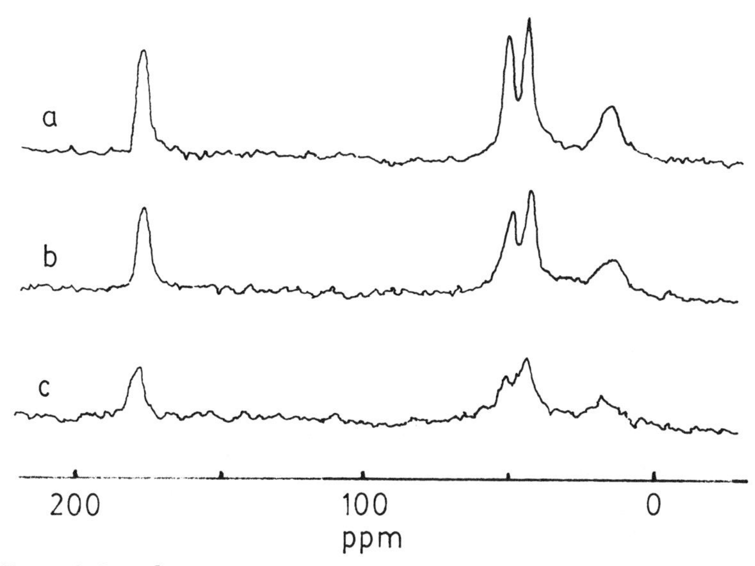

Figure 2. Signal intensity attenuations of PMMA in the PVF_2/PMMA blends. Each is a 15.0-MHz DD-MAS-CP ^{13}C NMR spectrum. (a) PMMA, (b) 50:50 PVF_2/PMMA blend (MEK), and (c) 50:50 PVF_2/PMMA blend (MEK, melted and quenched).

Table II. Calculations of the Residual NMR Signal Intensities

Samples	*PMMA*	*PVF_2/PMMA*	*PVF_2/PMMA (quenched)*
Weight, mg	463.2	411.9	390.6
Weight fraction of PMMA	1.0	0.50	0.50
Number of accumulations	640	3238	3600
Relative sensitivity (calculated)	1.0	1.000	1.000
Relative signal intensities (calculated)	1.0	2.250	2.372
Signal intensities (experimental) (arbitrary unit)			
carbonyl	3521	5935	3466
methoxyl	4294	5770	3572
quaternary	4844	7463	4948
α-methyl	1937	2999	2660
Residual signal intensities (%)			
carbonyl	100	75	42
methoxyl	100	60	35
quaternary	100	69	43
α-methyl	100	69	58

Table III. Residual NMR Signal Intensities of PMMA in the PVF_2/PMMA Blends

Compositions (wt. % of PVF_2)	*Methods of Blending*	*Residual ^{13}C Peak Intensities (%)*				T_m (°C)
		Carbonyl	*Methoxyl*	*Quaternary*	*α-Methyl*	
70	MEK	57	53	58	73	164
50	MEK	75	60	69	69	165
25	MEK	76	59	68	85	broad
70	MEK, quenched	43	31	38	57	161
50	MEK, quenched	42	35	43	58	none
25	MEK, quenched	69	59	72	68	none
70	DMF	109	118	116	94	173
50	DMF	113	111	116	94	174
25	DMF	114	101	104	112	174
70	DMF, quenched	44	29	35	41	163
50	DMF, quenched	62	40	59	69	163
25	DMF, quenched	73	60	68	84	none
70	extruded	41(44)[a]	30(30)[a]	43(42)[a]	50(58)[a]	159
50	extruded	35(42)[a]	28(26)[a]	37(30)[a]	51(49)[a]	148
25	extruded	52(50)[a]	45(46)[a]	47(52)[a]	66(71)[a]	none
100	—	—	—	—	—	171

[a] Results obtained at a higher MAS rate (2.3 kHz).

Table IV. Residual NMR Signal Intensities of PVAc in the PVF_2/PVAc Blends

Composition (wt. % of PVF_2)	*Methods of Blending*	*Residual ^{13}C Peak Intensities (%)*				T_m (°C)
		Carbonyl	*Methine*	*Methylene*	*α-Methyl*	
69.8	DMF	72	100	122	73	175
49.9	DMF	32	51	58	29	157
24.8	DMF	19	18	28	15	153
13.7	DMF	16	—	—	18	152
8.5	DMF	17	—	—	15	—
4.1	DMF	62	49	58	59	—
100.0	—	—	—	—	—	171

Table V. Residual NMR Signal Intensities of PVME in the PVF_2/PVME Blends

Composition (wt. % of PVF_2)	*Methods of Blending*	*Residual ^{13}C Peak Intensities (%)*			T_m (°C)
		Methine	*Methyl*	*Methylene*	
75	DMF	110	119	109	177
50	DMF	123	110	124	175
100	—	—	—	—	171

residual intensities of the various solid state ^{13}C signals of PMMA, PVAc, and PVME, respectively, when each was blended with PVF_2. The calculations were based on peak heights. Essentially the same results were obtained when peak areas were substituted for peak heights. The melting points of PVF_2 in the blends as determined calorimetrically are listed in the tables. Also, various compositions and thermal and solvent histories are indicated in the first two columns.

Values in these three tables that exceed 100% may do so for several reasons. Minor mass balance errors undoubtedly account for some scatter. For PMMA and its blends, repeated experiments indicated agreement in the 5–10% range as is shown in Tables VI and VII. In other cases weak, incompletely resolved peaks (e.g., the methylene carbon of PVAc) are estimated to have somewhat larger uncertainty. Some of the relevant instrument pulse sequence parameters that had to be considered and may be relevant in this regard will be discussed in conjunction with Tables VII and VIII. The values greater than 100% are generally observed on all of the carbon peaks in the DMF cast films, regardless of blend; melting points of PVF_2 were also found to be higher in each case. This behavior was quite consistent and may be a result of a modification of the crystal form of the PVF_2 (at least four are known) that occurred on casting from DMF. Perhaps a modification of relaxation times, particularly $T_{1\rho}^{H}$, might then result in larger initial ^{13}C magnetization in each recorded pulse, which would lead to enhanced peak intensities for the case materials at equal calculated S/N ratios.

Discussion

Shifts of NMR Frequencies. As a result of the interactions between PVF_2 and the other polymers in the miscible blends, some shifting of various NMR peaks away from their initial frequencies as is observed in FTIR studies might be expected (*15*). However, because of the inherent resolution limitation of NMR instruments in addition to the decreased sensitivity obtained with the miscible blends (signal attenuations), no significant NMR chemical shifts were induced by blending. The stereochemical microstructure of the vinyl polymers would help to obscure any such shift as well.

NMR Signal Attenuations and Degrees of Intermixing. Tables III, IV, and V, show that the magnitudes of ^{13}C signal attenuations of PMMA, PVAc, and PVME in the blends with PVF_2 were in very good agreement with the melting point depressions of PVF_2 in the blends. The method of melting point depression has become accepted as an indicator of the degree of blend miscibility; for example, this number can be directly related to the polymer–polymer free energy interaction parameter (*14*, *24*). Clearly, a

Table VI. Carbon-13 NMR Signal Intensities of PMMA Obtained at Various Experimental Conditions

Run No.	*CP Contact Time (ms)*	*Pulse Delay Time (s)*	*Peak Intensities (arbitrary unit)*			
			Carbonyl	*Methoxyl*	*Quaternary*	*α-Methyl*
1	3	2	4833	6301	7190	2191
2	1	3	5227	7731	8853	2694
3	3	3	4941	6383	7383	2159
4	5	3	3734	4982	6124	1816
5	3	2	4788	6510	7617	2263

Table VII. Carbon-13 NMR Signal Intensities of the Melt-Extruded 25:75 PVF_2/PMMA Blend Obtained at Various Experimental Conditions

Run No.	*CP Contact Time (ms)*	*Pulse Delay Time (s)*	*Peak Intensities (arbitrary unit)*			
			Carbonyl	*Methoxyl*	*Quaternary*	*α-Methyl*
1	3	2	3882	4031	5121	2113
2	1	3	4939	6190	8370	2894
3	3	3	4296	4468	5588	2095
4	5	3	2703	2808	3829	1524
5	3	2	4068	4307	5676	2301

Table VIII. Residual NMR Signal Intensities of PMMA in the 25:75 PVF_2/PMMA Blend, Calculated from the CP Spectra with Various Contact Times

CP Contact Time (ms)	*Residual ^{13}C NMR Peak Intensities (%)*			
	Carbonyl	*Methoxyl*	*Quaternary*	*α-Methyl*
1	68	59	68	77
3	63 (52)[a]	50 (45)[a]	54 (47)[a]	70 (66)[a]
5	52	60	45	60

Note: Pulse delay time = 3 s.
[a] Calculated from Table II.

macroscopic, thermodynamic observation is all that may be obtained in this case. In comparison, NMR signal attenuations of any polymer blended with PVF_2, as observed in our work, indicate to some extent the spatial proximity of that polymer to PVF_2. We will show that intermolecular distances for ^{13}C–^{19}F dipolar coupling as observed by the present method were in the range of 3 to 4 Å. At least for the blends that were examined, the methods of melting point depression and NMR signal attenuations provided consistent indications of the extent of compatibility from very different experimental approaches. The data in Tables III, IV, and V for the three series of blends are discussed individually.

PVF_2/PMMA Blends. Without any thermal treatment, MEK was a better common solvent for blending than DMF; attenuation of ^{13}C resonances was nil in the latter case. Although the mixtures of PVF_2 and PMMA in DMF were clear, the dried samples of all compositions cast from DMF were totally immiscible as measured by both the NMR and DSC studies.

Melting and quenching the blends appeared to improve miscibility according to the NMR work; PVF_2 crystallinity was undetected by DSC in some of these samples. The differences in NMR signal attenuations between the MEK-blended and the DMF-blended samples with the same composition became smaller after the melting and quenching process, although the miscibility remained poorer than that of the melt-extruded blends with the same compositions.

For solvent-blended samples in the composition ranges studied, as the PVF_2 content increased, the PMMA ^{13}C signal attenuation became greater. For the melt-extruded blends, the one with medium PVF_2 content (50% by weight) showed the highest degree of ^{13}C signal loss.

PVF_2/PVAc Blends. Compared to the PVF_2/PMMA blends that used DMF as common solvent, the PVF_2/PVAc blends made from DMF solutions had much better miscibility, as indicated by the NMR results in Table IV. Carbon-13 NMR signal attenuations of PVAc passed through a minimum as the PVF_2 content was increased. The loss of miscibility at the higher PVF_2 content probably was a result of the higher degrees of PVF_2 crystallinity in those blends.

PVF_2/PVME Blends. In Table V, a summary of observations on the PVF_2/PVME blends is presented. Although both pure materials dissolve readily in DMF, the cast films were concluded to be totally immiscible by both the NMR and DSC results; this finding confirms previously observed data (*19*). The coincidence of signal intensities that are enhanced rather than diminished by the DMF casting sequence and elevated PVF_2 melting point is especially noticeable in Table V.

NMR Signal Attenuations of Individual Carbons. The nature of the miscibility of both the PVF_2/PMMA and the PVF_2/PVAc blends previously has been attributed to a weak intermolecular "acid–base" association. In particular, an interaction was postulated between the acidic protons of the PVF_2 and the basic carbonyl groups for both PMMA and PVAc (*15*, *19*). Knowing that the definition of miscibility in these NMR experiments has to be confined to approaches of dissimilar nuclei to within about 4 Å, we might observe differing extents of NMR peak attenuations for each carbon. Depending on how much less than 4 Å the contact actually was on the average, we could calculate an estimate of the specificity of interatomic mixing.

Table II shows that, for the PVF_2/PMMA blends, four resolvable peaks occurred in the spectra of PMMA; the methoxyl carbon signal was the most attenuated, and the α-methyl carbon signal was the least attenuated. This difference in signal attenuation became more pronounced in the melt-and-quenched and the melt-extruded blends.

In the PVF_2/PVAc blends, the signal intensities of the two main chain carbons of PVAc seemed to be less attenuated than those of the two side chain carbons of PVAc, especially for the samples with higher PVF_2 contents. However, the initial peak intensities of the two main chain carbons were much weaker than those of the two side chain carbons (*see* Figure 1), partially because of the effects of tacticity. Thus, in view of the accuracy of the measurements, the uncertainty of any conclusion concerning the two main chain carbons is higher.

NMR Signal Attenuations and Magic Angle Spinning Rates. The direct nuclear dipolar coupling constant, D, between a ^{19}F nucleus and a ^{13}C nucleus in a rigid lattice depends on both the internuclear distance r and the orientation of r with respect to the external magnetic field B_o:

$$D(^{13}C - ^{19}F) = \frac{h\gamma_C\gamma_F(3\cos^2\theta - 1)}{4\pi^2 r^3}$$

where γ_C and γ_F are the magnetogyric ratios, and θ is the angle between r and B_o.

If r is considered in angstrom units, then

$$D(^{13}C - ^{19}F) = \frac{56815.2}{r^3}\left(\frac{3\cos^2\theta - 1}{2}\right) \text{ Hz}$$

that is,

$$D = 56815.2\left(\frac{3\cos^2\theta - 1}{2}\right) \text{ Hz, if } r = 1 \text{ Å}$$

$$D = 7101.9\left(\frac{3\cos^2\theta - 1}{2}\right) \text{ Hz, if } r = 2 \text{ Å}$$

$$D = 2104.2\left(\frac{3\cos^2\theta - 1}{2}\right) \text{ Hz, if } r = 3 \text{ Å}$$

$$D = 887.7\left(\frac{3\cos^2\theta - 1}{2}\right) \text{ Hz, if } r = 4 \text{ Å}$$

Both spin diffusion (among ^{19}F) and molecular motions can reduce D to some degree. The maximum value of D would be realized when the angle dependent term in brackets is unity.

Distance influences dipolar coupling, and this fact shows that the MAS itself can decouple long-range weak dipolar interactions; faster MAS decouples those of shorter range (i.e., estimations of miscibility in this study are limited by the chosen MAS rate). On the other hand, an investigation of NMR signal attenuations of polymer blends containing PVF_2 over a wide range of MAS rates would generate an average concentration profile of the spatial distribution of those ^{19}F nuclei in the vicinity of any resolvable ^{13}C nucleus. Unfortunately, because of the limitations of our equipment, the only comparison was at the spinning rates of 2.0 and 2.3 kHz. Some results for the melt-extruded PVF_2/PMMA blends are shown in Table II. No significant spectral differences were seen at these two spinning rates. Because of the inverse third power dependence of the dipolar coupling on the internuclear distance, the expected effect is only $(2.3/2.0)^{1/3} = 1.048$ in magnitude. Consequently, the results are not surprising, given the limited frequency range. MAS rates as high as 15 kHz have been achieved by using helium as the driving gas (*25*). This result suggests that if the NMR signal attenuations were compared at, for example, the MAS rates of 16 and 2 kHz, then a factor of two in the decoupling–distance relationship might be achieved to aid in the understanding of the internuclear distance questions.

NMR Signal Attenuations and Relaxation Times. The technique of ^{13}C–^{1}H cross polarization is not only used to rapidly polarize carbon magnetization (by $\sim T_{1H}/T_{1C}$) but also to achieve greater total nuclear magnetic alignment with the applied field than would otherwise be possible.

Maximum enhancement of ^{13}C polarization at the end of the CP contact is by a factor of $M_c = \gamma_H/\gamma_C M_{oc}$, where M_{oc} is the polarization that would be generated in B_o after waiting several relaxation times, T_{1C}. However, this optimal enhancement of M_{oc} can be achieved only if $T_{1\rho}^{H} \gg$ CP contact time $\gg T_{CH}$ (where T_{CH} is the cross-relaxation rate) (*10*, *26*). If ^{13}C nuclear spin alignment varied with the composition of PVF_2 polymer system, then unfair comparisons of attenuation might result. An equivalent statement would be that the ^{13}C relaxations might begin at different spin temperatures and attenuation calculations thus might be mistaken.

Both the $T_{1\rho}^{H}$ and the T_{CH} are modulated by molecular motion. Thus, average molecular mobility for a component in a pure polymer and in a polymer blend (especially highly miscible examples) conceivably might be different. Therefore, when we compare the NMR signal intensities from the CP spectra of a polymer and of its blends, these relaxation parameters have to be considered. In fact, any change of these relaxation times is an indication of the miscibility. The choice of CP time is especially important; comparing NMR intensities of a polymer and its blends from CP spectra in which the same contact time was specified could be misleading depending on $T_{1\rho}^{H}$ and T_{CH}. Further investigation of this point was required to have confidence in the results.

Tables VI and VII list the ^{13}C-NMR peak intensities of the PMMA and the melt-extruded 25:75 PVF_2/PMMA blends, respectively, obtained from spectra in which various CP contact times and different pulse delays were employed. Runs 1 and 5 were carried out at exactly the same conditions to show the reproducibility of our experiments. Run 3 used the same CP contact time as runs 1 and 5 but employed a longer pulse delay time. The results show that for both the PMMA alone and the blends, 3 s of pulse delay time is certainly long enough to ensure full repolarization of the protons between the scans.

From Tables VI and VII, the spin-locked magnetization relaxed faster in the blends than in the pure PMMA polymer (i.e., $T_{1\rho}^{H}$ is shorter in the blend than in the homopolymer). The CP contact time we used in all the examples for the PMMA and its blends was 3 ms, which is not the optimum for all systems investigated. Only when there are regions of CP contact times such that the optimum CP sensitivity enhancement can be achieved ($T_{1\rho}^{H} \gg$ CP time $\gg T_{CH}$) for both the homopolymer and its blends, will the ^{13}C–^{19}F coupling be the exclusive factor that would cause NMR signal attenuations. Further investigation of this point has shown that 3-ms contact time is close enough to optimum so that none of our conclusions are altered. Details of these new experiments will be published elsewhere. The residual NMR signal intensities (Table VIII) of PMMA in the 25:75 melt-extruded PVF_2/PMMA blend were calculated from Tables VI and VII for the CP spectra obtained at various CP contact times. Table VIII reveals that the residual ^{13}C NMR peak intensities calculated from the spectra with

3-ms contact time are greater than those found at an earlier date on the samples under the same experimental conditions (shown in Table II). Information in Table VIII came from the spectra recorded for the blends about 5 weeks after their preparation, whereas the facts reflected in Table III were calculated from spectra obtained for the blends within 2 days of their preparation. The aging of the blends is apparent from this data.

Of further interest in Table VIII is that although the absolute values of the numbers are different, the same trend was found in each contact time series when various carbons were compared. Further exploration of the kinetics of the events suggested by Table VIII is warranted.

Acknowledgment

This work was supported with help from NSF Grant no. DMR 76–11963 A02.

Literature Cited

1. Olabisi, O.; Robeson, L. M.; Shaw, M. T. "Polymer–Polymer Miscibility"; Academic Press: New York, 1979.
2. "Polymer Blends"; Paul, D. R.; Newman, S., Eds.; Academic Press: New York, 1978.
3. Morawetz, H.; Amrani, F. *Macromolecules* **1978**, *11*, 281.
4. Frank, C. W.; Gashgari, M. A. *Macromolecules* **1979**, *12*, 163.
5. Nishi, T.; Wang, T. T.; Kwei, T. K. *Macromolecules* **1975**, *8*, 227.
6. Kwei, T. K.; Nishi, T.; Roberts, R. F. *Macromolecules* **1974**, *7*, 667.
7. McBrierty, V. J.; Douglas, D. C.; Kwei, T. K. *Macromolecules* **1978**, *11*, 1265. Also Douglas, D. C.; McBrierty, V. J. *Macromolecules* **1978**, *11*, 766.
8. *Chem. Eng. News*, October 16, 1978, p. 23.
9. *JEOL News* **1980**, *16A(1)*, 2.
10. Yannoni, C. S. *Acc. Chem. Res.* **1982**, *15*, 201.
11. Lyerla, J. R. "Contemporary Topics in Polymer Science;" Plenum Press: New York, 1979; Vol. 3, pp. 143–213.
12. Schaefer, J.; Sefcik, M. D.; Stejskal, E. O.; McKay, R. A. *Macromolecules* **1981**, *14*, 188.
13. Stejskal, E. O.; Schaefer, J.; Sefcik, M. D.; McKay, R. A. *Macromolecules* **1981**, *14*, 275.
14. Nishi, T.; Wang, T. T. *Macromolecules* **1975**, *8*, 909.
15. Coleman, M. M.; Zarian, J.; Varnell, D. F.; Painter, P. C. *J. Polym. Sci. Polym. Lett. Ed.* **1977**, *15*, 745.
16. Wahrmund, D. C.; Bernstein, R. E.; Barlow, J. W.; Paul, D. R. *Polym. Eng. Sci.* **1978**, *18*(9), 677.
17. Bernstein, R. E.; Paul, D. R.; Barlow, J. W. *Polym. Eng. Sci.* **1978**, *18*(9), 683.
18. Bernstein, R. E.; Wahrmund, D. C.; Barlow, J. W.; Paul, D. R. *Polym. Eng. Sci.* **1978**, *18*(*16*), 1220.
19. Paul, D. R.; Barlow, J. W.; Bernstein, R. E.; Wahrmund, D. C. *Polym. Eng. Sci.* **1978**, *18*(*16*), 1225.
20. Lyerla, J. R., Jr.; Horikawa, T. T.; Johnson, D. E. *J. Am. Chem. Soc.* **1977**, *99*(*8*), 2463.

21. Wu, T. K.; Ovenall, D. W. *Macromolecules* **1974**, *7*, 776.
22. Johnson, L. F.; Heatley, F.; Bovey, F. A. *Macromolecules* **1970**, *3*, 175.
23. Cooper, J. W. *Comput. Chem.* **1976**, *1*, 55.
24. Olabisi, O.; Robeson, L. M.; Shaw, M. T. "Polymer–Polymer Miscibility"; Academic Press: New York, 1979; p. 1.
25. Zilm, K. W.; Alderman, D. W.; Grant, D. M. *J. Magn. Res.* **1978**, *30*, 563.
26. Pines, A.; Gibby, M. G.; Waugh, J. S. *J. Chem. Phys.* **1973**, *59(2)*, 569.
27. Andrew, E. R. *Philos. Trans. R. Soc. London, Ser. A* **1981**, *299*, 505.

RECEIVED for review February 3, 1983. ACCEPTED October 26, 1983.

Effect of Molecular Weight on Blend Miscibility

A Study by Excimer Fluorescence

STEVEN N. SEMERAK and CURTIS W. FRANK

Department of Chemical Engineering, Stanford University, Stanford, CA 94305

Excimer fluorescence from a guest polymer blended with a nonfluorescent polymer host is used to study the influence of the molecular weight of the two components on the miscibility. The ratio of the excimer to monomer emission intensities, I_D/I_M, *for poly(2-vinylnaphthalene) (P2VN) is measured in blends with polystyrene (PS). Differential scanning calorimetry (DSC) is employed to establish conditions for miscibility in 35% P2VN/PS blends from which a criterion for miscibility in low concentration systems may be inferred. Increases in* I_D/I_M *above the values expected for miscible systems that occur with increased host molecular weight or guest concentration are analyzed using Flory–Huggins theory. An interaction parameter of 0.020 ± 0.004 is consistent with the fluorescence results over a relatively narrow concentration range.*

THE STUDY OF POLYMER BLENDS has received considerable attention (*1–5*) because of the variety of bulk properties that may potentially be obtained at a lower cost than the production of a new homopolymer or copolymer (*6*). As a consequence, existing characterization procedures for polymer blends have been improved and new techniques have been developed. These techniques have been extensively reviewed elsewhere (*1–5*) and will not be considered again.

Criteria to be satisfied by any new method are the following:

1. Concentration sensitivity sufficient to allow the study of blends containing 0.5% or less of the minor polymer component. The study of low concentration polymer solutions has formed an invaluable part of polymer physics. Similar studies of low concentration polymer blends most likely will be equally fruitful.

0065-2393/84/0206-0077$07.00/0

2. Detectability of phase sizes of 10 nm. This small-scale morphology may significantly influence the mechanical properties of a blend (*7*).
3. The ability to characterize both miscible and immiscible polymer blends. The former case requires a measure of the coil size of the dispersed polymer or the second virial coefficient. In the latter case, such quantities as the phase size, volume fraction of each phase, and the nature of the interface between phases are desired.
4. Simplicity of sample preparation, compatibility of preparation and measurement techniques with the processing and environmental conditions encountered normally, and applicability of the characterization technique to a broad range of polymer blends.

Growing evidence indicates that excimer fluorescence provides a tool that can meet all of these requirements. An excimer may be formed between two identical aromatic rings in a coplanar sandwich arrangement, if one of the rings is in an electronically excited singlet state (*8–10*). Excimers have a lifetime in the range of 10–100 ns, and their fluorescence spectra exhibit a characteristic broad Gaussian shape at lower energy than the fluorescence of an isolated aromatic ring. The fluorescence of such an isolated chromophore is called monomer fluorescence, to distinguish it from excimer fluorescence. A convenient experimental measure of excimer fluorescence is the ratio of excimer to monomer emission intensities, I_D/I_M, obtained under photostationary state conditions.

The process of intermolecular excimer formation may be directly used in the study of low concentration polymer blends by mixing a small amount of a fluorescent aryl vinyl polymer (called the guest) with a nonfluorescent polymer (called the host) that is transparent to the wavelength of UV light required to excite the guest. Because aryl vinyl polymers contain one chromophore for each repeat unit in the polymer, they are readily detectable at low concentration. Furthermore, for a given extent of phase separation of the guest polymer, the largest number of intermolecular excimer-forming sites (EFS) will be generated when the guest is an aryl vinyl polymer. The minimum detectable phase size in such an experiment is of the order of the intermolecular distance in the pure chromophore-bearing compound. Finally, a variety of aryl vinyl polymers can be readily synthesized.

The objective of this chapter is to examine the dependence of I_D/I_M for a particular guest polymer upon the molecular weight of the host. By lowering the host molecular weight, it should be possible, at some point, to reduce or even eliminate immiscibility with a particular guest.

This chapter complements and extends two previous studies in which fluorescence spectroscopy (*11*) and differential scanning calorimetry (DSC) (*12*) were used to examine the blend thermodynamics. In these studies, three different poly(2-vinylnaphthalene) (P2VN) guests (molecular weights 21,000, 70,000, and 265,000) were blended with eight different polystyrene (PS) hosts (molecular weights 2200–390,000). In the earlier work, the guest concentration was quite low, 0.3% or less. In the present study, however, concentrations up to 30% are considered. The significant contribution of this work is the demonstration that, for all concentrations examined, a single interaction parameter for the P2VN/PS pair may be extracted from the fluorescence data. Also, by combining the results of DSC and fluorescence spectroscopy, the absolute phase behavior of all the P2VN/PS blends has been determined.

Experimental

The polymer samples, solvent casting procedure, and spectrofluorimeter have been described previously (*11*). The only exceptions in the present work were that blends containing 3 and 30 wt% P2VN (21,000) were examined under frontface illumination to minimize self-absorption. In addition, the 3 and 30 wt% blends were dried for only 9–12 h, rather than the standard 16-h time, to eliminate crazing.

Blend samples that were examined by DSC were prepared in the same manner as those prepared for fluorescence work. All DSC samples were annealed at 453 K for 20 min to remove any residual toluene casting solvent and to allow the sample to flow into the bottom of the pan. DSC thermograms for the samples were obtained with a Perkin–Elmer DSC-2 calorimeter operating at a heating rate of 20 K/min.

Review of Fluorescence Studies of Blends Containing Aryl Vinyl Polymers

We present a brief review of earlier blend work to place the excimer fluorescence technique in perspective and also justify the selection of P2VN as the guest polymer for detailed study.

Before 1979, only a few systematic studies of the excimer fluorescence of blends containing aryl vinyl polymers were reported. Although PS was the first polymer to be studied in pure films or in solution (*13–16*), no studies of PS in polymer blends were made during this period. Poly(1-vinylnaphthalene) (P1VN) was the next polymer to be studied in pure films (*17, 19*) and in solution (*14, 18, 19*). Fox et al. (*19*) gave brief qualitative results of the excimer fluorescence of poly(methyl methacrylate) (PMMA) blends containing 0.01–40% P1VN. At about the same time, pure films (*19, 20*) and solutions (*19–21*) of P2VN were first studied.

The P2VN excimer fluorescence of several P2VN/PS blends was also recorded (*20, 22–24*). These blends were prepared by polymerization at 393 K of styrene that contained about 0.1% P2VN (*20*) or by film casting from a benzene or chloroform solution containing P2VN and PS in the ratio 0.2:100 (w/w) (*22–24*). Frank (*23*) noted that solvent-cast P2VN/PS blends containing 1% or more P2VN were visually immiscible. Such immiscible systems were not included in this study of P2VN excimer fluorescence from P2VN/PS blends at temperatures near the glass transition temperature of PS. In a study of the effect of P2VN molecular weight on the P2VN excimer fluorescence of solvent-cast P2VN/PS blends, Nishijima et al. (*24*) found anomalously high excimer fluorescence for P2VN molecular weights above 140,000 that they attributed to phase separation.

Preliminary studies on the pure-film and solution fluorescence of poly(4-vinylbiphenyl) (PVBP) (*25*), poly(2-vinylfluorenone) (*26*), poly(*N*-vinylcarbazole) (PVK) (*27, 28*), poly(acenaphthylene) (PAN) (*19, 21, 29*), and poly(vinylpyrene) (*30–32*) were reported. Of these polymers, only PVBP (*25*) was studied in polymer blends. These benzene-cast blends contained 0.2% PVBP in PS or PMMA, and the PVBP excimer fluorescence was unaffected by the polymer host. However, the PVBP molecular weight was not given in this study (*25*).

Except for the comments by Frank (*23*) and Nishijima et al. (*24*) on the possibility of phase separation in the polymer blends, all the pre-1979 studies of aryl vinyl polymers in blends (*19, 20, 22–25*) tacitly assumed that the fluorescent guest was miscible with the host. This assumption also appeared in related studies of chromophore-bearing polymers in blends with nonfluorescent hosts. A typical example of a copolymer guest study is the work of Reid and Soutar (*33*) on the depolarization of the fluorescence of 1-vinylnaphthalene-*co*-methyl methacrylate copolymers dispersed at low concentration in PMMA. North and Treadaway (*34*) examined the interaction between a polymeric guest (PVK) and a low molecular weight guest (anthracene), both dispersed at low concentration ($<0.1\%$) in a PS or PMMA host. Despite the importance of the "miscible-blend" assumption in these reports, the morphology of the blends was generally not verified by other experimental or theoretical means.

By 1979 the focus of studies of aryl vinyl polymer blends had shifted away from the details of the photophysics of the blends and moved toward the morphology of the blends. Frank et al. (*35, 36*) studied P2VN excimer fluorescence from a series of poly(alkyl methacrylate) hosts containing 0.2% P2VN. When the fluorescence ratio I_D/I_M was plotted versus the solubility parameter of the methacrylate hosts, a smooth curve possessing a distinct minimum was observed. Similar results for methacrylate blends containing PVBP and PAN were also reported (*36*).

The fluorescence behavior of 0.2% PVK dispersed in PS, PMMA, and poly(isobutylene) was studied by Chryssomallis and Drickamer (*37*). The spectrum of PVK, in contrast to P2VN, consists of the emission from two excimers and does not show any monomer (i.e., *N*-ethylcarbazole) emission. The ratio of the low energy excimer intensity to the high energy intensity, I_{D2}/I_{D1}, declined in the following order at atmospheric pressure: PVK/

poly(isobutylene), neat PVK, PVK/PMMA, and PVK/PS. All these blends were reported to be clear.

In a similar study of PVK/PS and PVK/PMMA blends, Johnson and Good (*38*) found that I_{D2}/I_{D1} for PVK/PMMA was lower than for PVK/PS, a contradiction of the previous report. Moreover, for a PVK molecular weight of 84,000, Johnson and Good (*38*) reported that the 0.2% PVK/PMMA blend was optically hazy. They concluded that I_{D2}/I_{D1} should increase with increasing miscibility for blends containing less than 10% PVK.

In spite of the disagreement, both PVK blend studies (*37*, *38*) show that the spectral parameter I_{D2}/I_{D1} is only about two or three times larger for neat PVK compared to low-concentration PVK blends. By comparison, I_D/I_M for neat P2VN was found (*36*) to be 10–20 times larger than I_D/I_M for low concentration P2VN blends. A large spectral change between neat guest polymers and their low concentration P2VN blends. A large spectral change between neat guest polymers and their low concentration blends with miscible hosts is essential for good sensitivity to phase separation. P2VN is superior to PVK in this regard.

Results

Fluorescence Spectroscopy and Optical Clarity. SELECTION OF BLEND SYSTEMS FOR STUDY BY THE EXCIMER FLUORESCENCE TECHNIQUE. A general goal of this work was the continued development of the excimer fluorescence technique for the study of low concentration (0.5% or less) blends. The ideal blend system for this study should be miscible over some part of the low concentration range and immiscible at other concentrations. At the outset, however, it was not known how low the experimental detection limit of the excimer fluorescence technique would be.

The first task was to select the guest polymer from P2VN and PS, the most widely studied aryl vinyl polymers. Despite the disadvantage of being relatively exotic, P2VN was chosen as more suitable for a study of phase-separated blends. In addition to the potentially large spectral changes possible upon phase separation, P2VN (*39*) possesses a larger total quantum yield (0.13) in solution than the alternative polymer, PS (*40*, *41*) (quantum yield, 0.025). Moreover, the longest wavelengths at which P2VN and PS may be excited efficiently are 290 and 260 nm, respectively. This difference permits the use of a wider variety of host polymers and solvents with P2VN relative to PS. Finally, recent calculations of conformational statistics (*42*, *43*) have shown that the rotational dyad structure of P2VN is similar to PS, so that there is no disadvantage in choosing P2VN.

The host polymers were selected as follows: First, all host polymers were required to be in the glassy state at room temperature, to minimize the complications induced by different local viscosities on the photophysics of the guest. Second, host polymers that were shown earlier (*35*) to be grossly incompatible with P2VN were ruled out. At this point, a major question arose on the absolute miscibility of the P2VN blends. Clearly, if the excimer fluorescence technique alone were to be used, the results could only be interpreted in a relative way. However, the only way that an additional technique such as DSC would be helpful in determining the miscibility of low concentration blends requires a P2VN blend with a large margin of miscibility, so that a 35% blend would still be miscible. Detection of miscibility for the high concentration blend by DSC would imply that the low concentration blend was also miscible. The problem with this absolute procedure is that it requires an extraordinarily miscible blend system, which is generally nonexistent among commercially available polymers having molecular weights above 10,000.

This problem was solved by using the experimental parameter of host molecular weight. By lowering the host molecular weight to about 1000, a number of high concentration P2VN blends made with these host oligomers were found to be miscible by DSC. As a result, the miscibility of all the P2VN blends studied has been placed on an absolute basis, in contrast to the earlier studies. (*11*, *12*).

The choice of PS as the host polymer was dictated primarily by the availability of well-characterized, low molecular weight samples. Furthermore, PS is commonly chosen to provide a glassy matrix at room temperature for studies of the luminescence of small molecules. PS absorbs light very weakly at 290 nm (*44*): the optical density, $\log_{10}(I_0/I)$, of PS relative to P2VN (*45*) at 290 nm is $10^{-4}:1$, so that only about 3% of the exciting light is absorbed by PS in a 0.3 wt% P2VN/PS blend. Finally, the glass transition temperatures of the PS hosts lie between 368 and 388 K for $M \geq$ 100,000 and are always greater than 298 K even for the lowest molecular weights.

As in the majority of previous fluorescence studies of polymer blends (*19*, *22–25*, *33–38*), our samples were prepared by solution casting. In this procedure, a film of a solution of the two polymers in a common solvent is spread onto a flat surface and the solvent is allowed to evaporate. Solution casting avoids the harsh conditions of shear mixing and allows the preparation of small (9 mg) samples. A drawback of solution casting, however, is that the blend morphology may be affected by the casting solvent. For example, different casting solvents can produce both miscible and immiscible blends of PS with poly(vinyl methyl ether) (PVME) (*46*, *47*).

Gelles and Frank (*48*) used the excimer fluorescence of PS in such blends to observe a substantial increase in the ratio I_D/I_M for visibly phase-separated blends relative to miscible blends. The casting solvents employed were tetrahydrofuran and toluene, respectively. On the other hand, the previously mentioned studies of PVK blends revealed no clear-cut difference between the casting solvents dichloromethane (*37*) and benzene (*38*). Similarly, a fluorescence study on the relaxation behavior of solvent-cast P2VN/PS blends (*23*) reported no difference between the casting solvents chloroform and benzene.

Consequently, all of the blends we used were prepared from a single casting solvent, toluene. Toluene was selected as the casting solvent because this choice permitted a direct comparison of our results with a number of previous P2VN blend studies (*22–24, 35, 36*) in which the casting solvent was benzene or toluene.

EFFECT OF MOLECULAR WEIGHT AT LOW P2VN CONCENTRATION. Figure 1 presents the fluorescence results obtained previously (*11*) on the effect of PS host molecular weight on I_D/I_M for the three P2VN samples examined. These results are reported in a modified form to emphasize the relationship between the fluorescence behavior and the visual appearance. The interesting features of these results are the independence of guest I_D/I_M on host molecular weight for P2VN (21,000) and the significance of the increase in I_D/I_M in both P2VN (70,000) and P2VN (265,000) for host molecular weights greater than 4000.

The fluorescence intensity ratio is given by

$$\frac{I_D}{I_M} = \frac{Q_D}{Q_M}\left[\frac{1 - M}{M}\right] \tag{1}$$

where Q_D is the ratio of the fluorescence decay constant to the total decay constant for the excimer, and Q_M is the analogous ratio for the monomer species (*48, 49*). Also, in the equation, M is the probability that a photon absorbed by the P2VN guest will decay along a monomer pathway and is a function of the concentration and segregation of the P2VN in the blend. In separate work (*50*) we have shown that the factor Q_D/Q_M for the P2VN/PS pair can be assumed to be independent of the molecular weight of the PS host. Thus, the quantity $(1 - M)/M$ is the only factor that influences I_D/I_M in the blend studies reported in this chapter.

An immiscible blend will have a larger value of I_D/I_M than a miscible blend of the same bulk composition (*48*). In an immiscible blend most of the chromophores are located in the guest-rich phase; therefore the average value of M is low, and I_D/I_M is large. Thus, the fluorescence data for P2VN (70,000) and P2VN (265,000) indicate that blends with PS (2200) are more miscible than those with PS (390,000). The absolute miscibility of PS (2200)-containing blends must be determined by another technique, however, because the a priori calculation of the value of I_D/I_M for a miscible blend cannot yet be made with sufficient accuracy.

Optical clarity is one such alternative technique that is simple and that may be directly applied to blends prepared for fluorescence studies. The appearances of the members of the three series of blends are indicated by

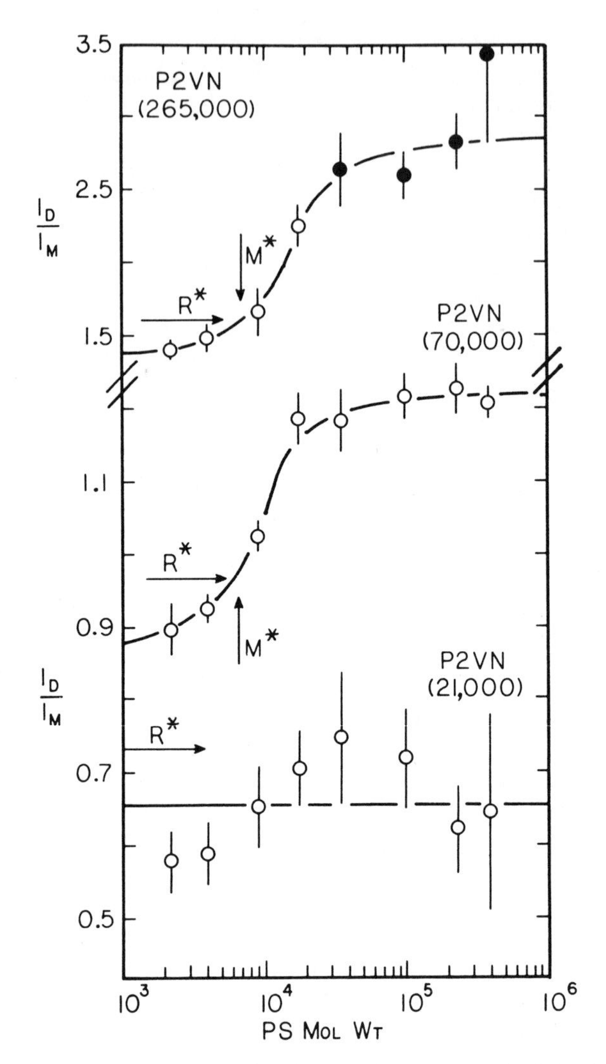

Figure 1. Effect of PS host molecular weight on I_D/I_M *for blends containing 0.3 wt % P2VN. The molecular weights of the P2VN samples are given in parentheses. The excimer and monomer intensities are measured at 398 and 337 nm, respectively.* I_D/I_M *is uncorrected for spectral overlap and was measured under backface illumination. Key: ○, optically clear blends; and ●, optically cloudy blend. (Reproduced from Ref. 11. Copyright 1981, American Chemical Society.)*

the plotting symbols in Figure 1: open circles for films that were optically clear and had no apparent phase separation, and filled circles for films that ranged from slightly bluish to those with regions of white. Phase separation has definitely occurred for blends of P2VN (265,000) with PS molecular weights of 35,000, 100,000, 233,000, and 390,000. The increase in I_D/I_M for P2VN must be associated with this phase separation.

The P2VN (70,000) results are ambiguous, however. Although the increase in I_D/I_M is similar to the increase for P2VN (265,000), all the blends are optically clear. Possibly the fluorescence behavior is caused by P2VN coil expansion in the low molecular weight PS. However, experiments at very low P2VN concentrations (0.003%) indicated that I_D/I_M was constant at 0.9 ± 0.3 for all eight PS hosts. Thus, the increase in I_D/I_M for P2VN (70,000) in Figure 1 must be a result of phase separation into domains that are very small.

Finally, all of the P2VN (21,000) blends are optically clear and I_D/I_M does not significantly change with change in host molecular weight. Thus, these blends are assumed to be miscible.

In spite of the convenience of the use of appearance as a characterization tool, another technique must be applied to determine the miscibility of the optically clear blends. Later, we will show by DSC that all P2VN/PS (2200) blends are miscible. For the moment, these blends will be assumed to be miscible so that a procedure for quantifying the fluorescence data can be described. First, R_{misc}, the value of I_D/I_M for a miscible blend having the same P2VN concentration as the other blends under study, is recorded. Second, the average relative error in I_D/I_M, W, is determined from all points in the figure that contains the fluorescence data under consideration. Then, the value of R^*, which corresponds to I_D/I_M for a blend that has just become immiscible, is defined as

$$R^* = (1 + 2W)R_{misc} \tag{2}$$

The molecular weight of the host at which the blend becomes immiscible is given by M^*, which is located on the figure at the point where I_D/I_M equals or exceeds R^*.

The values of R^* and M^* obtained from Figure 1 are listed in Table I. In each case, R_{misc} was chosen to be the value of the ratio for the PS (2200) blend, because these blends were assumed to be miscible. Note that $M^* = 8000$ for the P2VN (265,000) guest, as compared to the PS molecular weight of 35,000 at the point of turbidity. This and other fluorescence studies will demonstrate the earlier conclusion that the fluorescence method is more sensitive to immiscibility than the method of optical clarity.

Table I. R^* and M^* for 0.3 wt% P2VN/PS Blends

P2VN Mol. Wt.	R_{misc}[a]	W, % *Relative Error*	R^*[a]	M^*
21,000	0.58	13	0.73	>390,000[b]
70,000	0.90	4	0.97	7,000
265,000	1.40	7	1.60	8,000

[a] I_{397}/I_{337} measured under backface illumination, uncorrected for overlap.
[b] Although the ratio of the PS (35,000) blend is slightly greater than R^*, this is not considered to be significant because I_D/I_M decreases for molecular weights greater than 35,000.

Of course, the fluorescence method described in this chapter is exceedingly more sophisticated than the use of simple appearance as a diagnostic tool. Therefore, the direct comparison of the two methods could be misleading. In fact, for systems in which the refractive index difference between the two components is sufficiently large, sensitive light scattering measurements may be used to determine the molecular weight, the radius of gyration, and the interaction parameter for polymers in dilute solution. Although scattering from dust particles presents a significant problem for solid state mixtures, light scattering has been used to follow the kinetics of phase separation (*51*). However, similar phenomena have also been studied by using excimer fluorescence with comparable sensitivity (*52, 53*).

In spite of the differences in power at the levels at which they are employed, comparison of the excimer fluorescence results with the appearance does provide a link, albeit crude, between the molecular level interactions and the bulk morphology. Any information of this type is useful in establishing the limits of applicability of a new method such as the excimer probe.

EFFECT OF P2VN CONCENTRATION FOR TWO EXTREME PS MOLECULAR WEIGHTS. The point of immiscibility for polymer blends having variable guest concentration can also be quantified by the fluorescence ratio technique. The critical concentration, defined as C^*, is located at the point where I_D/I_M equals or exceeds R^*. However, because the values of I_D/I_M for miscible blends may vary with guest concentration, miscible blends must be prepared in conjunction with the blends of interest over the same guest concentration range.

To explore the relationship between cloud points determined by light scattering and C^* given by the fluorescence ratio, we prepared two series of blends by using the P2VN (70,000) guest over the concentration range of 0.01–10 wt% P2VN. A polydisperse PS host with M_w 158,000 was used in the first series of blends (*11*), and the monodisperse PS (2200) host was used in the second series. We obtained values of R_{misc} from the latter series after assuming that these blends were miscible.

The fluorescence results for the PS (158,000) and PS (2200) blends are compared in Figure 2. The lower half of this figure includes the 0–1% P2VN (70,000) concentration range. Blends for both PS hosts are optically clear in this concentration range. However, the value of the ratio for PS (2200) blends containing 0.6 wt% P2VN or less is constant at 1.07, and the corresponding value for PS (158,000) is always above 1.2. The relative error in I_D/I_M, W, is about 6% in the concentration range below 1 wt% P2VN. If R_{misc} is assumed to be 1.07 for blends below 0.6 wt% P2VN concentration, the value of R^* is found to be 1.20. Thus, C^* for the PS (158,000) blends is found to be 0.01 wt% P2VN (70,000). By contrast, these blends do not become optically cloudy until a concentration of 3 wt% P2VN.

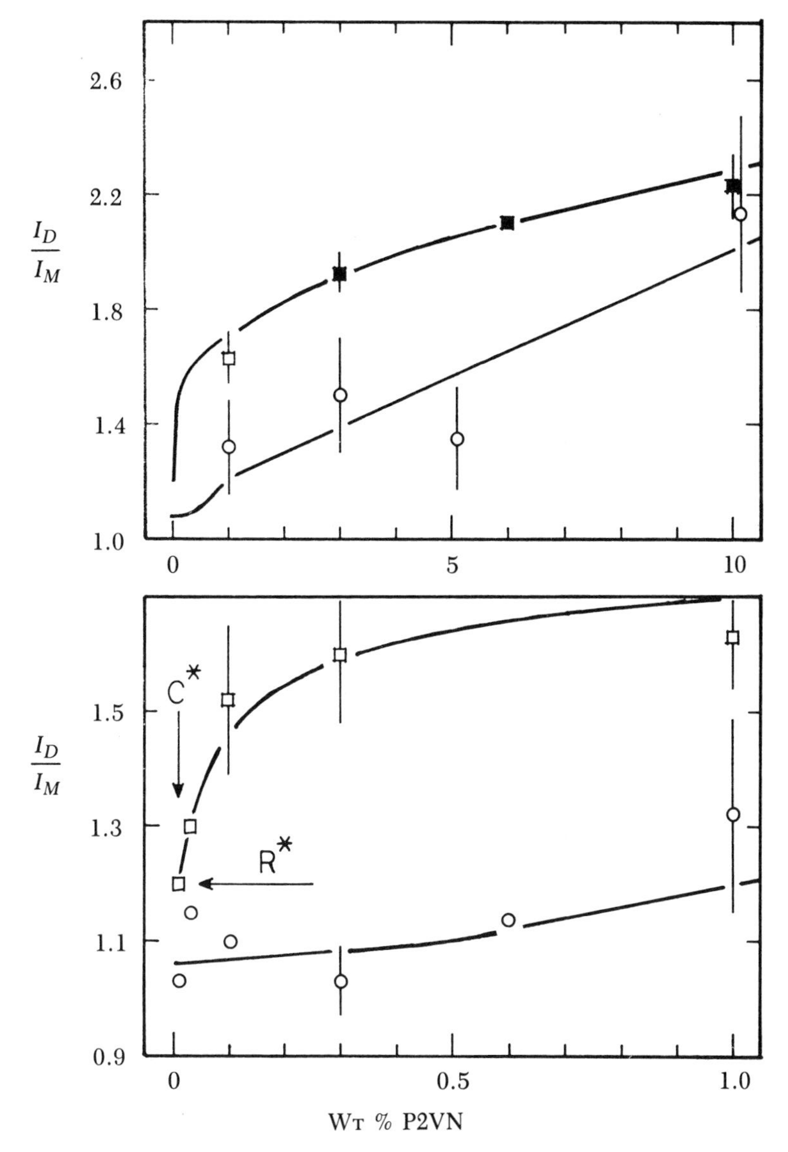

Figure 2. Effect of concentration on I_D/I_M *for P2VN (70,000) blended with PS (2200) (○) and PS (158,000) (□).* See *Figure 1 for details.*

The fluorescence ratio of the PS (158,000) blends rises smoothly between 1 and 10% P2VN concentration, as shown in the upper half of Figure 2. Although values of I_D/I_M for the PS (2200) blends lie below those for the PS (158,000) blends, the ratios become difficult to distinguish at 10 wt% concentration. The relative error in the ratio increases to nearly 15%

of I_D/I_M at 10 wt% P2VN. The distinction between the blends can be made visually at 10%, however, because the PS (2200) blends are optically clear.

EFFECT OF MOLECULAR WEIGHT AT MODERATE P2VN CONCENTRATION. From the fluorescence data and the appearance of the 0.3 wt% P2VN (21,000)/PS blends, we infer that they are miscible for all PS molecular weights. However, higher concentration P2VN (21,000) blends were expected to be immiscible and exhibit an increase in I_D/I_M with an increase in PS molecular weight. To check this prediction, 3 and 30 wt% P2VN (21,000) blends were prepared with the molecular weight series of polystyrenes. The fluorescence ratios of the 3- and 30-wt% blends are plotted against PS molecular weight in Figures 3 and 4, respectively.

The dependence of I_D/I_M on host molecular weight for the 3- and 30-wt% blends is somewhat different from that for the 0.3-wt% blends. Although the ratio is still lowest for very low host molecular weights, it increases almost linearly with log (M_{host}) between M_n 2200 and 390,000. Points of inflection in the ratio are barely discernible in Figures 3 or 4. Although values of I_D/I_M for the 0.3- and 3.0-wt% blends with PS (2200)

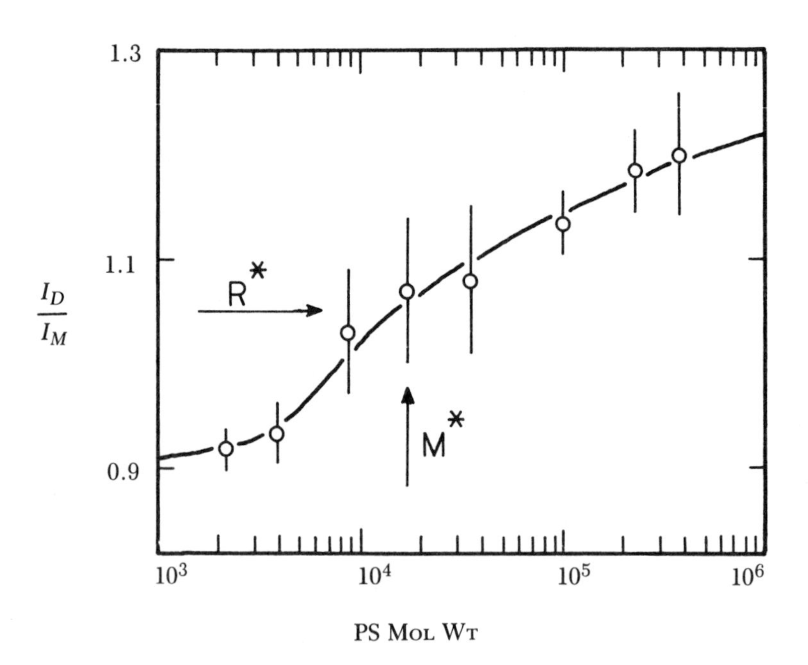

Figure 3. Effect of PS host molecular weight on I_D/I_M for blends containing 3.0 wt% blends of P2VN (21,000). The excimer and monomer intensities are measured at 398 and 337 nm, respectively. The ratio is uncorrected for spectral overlap and was measured under frontface illumination. The averages of two data points taken from one film of each blend are shown. Open circles denote optically clear blends.

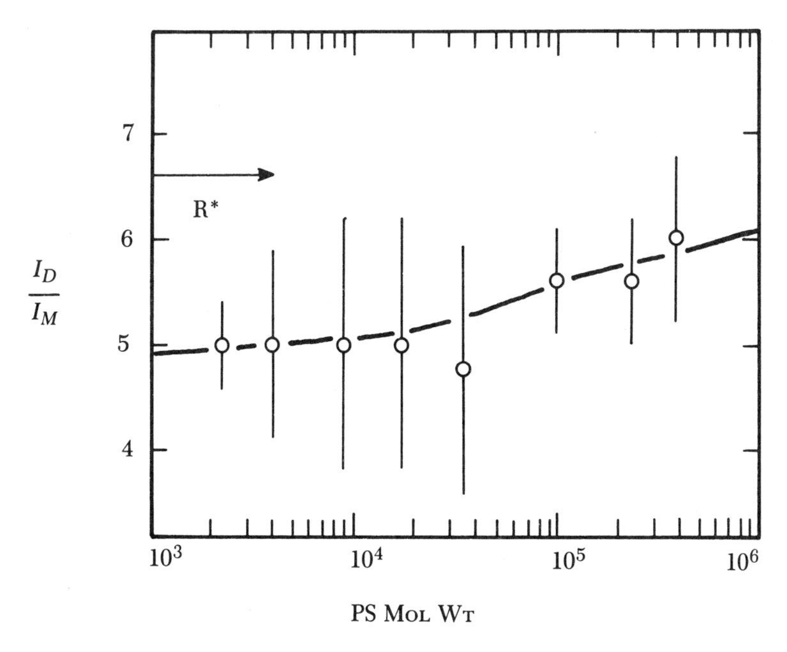

Figure 4. Effect of PS host molecular weight on I_D/I_M *for blends containing 30 wt% P2VN (21,000).* See *Figure 3 for details.*

are nearly equal, at 0.67 and 0.92, respectively, the value of 5.0 for the 30-wt% blend with PS (2200) is remarkably large.

The increase in the ratio with the increase in host molecular weight for the 3-wt% blends of Figure 3 can be attributed to phase separation. All these blends were optically clear, however, as was observed for the 0.3 wt% P2VN (70,000) blends. Thus, the domain size must be small. If the 3 wt% P2VN (21,000)/PS (2200) blend is assumed to be miscible, R^* is found to be 1.05 and M^* to be 18,000.

The data of Figure 4 for the 30 wt% P2VN (21,000) blends are interpreted in the same manner as Figure 3. All 30% blends were optically clear. Again, these results suggest that the domains in the phase-separated blends were very small. However, if the 30 wt% P2VN (21,000)/PS (2200) blend is assumed to be miscible (as confirmed by later DSC results), then R^* is 6.6. This value is larger than the fluorescence ratios for all 30% blends and indicates that all the blends are miscible, within experimental error. This result seems inconsistent, because the 3-wt% blends were found to be immiscible above M^* 18,000.

The inconsistency may be explained in two ways: First, the relative error of 16% for the ratio data in Figure 4 is quite large. This large relative error leads to a large value of R^*. This R^* value could probably be reduced with additional data, in which case a value of M^* less than 18,000 would be expected. Second, it is possible that the excimer fluorescence technique becomes less sensitive to differences in miscibility when the concentration of the guest polymer is large. This explanation is supported by the data of Figure 2, in which curves for miscible and immiscible blends appear to merge above the guest polymer concentration of 10%.

In a final test of miscibility of the 30 wt% P2VN (21,000) blends, the samples prepared for the fluorescence experiments were annealed for 20 min at 453 K and reexamined visually. After the heating, blends with PS hosts of molecular weights greater than or equal to 9000 appeared cloudy and bluish. Thus, an estimate of M^* 9000 was made for the unheated 30 wt% P2VN (21,000)/PS blends.

SUMMARY. All fluorescence and appearance results are collected in Table II. The PS molecular weight at which immiscibility first appears is listed for studies involving variable PS molecular weight, and the P2VN concentration where immiscibility first appears is listed for studies involving variable P2VN concentration. In every case, the fluorescence technique has superior (or at least equal) sensitivity to blend immiscibility relative to the visual technique. The absolute miscibility of the blends containing PS (2200) has been assumed in the past sections. In the next section we will use DSC to verify this assumption.

Differential Scanning Calorimetry (DSC). Two conditions limit the types of polymer blends that can be successfully studied by DSC. First, the T_g values of the constituent polymers must differ by at least 50 K to ensure sufficient resolution so that the two T_g values of an immiscible blend can be distinguished. Second, at least 20% of the minor component of the blend must be present, or its contribution to the heat capacity of the blend will be too small to be detected.

Table II. Immiscibility of P2VN/PS Blends as Determined by Appearance and Fluorescence Techniques

Blend	*Appearance*	*Fluorescence*
Variable PS Molecular Weight	M	M^*
0.3 wt% P2VN (21,000)	>390,000	>390,000
3.0 wt% P2VN (21,000)	>390,000	18,000
30.0 wt% P2VN (21,000)	>390,000	>390,000
30.0 wt% P2VN (21,000), heated to 453 K	9,000	—
0.003 wt% P2VN (70,000)	>390,000	>390,000
0.3 wt% P2VN (70,000)	>390,000	7,000
0.3 wt% P2VN (265,000)	35,000	8,000
Variable P2VN Concentration	C	C^*
P2VN (70,000)/PS (2200)	>10%	>10%
P2VN (70,000)/PS (158,000)	3%	0.01%

Abbreviations: M, blend cloudy at and above the stated PS molecular weight; M^*, blend has ratio > R^* at and above the stated PS molecular weight; C, blend cloudy at and above the stated P2VN concentration; C^*, blend has ratio > $R^*(C)$ at and above the stated P2VN concentration.

Typical DSC thermograms for the unblended polymers PS and P2VN are displayed in Figure 5. The temperature at which the heat capacity first begins to increase sharply is T_1; the local maximum in heat capacity is denoted T_2. The thermogram above T_2 has a slope slightly lower than the slope of the thermogram below T_1. The difference between T_2 and T_1, ΔT, is characteristic of the particular polymer under study. The value of T_g is obtained as the average of T_1 and T_2. These temperatures for PS and P2VN are listed in Table III.

Because the T_g values of pure P2VN and PS (2200) differ by 78 K, DSC would seem to be an ideal independent method for determining the state of miscibility of the P2VN/PS (2200) blends. Unfortunately, only one of the many blends previously studied contains more than 20% P2VN. Moreover, any P2VN/PS blend to be studied by DSC must be annealed for 20 min at 453 K. These conditions prohibited the direct study of any blend listed in Table II.

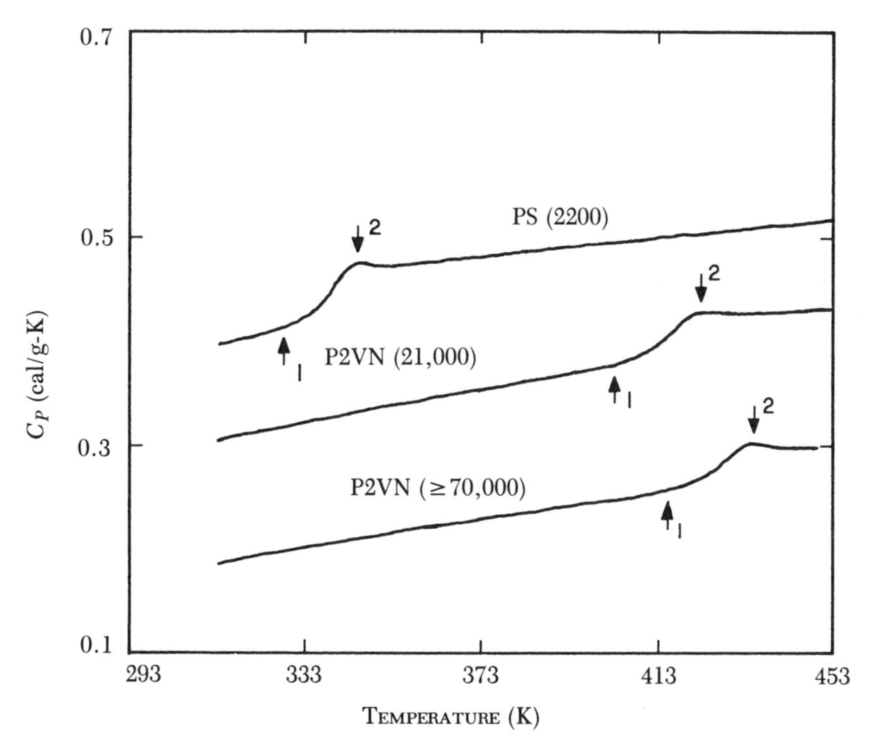

Figure 5. DSC thermograms for P2VN and PS (2200). Samples were heated at 20 K/min. The bottom curve is true; each additional curve has been displaced + 0.1 cal/g-K from the curve immediately beneath it. The upward and downward arrows indicate the points of deviation from the glassy and rubbery regimes, respectively. See Table III for temperatures. (Reproduced from Ref. 12. Copyright 1983, American Chemical Society.)

Table III. Transition Temperatures for Pure Polymers and Polymer Blends

Polymer	T_1	T_2	ΔT	T_g[a]
PS (2200)	328	345	17	337
P2VN (21,000)	403	426	23	415
P2VN (70,000)[b]	415	435	20	425
35% P2VN (21,000)/65% PS (2200)	343	383	40	363
35% P2VN (70,000)/65% PS (2200)	344	383	39	364
35% P2VN (265,000)/65% PS (2200)[c]	343	358	15	351

[a] $T_g = (T_1 + T_2)/2$.
[b] The temperatures for P2VN (265,000) are nearly the same as for P2VN (70,000).
[c] $T_3 = 366$, $T_4 = 400$, $\Delta T' = 34$, and $T_g' = (T_3 + T_4)/2 = 383$.

Indirect measurements are possible, however, because previous visual examination of P2VN/PS blends had shown that either heating or an increased P2VN concentration caused increased immiscibility. Thus, if a 35 wt% P2VN blend were found to be miscible by DSC, then this finding would imply that an unheated blend containing less P2VN would also be miscible. Consequently, a series of 35 wt% P2VN/PS (2200) blends was prepared and examined by DSC to determine indirectly the state of miscibility of the P2VN/PS (2200) blends at lower concentration.

DSC thermograms for the 35 wt% P2VN/PS (2200) blends are shown in Figure 6. The P2VN (265,000) guest will be considered first. Two transitions are observed: the lower is attributed to the PS (2200) host and the upper is attributed to the P2VN (265,000) guest. In this figure, T_1 and T_2 were determined in the same manner as for the pure polymers, and T_3 denotes the first point on the curve above T_2 at which the slope is the same as that of the curve below T_1. Because no local maximum occurs in the heat capacity above T_3, T_4 was located at the point where the high temperature heat capacity showed a downward deviation. As shown by the values of T_1–T_4 listed in Table III, the lower temperature transition is observed to be 14 K above the T_g of pure PS, and the higher temperature transition is 42 K below the T_g of pure P2VN. Moreover, the 34-K spread between T_3 and T_4 in the P2VN (265,000) blend is larger than the 20-K spread between T_1 and T_2 for pure P2VN.

Because two T_g values were recorded, we conclude that the 35 wt% P2VN (265,000)/PS (2200) blend is immiscible. However, the shift in the values of T_g from the pure-polymer values and the broadening of the glass transition interval for the P2VN component suggest that the immiscibility is not severe. This conclusion is confirmed by the appearance: although the blend was clear when first solvent cast, it became cloudy and bluish when it was heated in preparation for DSC.

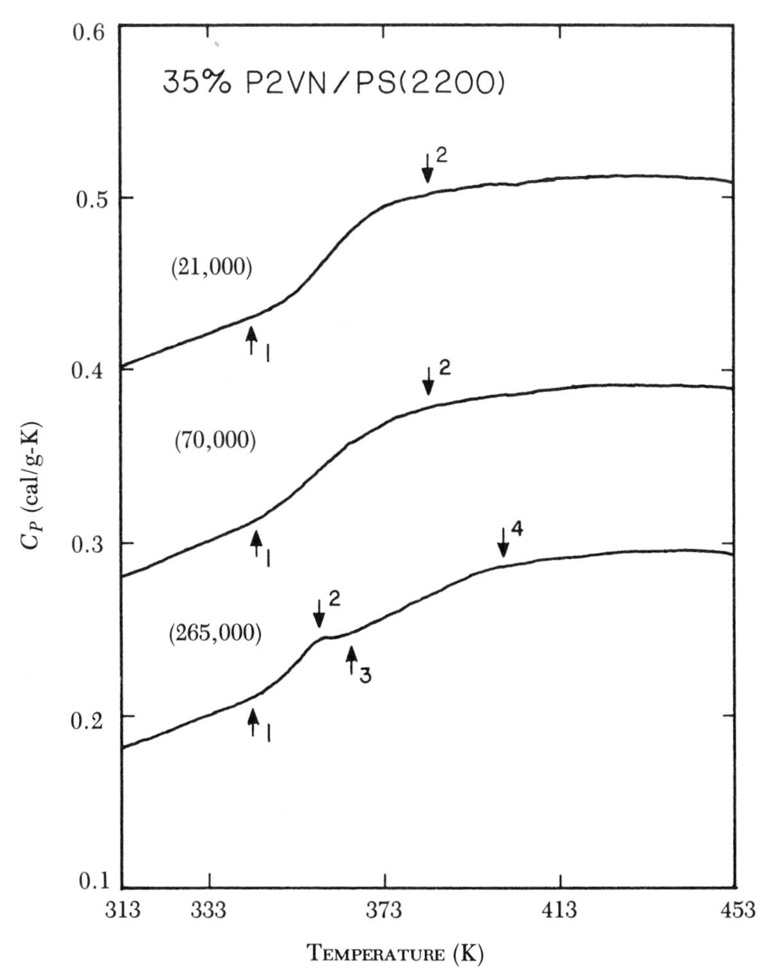

Figure 6. DSC thermograms for blends of PS (2200) containing 35 wt % P2VN. The molecular weight of the P2VN in each blend is given in parentheses. See *Table III for temperatures. (Reproduced from Ref. 12. Copyright 1983, American Chemical Society.)*

The upper two thermograms in Figure 6 for the P2VN (21,000) and P2VN (70,000) blends are nearly identical. Because no local maximum occurred in the heat capacity above T_1, T_2 was located at the point where the high temperature heat capacity showed a downward deviation. Values of T_1 and T_2 for the P2VN (21,000) and P2VN (70,000) blends are identical, and both blends have the same value of T_g, 364 K. As a check, we have applied the empirical Fox equation commonly used to predict the T_g of copolymers and of miscible blends.

$$\frac{1}{T_g} = \frac{w_A}{T_{gA}} + \frac{w_B}{T_{gB}} \tag{3}$$

where w_A and T_{gA} are the weight fraction and glass transition temperature, respectively, of homopolymer A. This equation yields 363 K, in excellent agreement with the value obtained experimentally.

In summary, both the 35 wt% P2VN (21,000)/PS (2200) and P2VN (70,000)/PS (2200) blends are miscible because only one glass transition is apparent in the DSC thermograms. Moreover, these blends were optically clear even after being heated in preparation for DSC.

On the basis of DSC results for 35 wt% P2VN/PS (2200) blends, we conclude that the unheated blends containing $\leq$35% P2VN (21,000) or P2VN (70,000) in PS (2200) are miscible. These results confirm almost all the assumptions made about the miscibility of the PS (2200)-containing systems listed in Table II. The sole exception is the 0.3 wt% P2VN (265,000)/PS (2200) blend. Both it and the 35 wt% blend are optically clear before heating, but the 35 wt% blend becomes cloudy after heating and shows two transitions by DSC. If the 0.3-wt% blend were immiscible, contrary to assumption, then the value of M^* given in Table II would be too large. In our discussion, we will assume that M^* 8000 is an upper limit for the 0.3 wt% P2VN (265,000) blends.

Discussion

The simplest possible theory for the thermodynamics of mixing of two polymers is the Flory–Huggins lattice treatment (*54*, *55*). Although this approach has limitations (*11*), binodal calculations presented earlier were useful in correlating the fluorescence results for the low concentration blends (*11*). In our present study, we have again referred to these calculations for explanation of the moderate concentration results. For convenience, the calculated binodals are reproduced in Figures 7–9 for the P2VN guests having M_v 21,000, 70,000, and 265,000, respectively.

The usual method of presentation is to give temperature (or interaction parameter) vs. concentration and hold the host molecular weight constant. These binodals are plotted as P2VN concentration vs. PS host molecular weight and the interaction parameter is held constant. A range of interaction parameters from 0.005 to 0.025 was selected because these seemed most consistent with the appearance and fluorescence behavior of the blends. The dashed lines in Figures 7 through 9 represent the bulk concentration of guest polymer in the films studied. As the molecular weight of the polystyrene host increases, the concentration of guest in the guest-lean phase at the binodal decreases rapidly. The films in which the host molecular weight is sufficiently large may show phase separation if the guest concentration at the binodal is less than the bulk guest concentration.

We first consider the 0.3-wt% P2VN (21,000) blends that were earlier concluded to be miscible. Because none of the binodals calculated for the different interaction parameters intersects the dashed line indicating the

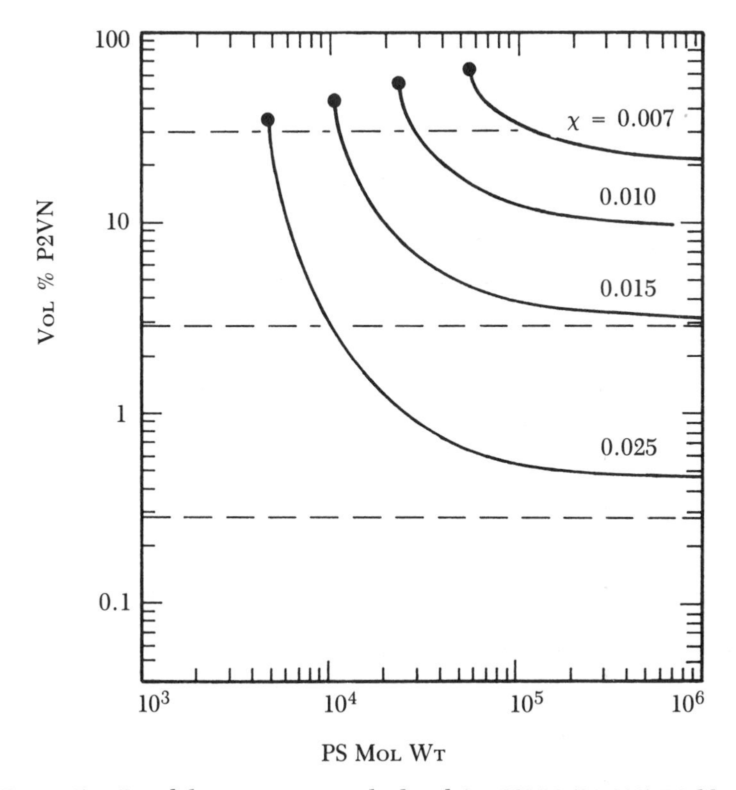

Figure 7. Binodal compositions calculated for P2VN (21,000)/PS blends. Results for the P2VN-lean phase are shown as volume percent P2VN vs. PS molecular weight. The interaction parameter for each curve is given, and the critical point is indicated by the filled circle. The dashed lines denote 0.3, 3, and 30 wt% P2VN. (Reproduced from Ref. 11. Copyright 1981, American Chemical Society.)

bulk film concentration, an upper limit of about 0.027 is established for χ_{AB} (*11*). Next, the 3-wt% P2VN (21,000) blends for which M^* is 18,000 are considered. By following the 3-wt% P2VN line in Figure 7 to the PS molecular weight equal to M^*, $\chi_{AB} = 0.021$ is obtained. Finally, the fluorescence data from the 30-wt% P2VN (21,000) blends in Figure 4 are considered. The observation that $M^* > 390{,}000$ is surprising, because the increase in P2VN concentration from 3- to 30-wt% was expected to reduce M^* below 18,000. Disregarding the fluorescence data for the 30-wt% blends in favor of visual observations made on heated 30-wt% blends, Table II shows that $M = 9000$ at the point of immiscibility. Reference to the 30-wt% P2VN line in Figure 7 yields $\chi_{AB} = 0.018$. This value is taken as an estimate for the interaction parameter of the unheated 30-wt% blends.

The approach just illustrated for the P2VN (21,000) blends is subject to several uncertainties. First, the method depends heavily on the accuracy and reliability of R_{misc}, the value of the fluorescence ratio of a miscible blend of the appropriate P2VN bulk concentration. Second, excessive overall error in I_D/I_M for a given set of blends will artificially increase M^* and lower χ_{AB}. Finally, if the P2VN sample is polydisperse, the initial increase in the fluorescence ratio above R_{misc} may be a result of the P2VN components of the highest molecular weight, which makes the proper P2VN molecular weight for the binodal calculations uncertain. Nevertheless, the uncertainty introduced by these factors does not obscure the question of whether a consistent interaction parameter can be extracted from the fluorescence data for the P2VN/PS pair.

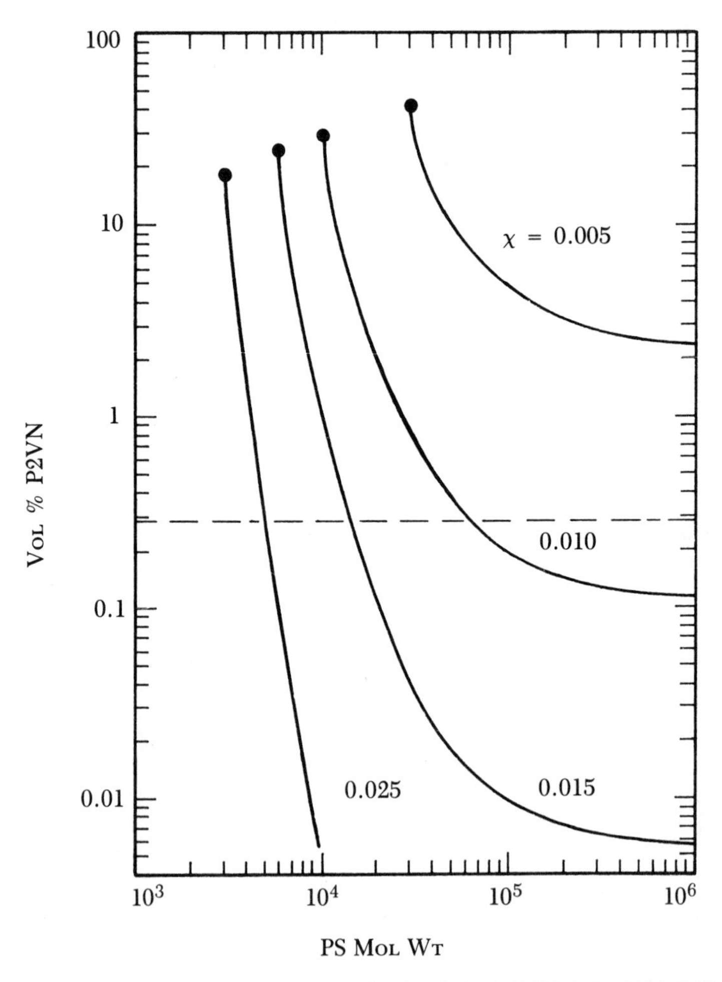

Figure 8. Binodal compositions calculated for P2VN (70,000)/PS blends. See *Figure 7 for details. (Reproduced from Ref 11. Copyright 1981, American Chemical Society.)*

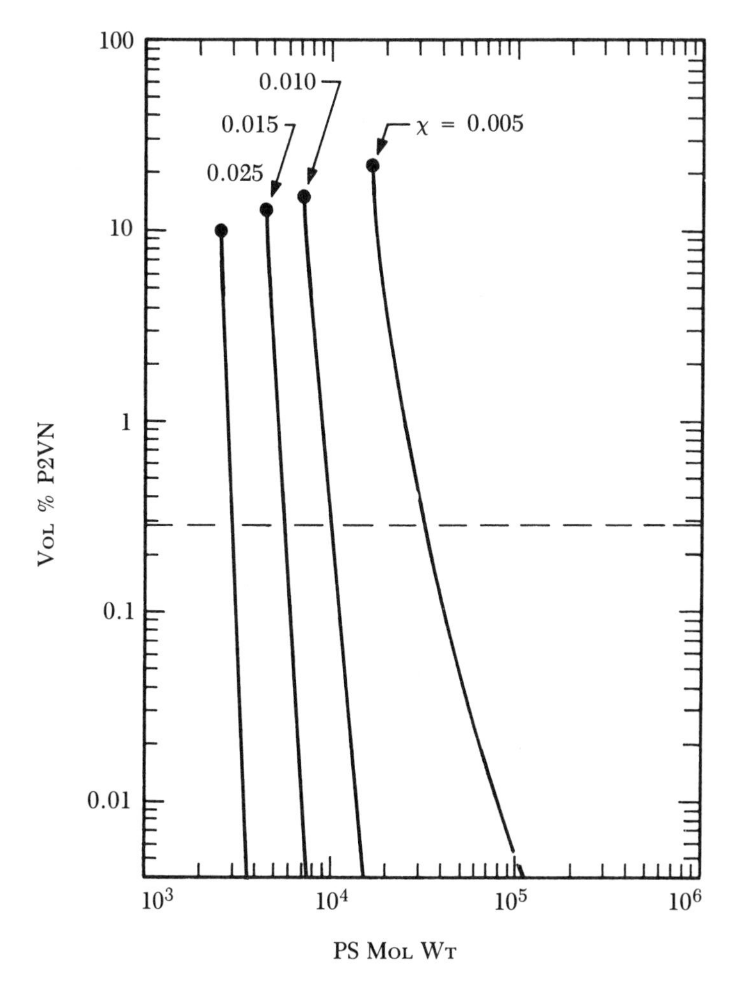

Figure 9. Binodal compositions calculated for P2VN (265,000)/PS blends. See *Figure 7 for details. (Reproduced from Ref 11. Copyright 1981, American Chemical Society.)*

Several sets of data are available for determination of χ_{AB} for the blends with P2VN (70,000). The first set consists of the host molecular weight dependence of I_D/I_M in Figure 1. The observed value of $M^* = 7000$ for 0.3-wt% P2VN (70,000) blends leads to an interaction parameter of 0.022 from Figure 8. The second set is the "cloud-point" concentration, C^*, determined from the concentration study in the PS (158,000) host to be 0.01 wt% or less. Figure 8 shows that χ_{AB} must be greater than 0.015. Finally, the data for 0.003-wt% P2VN (70,000) blends can be used to set an upper boundary on χ_{AB}. If these blends are all miscible as previously stated, then extrapolation of Figure 8 shows that χ_{AB} must be less than 0.017. This value is in good agreement with the restriction that $\chi_{AB} \geq 0.015$, which was derived from the cloud-point concentration study.

The last blends to be compared with binodal predictions are the 0.3 wt% P2VN (265,000) blends, for which fluorescence data are shown in Figure 1. The observed value of $M^* = 8000$ leads to an interaction parameter of 0.012 from Figure 9. This comparatively low value of χ_{AB} may be rationalized by noting that the DSC study of the P2VN (265,000)/PS (2200) blend did not conclusively show miscibility, as had been the case for the P2VN guests of lower molecular weight. If the miscible value of I_D/I_M for a 0.3-wt% P2VN (265,000) blend were less than 1.4 (e.g., 1.3) then the resulting interaction parameter would increase to 0.018. Regardless of the miscibility of the PS (2200)-containing blend, the calculations show the sensitivity of χ_{AB} in this case to the error in the miscible value of I_D/I_M.

The values of the interaction parameter χ_{AB} for seven P2VN/PS blend systems are compiled in Table IV. All these P2VN/PS blends are satisfactorily characterized by $\chi_{AB} = 0.020 \pm 0.004$. We have explained the low value of χ_{AB} for the P2VN (265,000) blends. However, an alternative proposal is that kinetic effects during solvent casting because of the slow diffusion of the P2VN (265,000) guest relative to other P2VN guests may cause χ_{AB} to be low for P2VN (265,000) blends. This discrepancy may also reflect differences in the polydispersity of the P2VN guests. If we consider all the possible grounds for error, it is remarkable that a single value of χ_{AB} explains the data for the variety of P2VN/PS blends studied. More extensive

Table IV. The Interaction Parameter χ_{AB} for P2VN/PS Blends Solvent Cast from Toluene at Room Temperature

P2VN Mol. Wt.	*P2VN Conc., wt%*	*PS Mol. Wt.*	χ_{AB}[a]
21,000	0.3	variable[b]	<0.027
21,000	3.0	variable[b]	$0.021^{+0.005}_{-0.004}$
21,000	30.0	variable[b]	≈0.018[c]
70,000	0.003	variable[b]	<0.017
70,000	variable	158,000[d]	>0.015
70,000	0.3	variable[b]	$0.022^{+0.003}_{-0.002}$
265,000	0.3	variable[b]	$0.012^{+0.006}_{-0.002}$[e]

[a] Determined by fluorescence ratio method; *see* Eqs. 1 and 2 in Ref. 11 for definition of χ_{AB}.

[b] 2200; 4000; 9000; 17,500; 35,000; 100,000; 233,000; 390,000; $M_w/M_n < 1.1$.

[c] Determined by appearance of heated blends.

[d] $M_w/M_n = 2.0$.

[e] $\chi > 0.012$ if P2VN (265,000)/PS (2200) blend is immiscible. *See* text.

measurements over a larger concentration range are necessary before any conclusions on the possible concentration dependence of the interaction parameter may be drawn.

Acknowledgments

This work was supported by the Polymers Program of the National Science Foundation.

Literature Cited

1. Platzer, N. A. J. "Copolymers, Polyblends, and Composites"; ACS ADVANCES IN CHEMISTRY SERIES No. 142. American Chemical Society: Washington, D.C., 1975.
2. Klempner, D.; Frisch, K. C., Eds., *Polym. Sci. Technol.* **1977**, *10.*
3. Sperling, L. H., Ed. "Recent Advances in Polymer Blends, Grafts and Blocks"; Plenum Press: New York, 1974.
4. Paul, D. R.; Newman, S., Eds. "Polymer Blends"; Academic Press: New York, 1978; Vol. 1.
5. Olabisi, O.; Robeson, L. M.; Shaw, M. T. "Polymer–Polymer Miscibility"; Academic Press: New York, 1979.
6. Rudin, A. *J. Macromol. Sci., Rev. Macromol. Chem.* **1980**, *C19(2)*, 267.
7. MacKnight, W. J.; Karasz, F. E.; Fried, J. R., Chapter 5 in Ref. 4.
8. Förster, Th. *Angew. Chem., Int. Ed. Engl.* **1969**, *8*, 333.
9. Birks, J. B. In "Photophysics of Aromatic Molecules"; Wiley-Interscience: New York, 1970; Chap. 7.
10. Klöpffer, W. In "Organic Molecular Photophysics"; Birks, J. B., Ed.; Wiley: New York, 1973; Vol. 1, Chap. 7.
11. Semerak, S. N.; Frank, C. W. *Macromolecules* **1981**, *14*, 443.
12. Semerak, S. N.; Frank, C. W. In "Polymer Characterization"; Craver, C. D., Ed.; ACS ADVANCES IN CHEMISTRY SERIES No. 203; American Chemical Society: Washington, D.C., **1983**; *p. 757.*
13. Yanari, S. S.; Bovey, F. A.; Lumry, R. *Nature* London **1963**, *200*, 242.
14. Vala, M. T.; Haebig, J.; Rice, S. A. *J. Chem. Phys.* **1965**, *43*, 886.
15. Longworth, J. W. *Biopolymers* **1966**,*4*, 1131.
16. Basile, L. J. *J. Chem. Phys.* **1962**, *36*, 2204.
17. David, C.; Demarteau, W.; Geuskens, G. *Eur. Polym. J.* **1970**, *6*, 1397.
18. Nishijima, Y.; Yamamoto, M.; Mitani, K.; Katayama, S.; Tanibuchi, T. *Rep. Prog. Polym. Phys. Jpn.* **1970**, *13*, 417.
19. Fox, R. B.; Price, T. R.; Cozzens, R. F.; McDonald, J. R. *J. Chem. Phys.* **1972**, *57*, 534.
20. Harrah, L. A. *J. Chem. Phys.* **1972**, *56*, 385.
21. Nishijima, Y., Yamamoto, M.; Katayama, S.; Hirota, K.; Sasaki, Y.; Tsujisaki, M. *Rep. Prog. Polym. Phys. Jpn.* **1972**, *15*, 445.
22. Frank, C. W.; Harrah, L. A. *J. Chem. Phys.* **1974**, *61*, 1526.
23. Frank, C. W. *Macromolecules* **1975**, *8*, 305.
24. Ito, S.; Yamamoto, M.; Nishijima, Y. *Rep. Prog. Polym. Phys. Jpn.* **1978**, *21*, 393.
25. Frank, C. W. *J. Chem. Phys.* **1974**, *61*, 2015.
26. Marsh, D. G.; Yanus, J. F.; Pearson, J. M. *Macromolecules* **1975**, *8*, 427.
27. Klöpffer, W. *J. Chem. Phys.* **1969**, *50*, 2337.
28. Klöpffer, W. *Ber. Bunsenges. Phys. Chem.* **1969**, *73*, 864.
29. Schneider, F.; Springer, J. *Makromol. Chem.* **1971**, *146*, 181.
30. McDonald, J. R.; Echols, W. E.; Price, T. R.; Fox, R. B. *J. Chem. Phys.* **1972**, *57*, 1746.

31. Yokoyama, M.; Tamamura, T.; Nakano, T.; Mikawa, H. *Chem. Lett. (Chem. Soc. Jpn.)* **1972**, 499.
32. Hirota, K.; Yamamoto, M.; Nishijima, Y. *Rep. Prog. Polym. Phys. Jpn.* **1973**, *16*, 509.
33. Reid, R. F.; Soutar, I. *J. Polym. Sci., Polym. Phys. Ed.* **1978**, *16*, 231.
34. North, A. M.; Treadaway, M. F. *Eur. Polym. J.* **1973**, *9*, 609.
35. Frank, C. W.; Gashgari, M. A. *Macromolecules* **1979**, *12*, 163.
36. Frank, C. W.; Gashgari, M. A.; Chutikamontham, P.; Haverly, V. J. In "Structure and Properties of Amorphous Polymers"; Walton, A. G., Ed.; Elsevier: New York, 1980; pp. 187–210.
37. Chryssomallis, G.; Drickamer, H. G. *J. Chem. Phys.* **1979**, *71*, 4817.
38. Johnson, G. E.; Good, T. A. *Macromolecules* **1982**, *15*, 409.
39. Ito, S.; Yamamoto, M.; Nishijima, Y. *Rep. Prog. Polym. Phys. Jpn.* **1976**, *19*, 421.
40. Reference 10, p. 381.
41. Gargallo, L.; Abuin, E. B.; Lissi, E. A. *Scientia (Valparaiso)* **1977**, *42*, 11.
42. Ito, S.; Yamamoto, M.; Nishijima, Y. *Bull Chem. Soc. Jpn.* **1982**, *55*, 363.
43. Seki, K.; Ichimura, Y.; Imamura, Y. *Macromolecules* **1981**, *14*, 1831.
44. Berlman, I. B. "Handbook of Fluorescence Spectra of Aromatic Molecules"; Academic Press: New York, 1965: p. 129.
45. Laitinen, H. A.; Miller, F. A.; Parks, T. D. *J. Am. Chem. Soc.* **1947**, *69*, 2707.
46. Bank, M.; Leffingwell, J.; Thies, C. *Macromolecules* **1971**, *4*, 43.
47. Bank, M.; Leffingwell, J.; Thies, C. *J. Polym. Sci., Part A-2* **1972**, *10*, 1097.
48. Gelles, R.; Frank, C. W. *Macromolecules* **1982**, *15*, 741 and 747.
49. Fitzgibbon, P. D.; Frank, C. W. *Macromolecules* **1982**, *15*, 733.
50. Semerak, S. N.: Frank, C. W. Submitted to *Macromolecules*.
51. Snyder, H. L.; Meakin, P.; Reich, S. *Macromolecules* **1983**, *16*, 715.
52. Gelles, R.; Frank, C. W. *Macromolecules* **1982**, *15*, 1486.
53. Gelles, R.; Frank, C. W. *Macromolecules* in press.
54. Flory, P. J. "Principles of Polymer Chemistry"; Cornell University Press: Ithaca, New York, 1953; Chapters 12 and 13.
55. Krause, S.; Chapter 2 in Ref. 4.

RECEIVED for review February 2, 1983. ACCEPTED July 11, 1983.

Structure–Property Relationships of Polystyrene / Poly(vinyl methyl ether) Blends

S. L. HSU, F. J. LU, and E. BENEDETTI

Department of Polymer Science and Engineering, University of Massachusetts, Amherst, MA 01003

Various incompatible and compatible poly(styrene) (PS) and poly(vinyl methyl ether) (PVME) blends have been prepared by varying composition, solvent, and thermal history. Fourier transform infrared spectroscopic analysis of these samples has revealed definite spectral features that are sensitive to compatibility. The vibrations most sensitive to change in molecular environment are the CH out-of-plane vibration in PS and the $COCH_3$ vibrations of PVME. Furthermore, by using a combination of vibrational and mechanical spectroscopy we have followed the segmental orientation of the two components when samples are stretched macroscopically. The orientation of chain segments achievable for the individual components depends strongly on the degree of compatibility, composition, and temperature of measurements.

THE PHASE SEPARATION BEHAVIOR OF POLYMER BLENDS and associated properties have been the subjects of a large number of theoretical and practical studies (*1*). We have primarily used vibrational spectroscopy to characterize microstructures of polymer blends and their changes when samples are deformed macroscopically. We have given particular attention to the structure–property relationship of the poly(styrene) (PS) and poly(vinyl methyl ether) (PVME) blends. The unusual compatibility behavior of this binary mixture is particularly interesting. Solvent, molecular weight, composition, and temperature can affect the compatibility of these two polymers (*2–9*).

In the first area of study, vibrational spectroscopy was used to search and characterize specific molecular interactions existing in compatible PS/

0065-2393/84/0206-0101$06.00/0

PVME blends. With exceptions, the favorable intermolecular interactions in most compatible blends are generally not clearly understood. Vibrational spectroscopy is an attractive technique for such studies. When properly assigned, the position, intensity, and shape of vibrational bands are useful in clarifying conformational and environmental changes of polymers at the molecular level. Because intrachain energy is so much higher than interchain nonbonded interactions, perturbation of spectroscopic features of a single molecule as a result of nonbonded interaction is generally quite small. In most cases, small differences in the vibrational spectra are difficult to observe directly and can only be shown in the difference spectra. The advent of Fourier transform infrared spectroscopy (FTIR) has made analysis of chemical mixtures by computer calculated difference spectra quite convenient and accurate.

In the second area of study, we have used a combination of vibrational and mechanical spectroscopy to follow the segmental orientation of the individual components when samples are deformed macroscopically. Undoubtedly, the overall properties of polymer blends are correlated to the sample morphology that, in turn, may vary with the composition or the degree of interpenetration of the components. In many polymers, the IR spectra exhibit absorption bands that can be assigned to vibrations of chain segments in the various phases. Furthermore, these vibrational bands may be characteristic of the details of local conformation. If band assignments are established and, additionally, if the directions of the transition moments with respect to the chain axis are known, vibrational spectroscopy offers the advantage that stress-induced changes in orientation, conformation, and packing can all be measured.

In this study, we have prepared compatible and incompatible films cast from various solvents. In addition, phase separated binary mixtures can be prepared at temperatures above the lower critical solution temperature (LCST). The IR data obtained from these samples have revealed specific features characteristic of localized structure and environment of individual components in compatible and incompatible binary mixtures. Furthermore, the degree and the rate of change of segmental orientation have been measured as a function of composition, temperature, strain rate, and compatibility.

Experimental

The atactic PS and PVME were obtained from commercial sources. PS of weight average molecular weight (M_w) 233,000 with polydispersity equal to 1.06 was obtained from Pressure Chemical Corporation. Other narrow molecular weight distribution atactic polystyrenes of M_w 175,000, 50,000, and 37,000 were obtained from Polysciences. PVME was also obtained from Polysciences. PVME molecular weights (M_w = 99,000 and M_n = 46,000) were determined by gel permeation chromatography (GPC) in our laboratory with poly(styrene) as the standard. Solvents of spectral grade were obtained commercially and used without further purification.

Thin films of the PS/PVME binary mixtures (15, 50, 64, and 84% of PS) were prepared from solutions (approximately 3–5% by weight) by casting onto potassium bromide windows. Compatible films were obtained from toluene solutions. Incompatible films were cast from chloroform or trichloroethylene (TCE). The solvents were completely removed by drying in vacuum ovens at 70 °C for at least 72 h. The morphology of the samples may be affected by differences in concentration, extreme thinness of film prepared, and solvent evaporation rate (*5*). Furthermore, vibrational band shape and relative intensities can depend on index of refraction or film thickness due to dispersion effects (*10*). We tried to remedy these errors in analysis by preparing films under identical conditions.

IR spectra were obtained with Nicolet 7199 FTIR instrument. Usually 300 scans of 2-cm^{-1} resolution were signal averaged and stored on a magnetic disk system. Entire or partial spectra can then be accessed for further analysis.

The hydraulic stretcher used to deform sample films and its interface to the spectrometer have been described elsewhere (*11*).

Results and Discussion

To date, qualitative and quantitative data from vibrational spectroscopy have been largely used in analyzing composition or for the elucidation of localized structures. As a result of the significant amount of conformational and packing disorder in polymers, even for those of highest crystallinity, detailed spectroscopic information related to the effects of intermolecular interactions in solid polymers is difficult to obtain. Therefore, with few exceptions, the vibrational spectra obtained thus far reflect regular or irregular structure of a repeat unit or a single chain. In a few systems, strong intermolecular interactions such as hydrogen bonding can significantly affect the band positions or intensity. In most cases, however, spectroscopic perturbations that arise from interchain forces are simply too small to be observed with a high degree of confidence. The initial goal in these blend studies is to search and characterize features sensitive to composition or degree of phase separation.

From previous studies (*12, 13*), vibrations involving the oxygen atom usually exhibit greatest sensitivity to blend compatibility. In the IR spectrum of PVME, a strong doublet at 1085 and 1107 cm^{-1} (room temperature) with a shoulder at 1132 cm^{-1} showed the greatest change in relative intensity when the sample was blended with PS. The intensities were also sensitive to changes in temperature. Although the exact assignment of these bands is not understood, one can obtain some information from normal vibrational analysis of model compounds. Snyder and Zerbi (*14*) carried out a normal vibrational analysis for a series of aliphatic ethers. They showed that model compounds, such as methyl ethyl ether, diethyl ether, and 1,2-dimethoxyethane, exhibit intense bands in this region. In those cases, only one component involves essentially pure C–O stretching. The other contains a significant amount of C–O stretching with contributions from methyl rocking and C–C stretching in the potential energy distribution.

Space-filling models show that considerable rotational freedom exists for the $-OCH_3$ group about the main chain. However, the exact relationship between structural differences and changes in the IR spectrum remains unclear. Therefore, at present we can only demonstrate that the relative intensity of this doublet is sensitive to the compatibility or incompatibility of PS/PVME blends.

Previously it had been established that high molecular weight PS/PVME binary mixtures containing ~20–80% PS ($M_w > 17,500$) dissolved in chloroform or TCE can be used to cast incompatible films (*2–4*). For the same binary mixtures dissolved in toluene, compatible blends can be prepared. Although the frequencies of the doublet in the 1100-cm^{-1} region do not change, the relative intensities of the two bands do differ in the two types of samples. In these cases we have removed the PS contribution at 1070 cm^{-1} (in-plane bending) from the mixture by difference spectroscopy. The 1028-cm^{-1} band of PS was used as an internal standard. Subtraction is accomplished when this band is completely eliminated. As can be seen in Figure 1, for compatible blends such as the binary mixtures obtained from toluene, the intensity of the 1085-cm^{-1} band is slightly greater than the 1107-cm^{-1} component. For incompatible blends obtained from

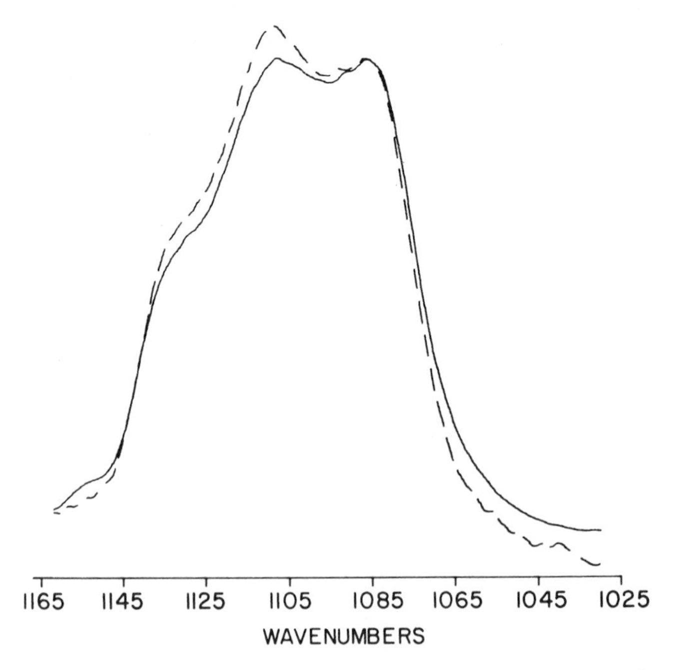

Figure 1. IR spectra of 50:50 PS/PVME blends in the 1100-cm^{-1} region. Key: —, film cast from toluene solution; and ---, film cast from TCE.

chloroform or TCE, the reverse is true. Allara (*10*) pointed out that when the index of refraction of the individual components differs significantly, spectral differences can arise from the changing bandshape that results from dispersion effects and may not be associated with the magnitude or specificity of intermolecular interactions. Because we could not easily determine optical coefficients of the individual components and their blends in our experiments, we used the same binary PS/PVME mixtures to prepare various samples. Only then could we be sure that the spectroscopic differences were dependent only on compatibility and not on composition.

In the IR spectrum of PS, a number of bands showed small changes in position or shape when PS was blended with PVME. The band most sensitive to phase compatibility is located near 700 cm^{-1}. This band is generally assigned to the CH out-of-plane bending vibration (*12*). It is found at 699.5 cm^{-1} for film cast from toluene solutions containing equal amounts of PS and PVME. This is to be compared to the peak maximum located at 697.7 cm^{-1} in neat PS. For incompatible blends of PS/PVME, such as the film cast from TCE solutions, the peak maximum is usually found at an intermediate position between the two extremes. These differences are shown in Figure 2. Similar changes in frequency for this vibration have also been observed for PS blended with poly(2,6-dimethylphenylene oxide) (PPO) (*12*).

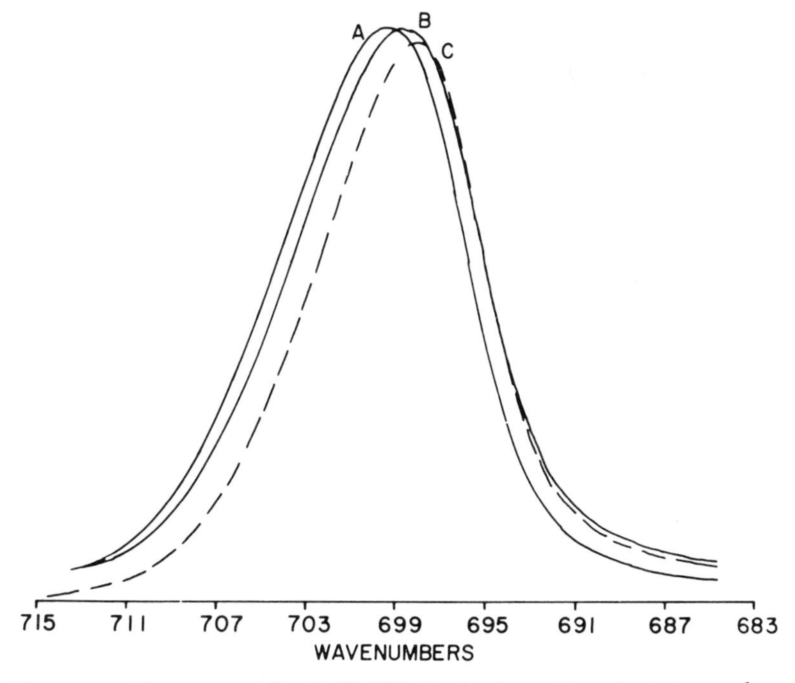

Figure 2. IR spectra of 50:50 PS/PVME blends or PS in the 700-cm^{-1} region. Key: A, film cast from toluene; B, film cast from TCE; and C, PS only.

We carried out additional experiments to substantiate that the spectroscopic differences found for PVME or PS are indeed associated with compatibility. Unlike the samples containing 50 wt% of PS, the binary mixtures containing 15% PS are compatible with PVME for all molecular weights and solvents. For binary mixtures containing 15 wt% of PS, the IR spectra obtained for films cast from toluene and TCE solutions are identical.

PS/PVME blends possess a lower critical solution temperature. Above this critical temperature a compatible blend can phase separate. For the sample containing equal weight parts of PS and PVME, the initially transparent film (compatible) will turn slightly bluish at ~ 130 °C and then optically opalescent at ~ 150 °C (incompatible). If quenched at this stage, the two phases (one rich in PS, the other rich in PVME) would remain separated. However, if the binary mixtures were allowed to cool slowly, the two phases would return to the compatible state. IR spectra were obtained for binary mixtures that contained equal parts of PS or PVME and had different thermal histories. The spectroscopic differences observed for either PVME or PS in compatible or incompatible samples are entirely consistent with the results presented in Figures 1 and 2. These experiments further substantiate that these vibrational bands can be used to characterize the compatibility of the two components. The molecular or structural basis for miscibility is a subject of intense scientific interest driven by the need to be able to predict a priori which polymer pairs will form miscible mixtures. This kind of spectroscopic information provides a basis to deduce the type and strength of intermolecular interactions in PS/PVME blends.

In the second area of study, we are more interested in the microstructural changes during deformation. The mechanical properties of polymer blends may depend on the conformation and internal organization of individual components. The properties may also depend upon the degree of interpenetration of the two components. In many cases, the structure–property relationship of polymer blends may be assessed by the relative orientation of the polymer segments associated with each phase. For an ideal polymer mixture, the property function is a linear combination of the properties of the pure components. In the other extreme case, when macrophase separation occurs, the blend may show very poor mechanical properties because it has large domains that have poor interfacial bonding.

As we have mentioned, the IR spectra obtained for PS or PVME contain a number of absorptions that are assignable to localized structure. Furthermore, if the transition moments of these absorptions can be assigned, IR spectroscopy provides an opportunity to measure microstructural changes when the samples are deformed macroscopically. In our experiment, macroscopic mechanical properties in terms of stress–strain curves can be measured simultaneously with microstructural response in terms of polarized IR spectra.

We chose to follow the four bands at 540, 700, 906, and 1028 cm^{-1} to characterize the orientation of polystyrene (*15*). Because the IR spectrum of PVME is not well understood, we chose the very localized CH stretching vibration at 2820 cm^{-1} as a standard to calculate the orientation of PVME segments. The two components oriented differently, even in the compatible blend, as shown in Figure 3. The polarization of polystyrene bands increased with sample elongation, as expected. However, it was difficult to capture and compare orientation behavior of PVME unless the samples were deformed at an extremely high rate. The different orientation function measured for the two components is unexpected because Cooper et al. (*16*) found the segmental orientation of both components in poly(ϵ-caprolactone)/poly(vinyl chloride) (PCL/PVC) blends to be similar. In fact, the identical orientation functions measured for the two components may be part of the criteria to determine compatibility at the molecular level. In a subsequent study of PS and PPO blends, Lefebvre et al. (*17*) demonstrated that the orientation of the two components may differ substantially, as we found for the PS/PVME blends. The fact that the relative orientation measured for the two components in the blend is strongly strain-rate dependent suggests the existence of different relaxation rates. This fact should be incorporated in the modeling to relate our macroscopic and microscopic measurements.

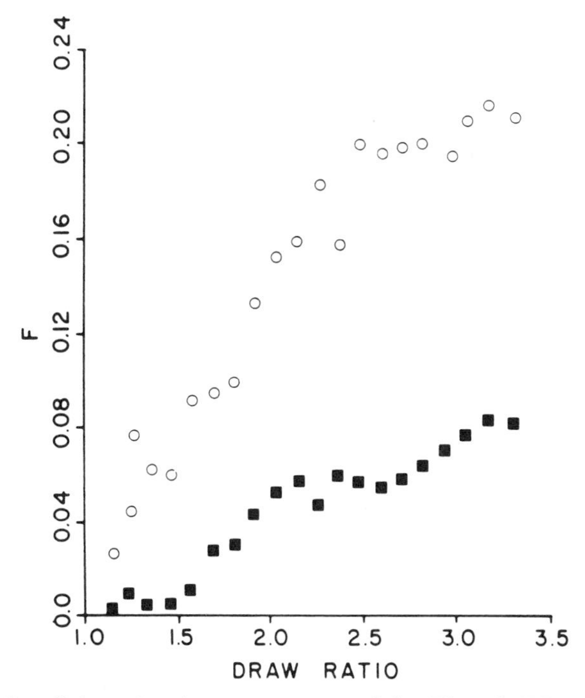

Figure 3. Orientation functions measured for PS and PVME in a 50:50 compatible blend. Key: ○, PS; and ■, PVME.

Usually PS is a brittle polymer when stressed. However, its deformation behavior in blends is quite interesting. As can be seen in Figure 4, highly oriented PS segments can be obtained at room temperature. However, the orientation behavior of PS greatly depends upon composition, molecular weight, draw rate, temperature of measurement, and the degree of compatibility of the two components. In Figure 4 we have shown the dichroic ratio measurements of the polystyrene bands for various samples. The orientation function of the polystyrene in blends measured at room temperature is similar to the orientation function measured for polystyrene drawn at 100 °C (5). Both sets of values are significantly higher than PS drawn at 110 °C (5). Undoubtedly, the orientation that can be achieved is strongly dependent on the temperature of measurement. Generally it is accepted that the lack of segmental orientation obtained for samples drawn or extruded above the glassy transition temperature is a result of the increasing chain mobility. The glass transition temperature (T_g) of PVME, is usually found at −25 °C. On the other hand, T_g of PS is approximately 100 °C. The T_g of compatible blends is usually found between these two extremes. For the samples we prepared, the T_g actually measured is somewhat below room temperature (55% blend, T_g = 18 °C). As in the case of

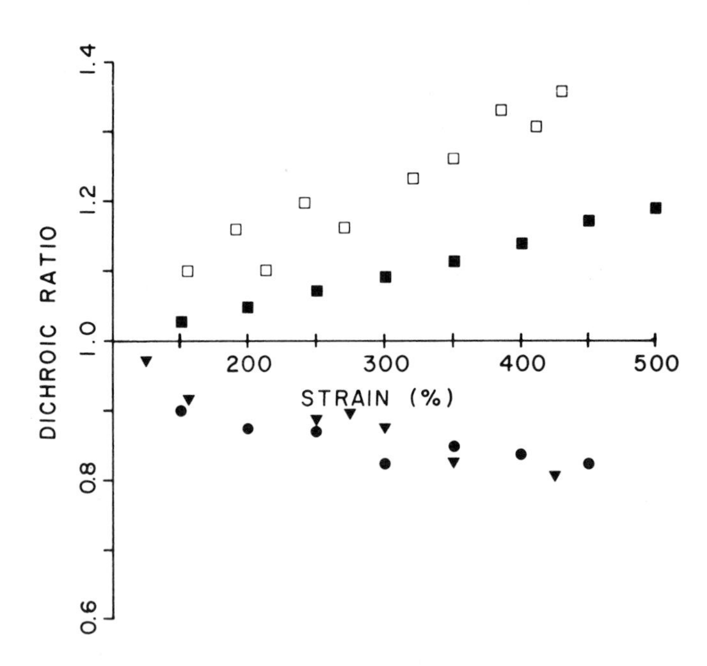

Figure 4. Orientation behavior of neat PS and PS in PS(55)/PVME(45) blends. Key: □, 540-cm^{-1} PS band in blends; ■, neat PS drawn at 110 °C; ▼, 1028-cm^{-1} band of PS in blends; and ●, 1028-cm^{-1} PS band for neat PS drawn at 100 °C.

neat PS, we found that the orientation function of the components in blends decreased significantly when the temperature was raised above the T_g.

Even though we could not compare the orientation obtained for PS or PVME segments easily, we were able to show dramatic differences between the orientation obtained for PS in compatible or incompatible blends. Because, as we have demonstrated earlier, compatible or incompatible blends can be prepared by varying solvent or thermal history of mixtures of the same composition, the orientation functions measured for the polystyrene bands shown in Figure 5 are most interesting. Clearly, the orientation of polystyrene is higher in compatible mixtures.

In this initial study we have demonstrated that vibrational spectroscopy can indeed be useful in characterizing conformation and environment of individual components in polymer blends. Furthermore, if we allow sufficient time resolution, FTIR can also be used to follow orientation changes of individual components when samples are deformed macroscopically.

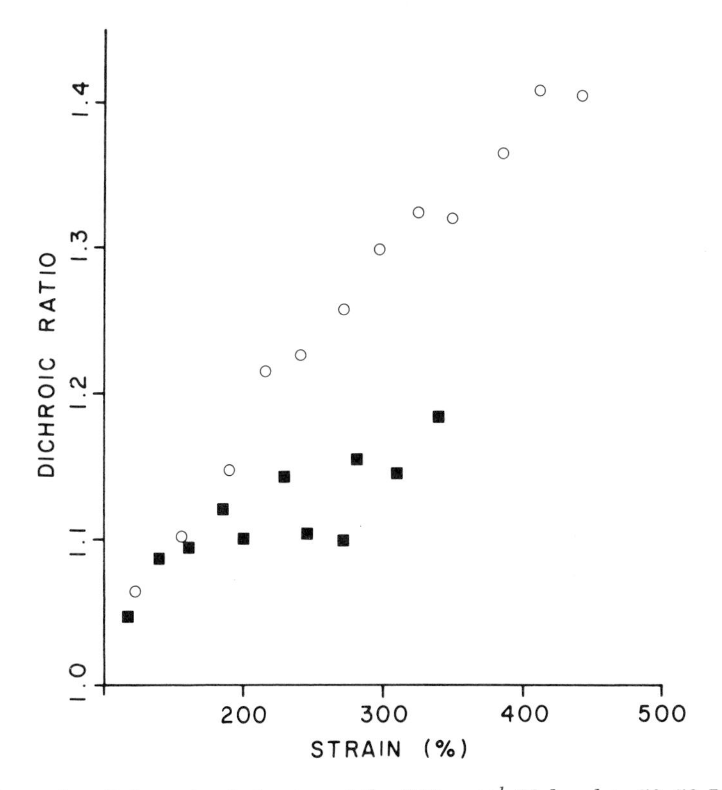

Figure 5. Orientation behavior of the 700-cm^{-1} PS band in 50:50 PS/PVME blends. Key: ■, *phase separated blends; and* ○, *compatible blends.*

Literature Cited

1. Barlow, J. W.; Paul, D. R. *Polym. Eng. Sci.* **1981**, *21*, 985.
2. Kwei, T. K.; Nishi, T.; Roberts, R. F. *Macromolecules* **1974**, *7*, 667.
3. Davis, D. D.; Kwei, T. K. *J. Polym. Sci. Polym. Phys. Ed.* **1980**, *18*, 2337.
4. Nishi, T.; Wang, T. T.; Kwei, T. K. *Macromolecules* **1975**, *8*, 227.
5. Reich, S.; Cohen, Y. *J. Polym. Sci. Polym. Phys. Ed.* **1981**, *19*, 1255.
6. Bank, M.; Leffingwell, J.; Thies, C. *Macromolecules* **1971**, *4*, 43.
7. McMaster, L. P. *Macromolecules* **1973**, *6*, 760.
8. Zeman, L.; Patterson, D. *Macromolecules* **1972**, *5*, 513.
9. Roberd, A.; Patterson, D.; Delmas, G. *Macromolecules* **1977**, *10*, 706.
10. Allara, D.L. *Appl. Spectrosc.* **1979**, *33*, 358.
11. Burchell, D. J.; Lasch, J. E.; Farris, R. J.; Hsu, S. L. *Polymer* **1982**, *23*, 965.
12. Wellinghoff, S. T.; Koenig, J. L.; Baer, E. *J. Polym. Sci. Polym. Phys. Ed.* **1977**, *15*, 1913.
13. Benedetti, E.; Hsu, S.L., in press.
14. Snyder, R. G.; Zerbi, G. *Spectrochim. Acta* **1967**, *23A*, 391.
15. Jasse, B.; Koenig, J. L. *J. Polym. Sci. Polym. Phys. Ed.* **1979**, *17*, 799.
16. Hubbell, D. S.; Cooper, S. L. *J. Polym. Sci. Polym. Phys. Ed.* **1977**, *15*, 1143.
17. LeFebvre, D.; Jasse, B.; Monnerie, L. *Polymer* **1981**, *22*, 1616.

RECEIVED for review January 20, 1983. ACCEPTED July 8, 1983.

8

Segmental Orientation in Multicomponent Polymer Systems

CARL B. WANG[1] and STUART L. COOPER

Department of Chemical Engineering, University of Wisconsin, Madison, WI 53706

The technique of IR dichroism was used to characterize segmental orientation in several multicomponent polymer systems. Compatibility at the molecular level usually leads to similar segmental orientation of the blend components. In contrast, different orientation behavior is observed for components in those incompatible blends that exhibit large-scale phase separation. Orientation characteristics for the blends of poly(vinyl chloride) (PVC) with poly(ϵ-caprolactone) (PCL), PCL with nitrocellulose (NC), PVC with a segmented polyurethane, and PVC with butadiene acrylonitrile rubber are described. Studies of the temperature dependence of segmental orientation show that the orientation achievable is strongly dependent upon the viscoelastic state of the blend. The orientation of the soft and hard segments of a polyether urethane–urea was shown to depend on domain morphology as affected by hard segment weight fraction, hard and soft segment length, and hard domain urea content.

THE USE OF ORIENTED PLASTICS AND FIBERS has grown recently because of the improved mechanical properties that orientation can impart to polymeric materials. The field of study in which strain is related to orientation of polymers has also grown to include a large number of experimental methods (*1*, *2*). Among the various techniques, IR dichroism can be used to measure the average orientation of particular chemical bonds or groups on the polymer chain by monitoring changes in the IR spectrum during stretching of the polymer. Additionally, in multiphase polymer systems, IR dichroism can be used to study the deformation of chain segments in each

[1]Current address: IBM Research Laboratory, San Jose, CA 95193.

0065–2393/84/0206–0111$06.00/0

domain by following the absorption of functional groups residing on segments in different phases.

Dichroism is based on the fact that IR absorption by molecules occurs only at a specific angle, represented by a transition moment vector **M**. The absorption A of plane polarized light by a chemical species is proportional to the square of the dot product of the transition moment vector and the electric vector **E**, which reside at an angle β with respect to each other.

$$A \sim (\mathbf{M} \cdot \mathbf{E})^2 = k(|\mathbf{M}|\,|\mathbf{E}|)^2 \cos^2 \beta \tag{1}$$

Thus, incident light that is perpendicular to **M** is not absorbed. If a polymer is elongated such that the **M** vectors are preferentially oriented, the amount of absorbance of plane polarized light parallel to the direction of stretch ($A_{\parallel}$) will differ from the absorbance of plane polarized light perpendicular to the direction of stretch ($A_{\perp}$). The ratio of these two absorbances defines the dichroic ratio D,

$$D = A_{\parallel}/A_{\perp} \tag{2}$$

which varies from zero to infinity (unity represents random orientation).

In the case of uniaxial stretching, the orientation of the polymer chain can be related to the stretch direction through an orientation function f (3),

$$f = \left(\frac{D_0 + 2}{D_0 - 1}\right)\left(\frac{D - 1}{D + 2}\right) \tag{3}$$

where D_0 is the dichroic ratio for perfect orientation, which is related to the angle α between the transition moment vector, **M**, and the chain axis by the following equation:

$$D_0 = 2 \cot^2 \alpha \tag{4}$$

In IR dichroism experiments, samples are typically stretched to a desired strain level and then allowed some degree of stress relaxation. Absorptions $A_{\parallel}$ and $A_{\perp}$ are then recorded by placing the sample in the sample beam of a spectrophotometer at a 45° angle. This angle minimizes machine polarization effects. A polarizer is placed between the sample and monochromator, and the parallel and perpendicular spectra are recorded by adjusting the angle between the polarized beam and the sample.

For dynamic experiments, a differential dichroism method may be used. The sample is strained in the common beam of a double-beam spectrophotometer and the transmitted radiation is split into two beams polarized parallel and perpendicular to the direction of stretch. The recorded output is the difference between the two absorptions. The orientation function is calculated by using Equation 5:

$$f = \left(\frac{D_0 + 2}{D_0 - 1}\right)\left(\frac{A_\parallel - A_\perp}{3A_0(l/l_0)^{-1/2}}\right) \quad (5)$$

where A_0 is the unpolarized absorption of the undeformed sample and l and l_0 are the stretched and unstretched sample lengths. The term $A_0(l/l_0)^{-1/2}$ equals the unpolarized absorption of any strain level assuming constant volume deformation.

Materials and Sample Preparation

Poly(ϵ-caprolactone) (PCL) and the poly(vinyl chloride) (PVC) were obtained from Union Carbide. Nitrocellulose (NC) (Hercules Inc.) contained 2.25 ± 0.60 nitro groups per anhydroglucose ring. Polyether polyurethane was prepared by the B. F. Goodrich Company (ET-38-1). The hard segment is composed of the 4,4′-(dicarbonylamino)diphenylmethane chain extended with 1,4-butanediol, and poly(tetramethylene oxide) (PTMO) is the soft segment. The sample code ET-38-1 represents the polyether soft segment (ET); the weight percent of 4,4′-(dicarbonylamino)diphenylmethane present (38%); and approximate soft segment molecular weight in thousands (*1*). The poly(butadiene-*co*-acrylonitrile) (BAN) used was from Scientific Polymer Products Co. The number after BAN indicates the weight percent of acrylonitrile in the copolymer. The segmented polyether polyurethane–ureas (PEUU) were synthesized in this laboratory by a two-step condensation reaction. The molar ratios of 4,4′-(dicarbonylamino)diphenylmethane, ethylenediamine (ED), and PTMO, and the soft segment (PTMO) molecular weight were altered in different syntheses to produce samples with systematically varied hard segment content and block length. A polymer made from 3 mol of 4,4′-(dicarbonylamino)diphenylmethane, 2 mol of ED, and 1 mol of 1000 molecular weight PTMO is designated as PEUU-46-1000; The first two digits represent the weight percent of the hard segment content and the second group of digits is the soft segment molecular weight.

All samples for this study were prepared by spin casting from polymer solutions and were about 75–125 μm thick. For each blend sample, the ratio specifies the weight percent composition of each component. Table I summarizes the sample preparation conditions and the type of apparatus used for the dichroism experiments.

Results and Discussion

NC/PCL Blends (*4*, *5*). In this blend system, dynamic differential dichroism was used to follow chain orientation while the sample was elongated at a constant strain rate. Measurements were performed by

Table I. Summary of Sample Preparation Method, IR Dichroism Technique, and IR Apparatus Used

Sample	*Casting Solvent*	*Casting Temperature (°C)*	*IR Dichroism Technique*[a]
PVC/PCL blends	THF	25	D
NC/PCL blends	THF	25	D
ET-38-1/PVC Series 1 blends	THF	25	C
ET-38-1/PVC Series 2 blends	THF/dioxane	60	C
BAN31/PVC blends	THF	25	C
BAN44/PVC blends	THF	25	C
PEUU copolymers	*N,N′*-dimethyl acetamide	70	C

[a] D = differential; C = conventional. The IR apparatus used was the Perkin-Elmer 180 IR spectrophotometer except for PEUU copolymers, for which the Nicolet 7199 FTIR was used.

using different IR peaks, which allowed a comparison of the molecular orientations for each blend constituent. In addition, a cyclic experiment was employed where the film was strained to a predetermined elongation, relaxed at the same strain rate until the stress was reduced to zero, and then elongated to a higher level of strain and so forth. The experimental data indicated an orientation response in the cyclic deformation similar to the single cycle process.

The orientation of the two components of a compatible, amorphous, 50% NC blend is shown in Figure 1. The films have been strained to 25, 50, and 75% in successive cycles; each strain increment is followed by a relaxation to zero stress. The orientation of the PCL and NC was followed by observing the carbonyl (1728 cm^{-1}) and NO_2 (1660 cm^{-1}) stretching peaks, respectively. Figure 1 shows that the orientation of the two chains follows similar paths. The similarity suggests that the local environment is very much the same for both blend constituents.

The orientation functions for the NO_2 and carbonyl of the 25% NC blend are shown in Figure 2. The PCL crystallinity is approximately 40%, and has a rod-like superstructure (*3*). The NO_2 orientation is similar to that observed for the NO_2 orientation in 50% NC, except that its magnitude is somewhat lower.

The orientation function of the PCL carbonyl in semicrystalline samples has both crystalline and amorphous components. If it is assumed that the orientation of the amorphous PCL can be represented by the orientation of a compatibly blended second polymer such as

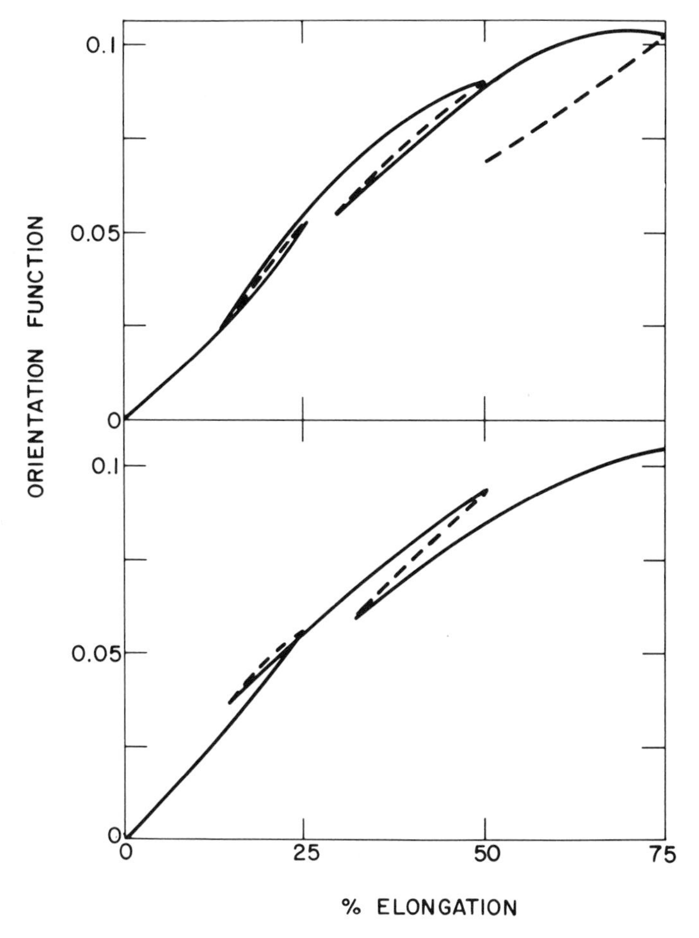

Figure 1. Carbonyl (top) and NO_2 (bottom) orientation function vs elongation curves for 50% NC/50% PCL in a cyclic IR dichroism experiment.

NC, then the orientation of the crystalline PCL can be calculated by making the following assumptions:

1. The amorphous and crystalline contributions to the dichroic difference ΔA and to the initial unpolarized absorption A_0 are additive.
2. The extinction coefficients are the same for the carbonyl in both phases. Therefore, the total orientation function is a linear function of the component orientations, i.e.,

$$f = \phi_c f_c + \phi_a f_a \quad (6)$$

where ϕ is the volume fraction, and the subscripts c and a represent the crystalline and amorphous phases, respectively.

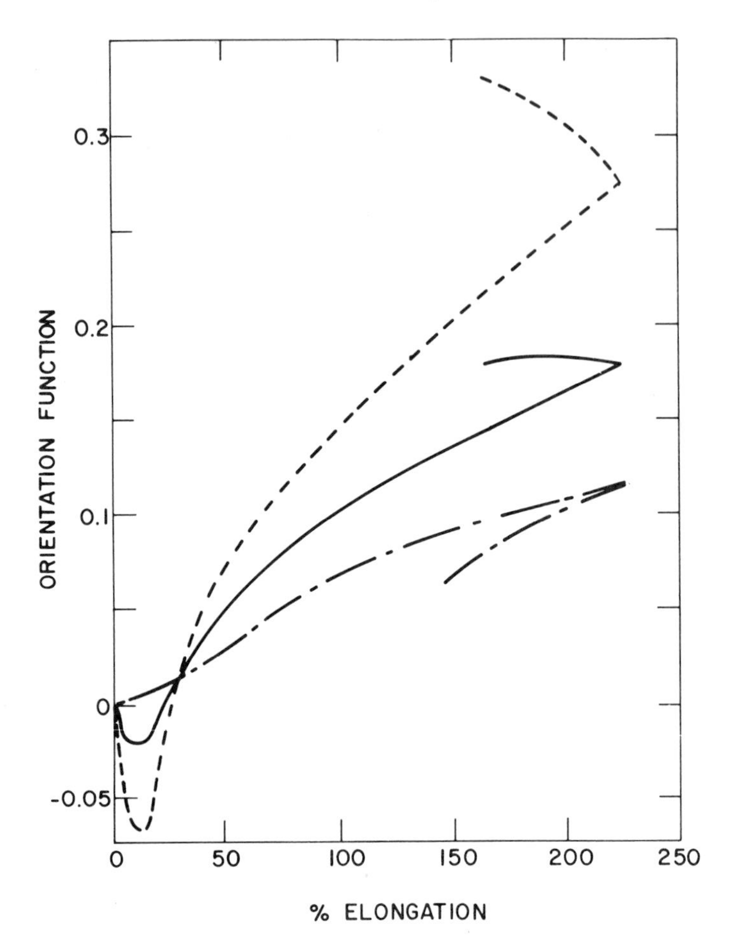

Figure 2. Carbonyl (—), NO_2 (—-—), and calculated crystalline (---) orientation function vs elongation curves for 25% NC/75% PCL.

3. The volume fraction ϕ is not affected by the applied deformation.

Equation 6 has been applied to single cycle 25% NC orientation data, and the calculated crystalline orientation is shown in Figure 2. The crystalline volume fraction was calculated on the basis of component densities and the differential scanning calorimetry (DSC) heat of fusion measurements (*4*). As expected, the crystalline PCL orientation was much higher than the orientation for the amorphous PCL and NO_2. As the stress is relieved, the amorphous chains recoil back to their high-entropy random-coil conformation. However, the crystallites have oriented as units, and the residual stress exerted on the crystals by tie molecules tends to maintain or increase their orientation.

The usefulness of Equation 6 goes beyond analyzing semicrystalline polymer blends. The NC may be viewed as a molecular probe. A small amount of a compatible polymer suitable for IR analysis may be mixed with a semicrystalline polymer and can give a direct measure of the amorphous orientation. A polymeric probe would also be useful in phase-separated systems such as block copolymers where suitable IR bands are not available because of band overlap or a lack of known transition moment vector directions. The probe molecule should be of comparable molecular weight, however, to equalize the effect of chain relaxation on the orientation function (discussion under "BAN/PVC Blends").

PVC/PCL Blends (*4*, *5*). Figure 3 shows that the PCL chains in the 75% PVC blend orient in essentially the same way as the isotactic segments and the other folded-chain PVC segments represented by the 693-cm^{-1} peak.

However, the syndiotactic segments of extended chains (613 cm^{-1}) are obviously situated in a different local environment. These more rigid segments form into microcrystalline phases that orient as units to much higher degrees than the amorphous isotactic sequences.

ET-38-1/PVC Blends (*6*). As determined from DSC and Rheovibron data (*5*), Series 1 samples are partially compatible and Series 2 samples are

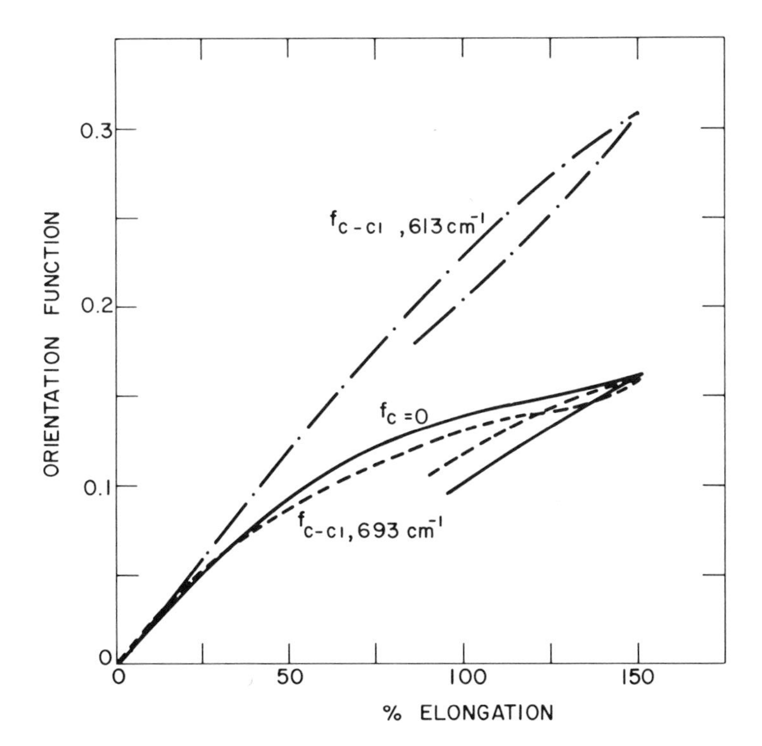

Figure 3. Carbonyl and C–Cl orientation function vs elongation curves for 75% PVC/25% PCL.

incompatible. In this study, the hydrogen bonded N–H(NH_B) and C=O(CO_B) groups were used to follow the orientation of the urethane hard segment, and the orientation of the soft segment was followed by using the nonhydrogen bonded or free C=O(CO_F) groups. We assumed that isolated 4,4′-(dicarbonylamino)diphenylmethane segments in the soft segment domains will orient in a similar fashion to the polyether segments. Both of the C–Cl stretching peaks at 637 and 691 cm^{-1} originate from amorphous phase PVC chain segments. Orientation of the PVC phase was followed by using the 637-cm^{-1} peak because it is both stronger and sharper than the 691-cm^{-1} peak.

Orientation functions for the NH_B and CO_B of each of the Series 1 samples are very similar (Figure 4). This similarity is a result of the fact that both groups reside in the hard segment domain. The CO_F orientation is observed to be greater than those of the NH_B and CO_B at low strain, whereas the inverse is true at high elongations (Figure 4). Similar behavior is also observed for Series 2 samples (not shown). This observation suggests that microphase separation in the polyurethane occurs in both blend systems. Initially, the stiffer hard segment domains are less affected by stretching and the coiled soft segments are easily oriented because they are above their glass transition temperature (T_g). As the film is elongated, the soft segments apply increasing shear stress to the hard segments and cause them to orient into the stretch direction. The dichroism data were taken

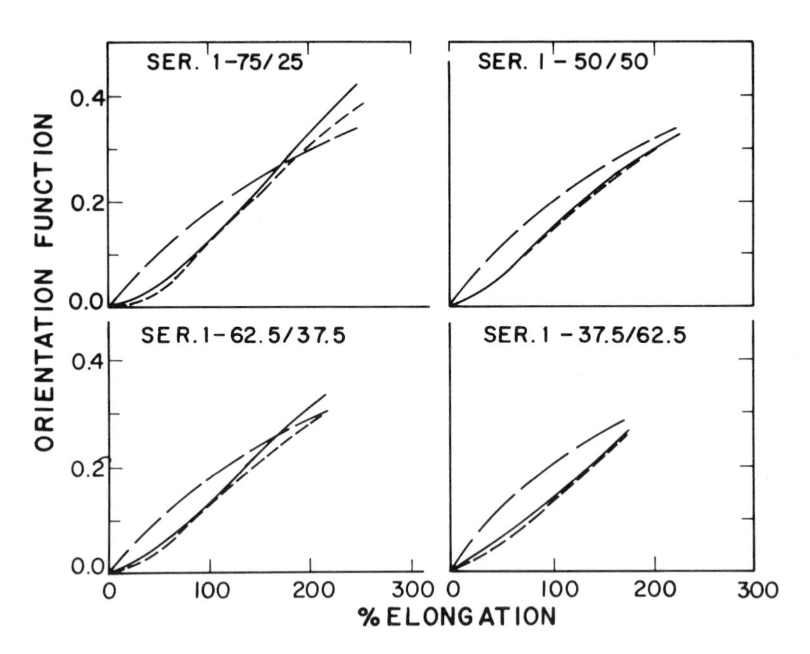

Figure 4. Orientation function vs elongation curves of Series 1 blends of ET-38-1/PVC. Key: —, NH_B; ---, CO_B; and --, CO_F.

after 10 min of stress relaxation. Consequently, the soft segments tended to disorient toward a more random conformation via an entropy-driven mechanism. The accompanying retractive force exerts a tension on the hard segments and causes them to become further aligned in the stretch direction. For Series 1 samples, at low strain, the C–Cl orientation is similar to that of the CO_F (Figure 5). However, the orientation functions of C–Cl and CO_F are well separated for the corresponding samples in Series 2. The similarity suggests that the local environment is very much the same for both PTMO and PVC in the Series 1 compatible blends.

For the 37.5/62.6 blends, the match of C–Cl and CO_F orientation for Series 1 samples at low strains indicates the segmental compatibility between PTMO and PVC phases (Figure 6). However, at higher strains the C–Cl orientation is observed to be higher than those of CO_F in both series of blends. For the Series 1 samples, the PTMO–PVC compatible matrix apparently is PVC-rich and shows similar orientation of C–Cl and CO_F only at low strains. In the Series 2 samples the PVC and polyurethane chains reside in two interconnecting matrices, and the more rigid PVC segments show higher orientation at all strain levels.

BAN/PVC Blends (7). DSC, Rheovibron, stress–strain, and transition electron microscopy (TEM) results indicate that BAN 31/PVC blends are substantially compatible but BAN 44/PVC blends are incompatible (*6*).

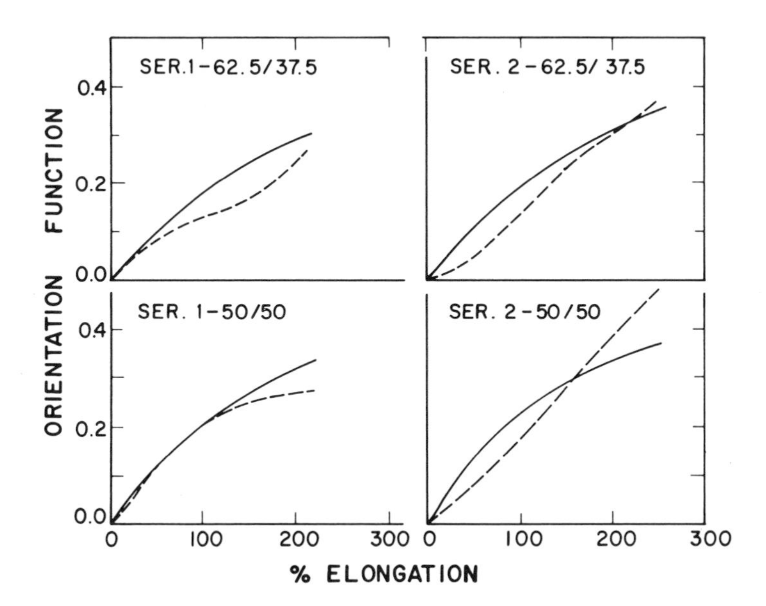

Figure 5. Orientation function vs elongation curves of Series 1 and 2 blends of ET-38-1/PVC (f_{CO_F} and $f_{C\text{-}Cl}$ for 62.5/37.5 and 50/50 compositions). Key: —, CO_F; and --, C-Cl.

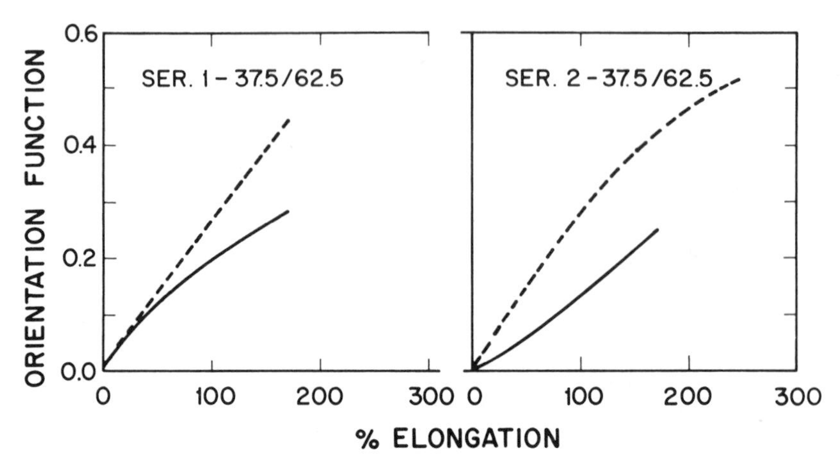

Figure 6. Orientation function vs elongation curves of Series 1 and 2 blends of ET-38-1/PVC (f_{CO_F} and $f_{C\text{-}Cl}$ for 37.5/62.5 composition). Key: —, CO_F; and --, C–Cl.

In this study the C≡N group (peak at 2237 cm^{-1}) was used to follow the orientation of the BAN chain segments. Orientation of the PVC phase was followed by using the C–Cl peak at 637 cm^{-1}.

During the stress relaxation period following sample stretching, the preferential orientation of the chain is influenced by segmental relaxation. For the PVC blends studied, separate experiments showed that the extent of segmental relaxation was determined by how close the testing temperature was to the T_g of the corresponding phase. The relaxation of chains in a compatible blend is affected by the molecular weight of the individual components. For example, in the deformation of a blend of different molecular weight polystyrenes, the lower molecular weight species may relax more rapidly and cause a lower orientation (*8*). However, for the polymer blends in this study, the influence of molecular weight on the segmental orientation is less prominent because components of comparable molecular weight were used.

Figure 7 is the schematic diagram showing the T_g range (between the onset and end temperature, as indicated by a slope change on the DSC thermogram) and the temperatures (vertical dashed lines) where the IR dichroism experiments were carried out for each BAN 31/PVC blend. In this series of nearly compatible blends, the unusually broad transition regions suggested that some degree of heterogeneity exists in the corresponding amorphous phase. Probably this phase contains a truly miscible BAN 31–PVC matrix as well as "microdomains" of either BAN 31 or PVC. Consequently, the measured chain orientation is a combination of the segmental orientation from both the matrix and the microdomains.

Figure 8 shows orientation functions (f) plotted against percent strain

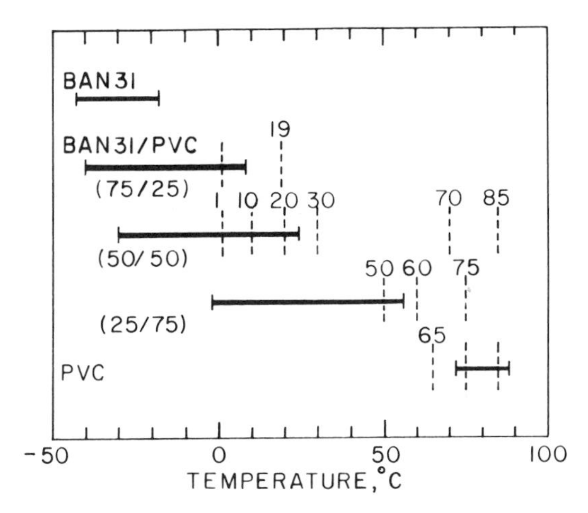

Figure 7. Relationship between the T_g range and IR dichroism measurement temperatures for the BAN 31/PVC blends.

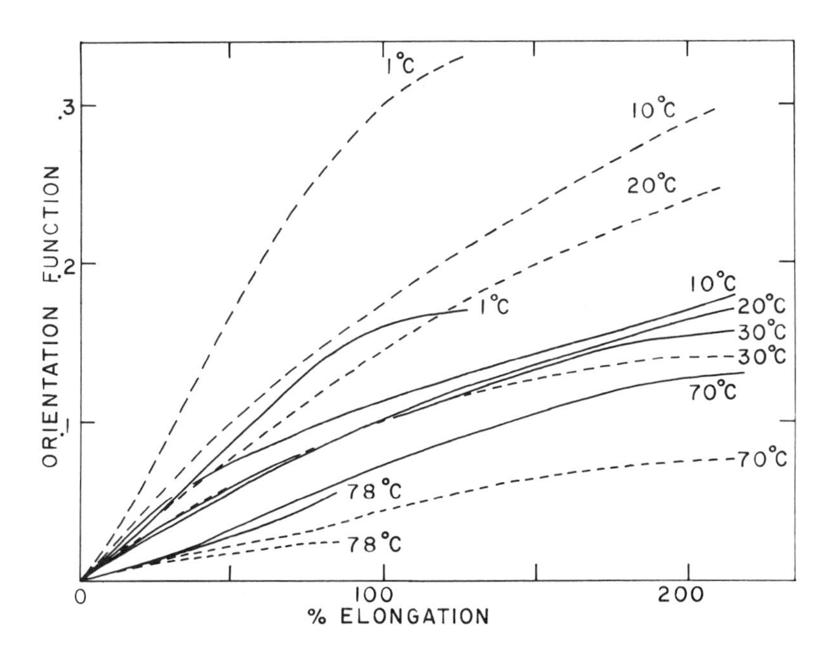

Figure 8. Orientation function vs elongation curves of the BAN 31/PVC (50/50) blends (the comparison between f_{CN} and f_{C-Cl}) at various temperatures. Key: —, C–Cl; and --, CN.

for the 50/50 blend of BAN 31/PVC. As the temperature was raised, the extent of segmental relaxation increased and lowered the f values. In the miscible BAN 31–PVC matrix, both components were considered to be similarly oriented because their local environments were indistinguishable. However, the contributions to the f values from the relatively pure PVC or BAN microdomains are rather different and depend upon the test temperature. For example, in the transition region (e.g., 1, 10, and 20 °C), the BAN 31 microdomains are easily aligned and not significantly disoriented within the experimental time scale. This ease of alignment tends to increase the overall BAN 31 chain orientation. In contrast, any PVC microdomains present are too rigid to undergo significant orientation. Hence, in the 1–20 °C range, f_{CN} is greater than $f_{C\text{-}Cl}$. At ~ 30 °C, the BAN 31 microdomains exhibit substantial disorientation within the given stress relaxation period so that f_{CN} is comparable to $f_{C\text{-}Cl}$. At somewhat higher temperatures (e.g., 70 and 78 °C) the PVC microdomains may be oriented so that f_{CN} is observed to be lower than $f_{C\text{-}Cl}$.

As DSC and Rheovibron data have shown, blends of BAN 44/PVC contain two amorphous phases; one is mainly BAN 44 and the other is a compatible BAN 44–PVC mixture. Because of poor interfacial bonding, efforts to uniformly deform the specimens failed and IR dichroism experiments could not be carried out at temperatures lower than ~ 50°C (Figure 9). The orientation function versus percent strain curves are shown in Figure 10 for the 50/50 composition. Because of the essentially relaxed state of the BAN 44 phase, which is well above its T_g, the PVC orientation as measured by $f_{C\text{-}Cl}$ is higher than the corresponding values of f_{CN}. The f_{CN} at 52°C, however, is greater than the data taken at higher temperatures because that sample was tested within the transition zone of the PVC-rich phase, which does contain some BAN 44.

The experimental results for NC/PCL, PVC/PCL, and ET-38-1 blends suggest than an amorphous compatible blend may exhibit the characteristics of a single homopolymer, not only in its glass transition behavior and mechanical properties, but also in the uniform way in which the polymer chains orient. In contrast, a phase-separated binary mixture of polymer segments will exhibit a different orientation for each component. For example, when the incompatible high-density polyethylene/isotactic polypropylene blends were deformed, the component that constitutes the continuous phase always orients to a higher degree than the component in the dispersed phase, regardless of which polymer dominates the composition. Furthermore, the polymers are not joined by covalent bonds in a polymer blend, and thus the discrete phases may experience different extents of applied deformation, depending upon the interfacial cohesion between the separated phases. An example has just been shown for the incompatible BAN 44/PVC blends.

However, some hypothetical polymer blend situations may not conform

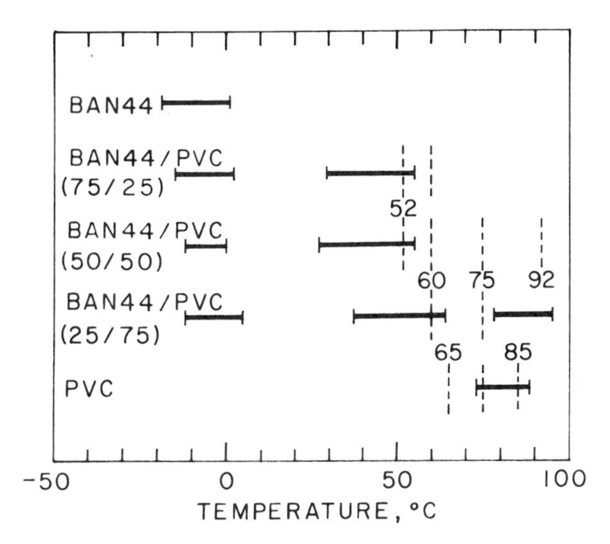

Figure 9. Relationship between the T_g range and IR dichroism measurement temperatures for the BAN 44/PVC blends.

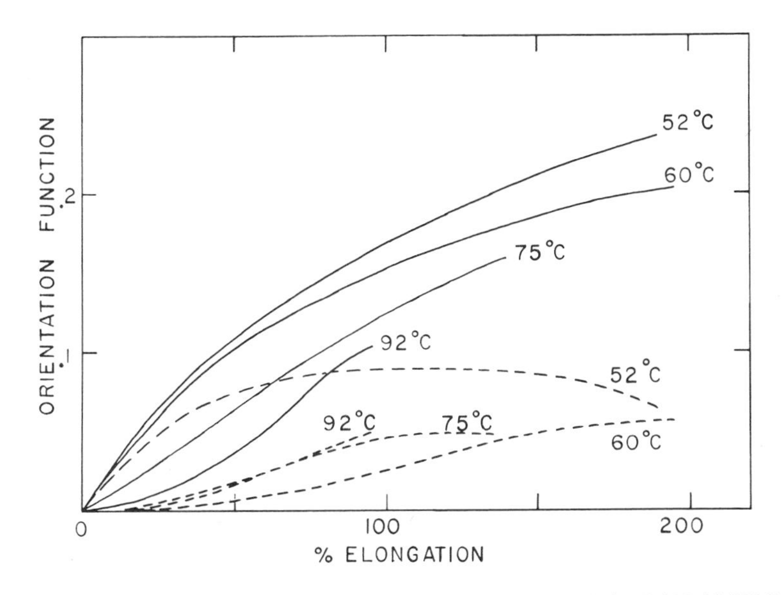

Figure 10. Orientation function vs elongation curves of the BAN 44/PVC (50/50) blends (the comparison between f_{CN} and f_{C-Cl}) at various temperatures. Key: —, C–Cl; and --, CN.

to the aforementioned model. First, two incompatible polymers might form an interpenetrating network such that the phases effectively act mechanically in parallel. Thus their orientation functions would be related to their respective moduli and could possibly be quite similar. Second, if the backbone motion in a compatible polymer blend is effectively restricted by entanglements, the segmental orientation of such a chain would vary inversely with the segment length between entanglement points. This theory predicts that the individual chain of a compatible blend will exhibit similar orientation only if the corresponding segment lengths are comparable. In a compatible poly(phenylene oxide) (PPO)/PS blend, the stiffer chain component (PPO) had a higher orientation function (*9*).

PEUU Copolymers (*10*). GENERAL ORIENTATION CHARACTERISTICS. Several functional groups were used to follow segmental orientation in the soft and hard segment domains of the PEUU samples. They include the following: (1) the average orientation of the symmetric and asymmetric CH stretching bands, which are a measure of soft segment orientation; (2) the C=O band ($CO_{F,UT}$) at 1731 cm^{-1}, which is a measure of the average orientation of the nonhydrogen bonded urethanes either at the hard domain interfacial zone or of hard segments solubilized in the polyether matrix; (3) the C=O absorption band ($CO_{B,UT}$) at ~1708 cm^{-1}, which follows the orientation of hydrogen bonded urethane at the interface; (4) the C=O band ($CO_{B,UA}$) at ~1635 cm^{-1}, which is a measure of the orientation of urea linkages within the hard domains; and (5) the NH band at ~3317 cm^{-1}, which characterizes the hard segment orientation.

Figure 11 shows the typical relationship between the orientation function (f) and the percent elongation for PEUU samples as represented by the two samples containing 46 wt% hard segment. Upon deformation, the soft segments, as monitored by f_{CH}, orient into the stretch direction. The orientation behavior of the hard segments is more complicated. A negative value of $f_{CO_{B,UA}}$ or f_{NH_B} indicates that the hard segments within the domains first orient transversely to the stretch direction. In contrast, both the urethane carbonyls orient positively; this orientation suggests that hard segments at the interface become aligned into the stretch direction. This phenomenon can be explained by suggesting a morphological model of polyurethanes consisting of interlocking hard and soft domains (*11*). In the unstrained state, the soft segments are essentially in the isotropic random coil conformation. Urethane segments are, however, aligned approximately perpendicular to the long direction of the hard domains, thereby making the hard domains locally anisotropic. Overall, the hard domains are randomly arranged so that the bulk sample appears mechanically and optically isotropic. In Estes' model (*8*) the individual hard segment may be initially oriented to different extents; this extent of orientation depends on the relative orientation of the long dimension of the domain to the stretch direc-

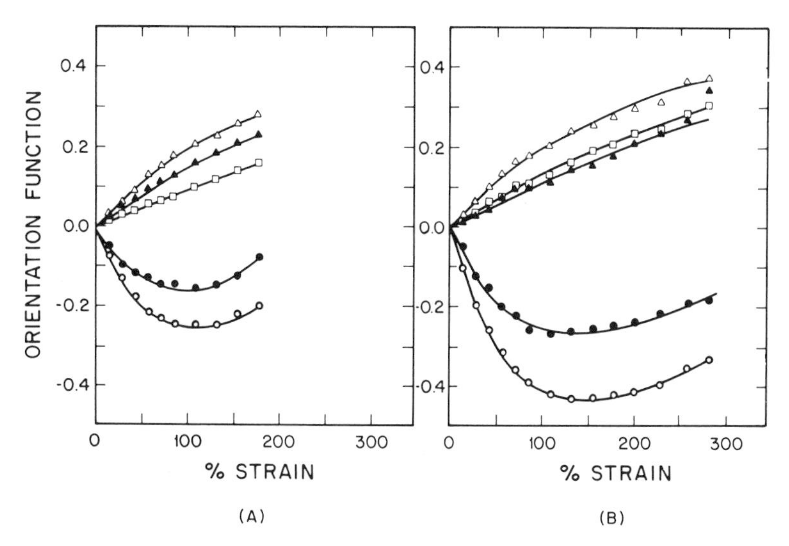

Figure 11. Orientation function vs elongation curves of (A) PEUU-46-1000, and (B) PEUU-46-2000. Key: □, CH; △, $CO_{F,UT}$; ▲, $CO_{B,UT}$; ●, NH_B; and ○, $CO_{B,UA}$.

tion. The orientation of such an assembly of randomly oriented lamellae gives rise to an average negative orientation of the urethane chain segments because the chain axis is perpendicular to the long axis of the domain. Because the NH_B absorption combines NH groups participating in either the interurea, interurethane, or hard-to-soft segment hydrogen bonding, it is not surprising that the average bonded N–H orientation function (f_{NH_B}) is less negative than the bonded urea carbonyl function, $f_{CO_{B,UA}}$ (Figures 11 and 12).

Soft Segment Orientation Studies. Figure 13 summarizes the soft segment orientation plotted against percent elongation for all the PEUU materials studied. As the hard segment content and block length were reduced, lower soft segment f values were observed. The storage modulus (E') at room temperature, as determined by DMA, also shows a trend consistent with the orientation behavior; that is, a higher E' correlates with higher domain rigidity and less segmental disorientation during the 5-min relaxation period. Both factors result in higher f values.

During the stress relaxation period following sample stretching, the orientation of the chain segments will be influenced by molecular relaxation processes. The extent of segmental relaxation will be affected by the relative rigidity of the corresponding phase at the testing temperature (room temperature for this work). Consequently, the strained soft segments with a T_g much lower than room temperature tend to disorient toward a more random conformation via an entropy-driven mechanism. The accompanying retractive force exerts a tension on the urethane junc-

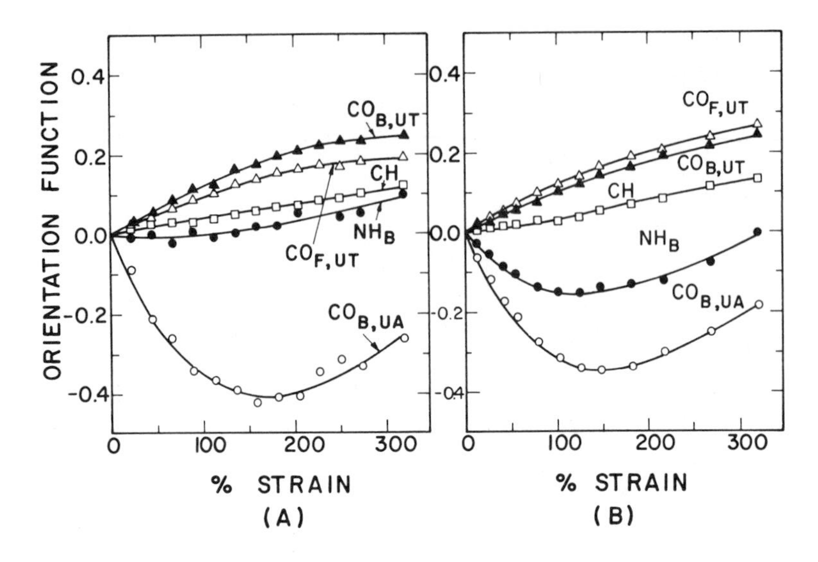

Figure 12. Orientation function vs elongation curves of (A) PEUU-25-1000 and (B) PEUU-25-2000. Key: □, CH; △, $CO_{F,UT}$; ▲, $CO_{B,UT}$; ●, NH_B; and ○, $CO_{B,UA}$.

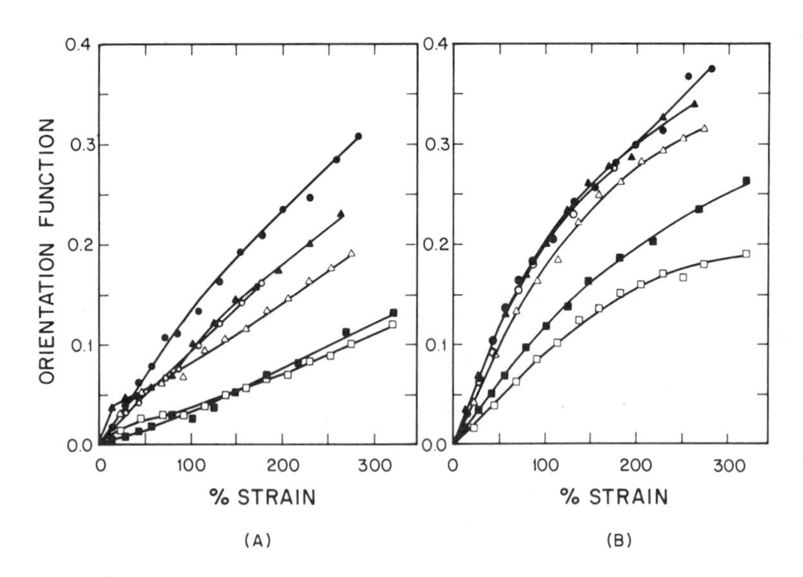

Figure 13. Orientation function vs elongation curves of (A) CH and (B) $CO_{F,UT}$ of polyether polyurethaneurea samples. Key: ○, PEUU-46-1000; ●, PEUU-46-2000; △, PEUU-36-1000; ▲, PEUU-36-2000; □, PEUU-25-1000; and ■, PEUU-25-2000.

tions at the interface between the hard domains and the soft segment matrix and causes them to become further aligned into the stretch direction. The hard domains are less affected by the PTMO relaxation, because of their relatively high rigidity. Thus, for each PEUU sample, the soft segment orientation (f_{CH}) was observed to be lower than the orientation of the nonhydrogen bonded urethanes ($f_{CO_{F,UT}}$) (Figure 13).

HARD SEGMENT ORIENTATION STUDIES. To study the influence of block length on segmental orientation, it is instructive to compare IR dichroism results on samples containing similar hard segment concentrations.

Stress hysteresis data showed that PEUU-46 samples contained highly interlocking hard segment domains. However, both the DSC and Rheovibron results also indicated that PEUU-46-2000 had higher phase purity and degree of order in both the hard domains and the polyether matrix phase. Figures 11A and 11B show that $f_{CO_{B,UA}}$ and f_{NH_B} of PEUU-46-1000 are less negative than those of PEUU-46-2000. The hard segment lamellae in the 2000 MW sample apparently have better domain cohesion. This improved cohesion arises because PEUU-46-2000 has a higher fraction of urea groups and degree of interurea hydrogen bonding. Therefore the domains in the longer segment material are not disrupted into microdomains containing bundles of hard segments oriented preferably along the stretch direction until higher strain levels.

Similar trends in orientation behavior for soft and hard segments within domains or at the interface are observed for sample pairs PEUU-36-1000 and 2000 (not shown) and PEUU-25-1000 and 2000 (Figures 12A and 12B). In contrast, an initially positive f_{NH_B} in PEUU-25-1000 is observed. This positive f arises from the relatively small contribution of the urea linkages to the average hard segment orientation. (PEUU-25-1000 has the lowest urea content among the polyurethaneureas studied.)

Conclusions

For amorphous compatible blends of PCL and PVC or NC, PCL oriented in essentially the same manner as NC and the isotactic segments of PVC. Syndiotactic PVC segments showed much higher orientation functions; these higher functions imply the existence of a microcrystalline PVC phase.

The PCL orientation functions for semicrystalline blends were substantially higher than for the amorphous component NC or isotactic PVC segments. The crystalline and amorphous PCL orientations for a 25% NC blend were determined by assuming that the NC orientation behavior was representative of the segmental orientation of the amorphous phase chain segments of PCL.

For the polyurethane (ET-38-1)/PVC blends, the IR dichroism results suggest that segmental compatibility between polyether (soft segment of

ET-38-1) and PVC depends on the details of the sample preparation method. Samples cast from tetrahydrofuran (THF) were more compatible than those cast from THE/dioxane. The aromatic urethane segments that exhibit microphase separation in the pure PU are not solubilized by blending with PVC by any of the sample preparation methods used.

For the blends of BAN/PVC, IR dichroism measurements are strongly dependent upon the viscoelastic state of the blends. The compatibility between BAN and PVC, the extent of structural heterogeneity, and the presence of ordered PVC microdomains significantly affect the orientation characteristics of the individual polymer chains.

The orientation behavior of segmented PEUU copolymers is dependent upon the hard domain morphology and the viscoelastic state of the soft segment matrix. The hard segments within rigid domains initially orient transverse to the stretch direction whereas the soft segments orient parallel to the stretch direction.

Acknowledgment

The authors acknowledge partial support of this work by the polymers section of the Division of Materials Research of the National Science Foundation through Grant DMR 81–06888 and the Naval Air Systems Command through Contract N00019–82–C–0246. PCL was supplied by J. Koleske of Union Carbide.

Literature Cited

1. Stein, R. S.; Finkelstein, R. S. *Ann. Rev. Phys. Chem* **1973**, 207.
2. Wilkes, G. L. *J. Macromol. Sci. Rev. Macromol. Chem.* **1974**, *C10(2)*, 149.
3. Read, B. E.; Stein, R. S. *Macromolecules* **1968**, *1*, 116.
4. Hubbell, D. S.; Cooper, S. L. *J. Appl. Polym. Sci.* **1977**, *21*, 3035.
5. Hubbell, D. S.; Cooper, S. L. *J. Polym. Sci. Polym. Phys. Ed.* **1977**, *15*, 1143.
6. Wang, C. B.; Cooper, S. L. *J. Appl. Polym. Sci.* **1981**, *26*, 2989.
7. Wang, C. B.; Cooper, S. L. *J. Polym. Sci. Polym. Phys. Ed.* **1983**, *21*, 11.
8. Monnerie, L., In "Proceedings of the 28th Macromolecular Symposium"; IUPAC: Amherst, Mass., 1982; p. 788.
9. Lefebvre, D.; Jasse, B.; Monnerie, L. *Polymer* **1981**, *22*, 1616.
10. Wang, C. B.; Cooper, S. L. *Macromolecules* **1983**, *16*, 775.
11. Estes, G. M.; Seymour, R. W.; Cooper, S. L. *Macromolecules* **1971**, *4*, 452.

RECEIVED for review January 20, 1983. ACCEPTED August 16, 1983.

9

Properties and Morphology of Poly(methyl methacrylate)/Bisphenol A Polycarbonate Blends

ZACK G. GARDLUND

General Motors Research Laboratories, Polymers Department, Warren, MI 48090-9055

Poly(methyl methacrylate) (PMMA) and bisphenol A polycarbonate (PC) blends prepared by melt extrusion techniques have been investigated by dynamic mechanical analysis and scanning electron microscopy. Partial miscibility was established by a comparison of the experimentally determined glass transition temperatures (T_g) *with those predicted by the Fox equation for a miscible pair of polymers. At PC concentrations less than 50 wt %, the PC is dispersed in a continuous PMMA matrix; at higher PC concentrations, both polymers form continuous phases. A correlation has been observed between the morphology and the storage modulus at 130 °C, a temperature between the* T_g *values of the pure polymers. The partial miscibility is discussed in terms of possible* n–π *complex formation between the* n *electrons of the ester group of the PMMA and the* π *electrons of the benzene rings of the PC.*

THE BLENDING OF BISPHENOL A POLYCARBONATE (PC) with numerous polymers has been the subject of some interest in recent years. The polymer blends have contained aliphatic polycarbonates (*1*, *2*), polyethylene (PE) (*3*, *4*), polypropylene (PP) (*5*), polystyrene (PS) (*3*), polysulfone (*6*), and a variety of other polymers and copolymers (*4*, *7*). The major effort has been in the area of PC–PE blends (*8*, *9*). Paul et al. (*10–14*) have done detailed studies of the effects of polyester structure on blend miscibility.

We have been interested in the interactions between PC and poly(methyl methacrylate) (PMMA), a polymer that although not a linear polyester does have a large density of pendent ester groups. A preliminary report on this research has been published (*15*). In this chapter, we will consider the interactions between bisphenol A PC and PMMA, over a wide range of compositions, as reflected by changes in dynamic mechanical response and morphology.

0065–2393/84/0206–0129$06.00/0

Experimental

Injection molding grade PMMA was obtained from Rohm and Haas Company, and an additive-free bisphenol A PC was obtained from General Electric Company. As determined by gel permeation chromatography in tetrahydrofuran with a PS standard, the PMMA had a number average molecular weight (M_n) of 99,000 and a molecular weight distribution (MWD) of 2.1, and the PC had M_n of 14,000 and MWD of 2.2. Mixtures of the resins were made on a weight percent basis and were dried in a vacuum oven at 90 °C. The mixtures were extruded under dry conditions and granulated before a second run through an injection molding machine in an extrusion mode. Molten resin was run onto a sheet of aluminum foil, covered with a second sheet of foil, and immediately compression molded. The plates were preheated to 250 °C and a 5-min warmup was used, followed by a 3-min press at 630 MPa. The plates were then cooled in a press at 210 MPa. Samples with cross-sectional dimensions of between 7×10^{-4} and 12×10^{-4} cm^2 and lengths of 5.3 cm were used for dynamic mechanical analysis (DMA) at 11 Hz. The sample was placed in the jaws of the tension arms and the sample was cooled to −150 °C before analysis was begun. Dynamic mechanical response was established by means of a Toyo Baldwin Co. Rheovibron DDV IIC. The Rheovibron was controlled by a Hewlett Packard 9825A desktop computer, Hewlett Packard 6940B multiprogrammer, and Princeton Applied Research Model 5204 Lock-In analyzer. The control program was developed in cooperation with IMASS, Inc. Data were recorded on tape and plotted by a Hewlett Packard 9245A plotter printer. Scanning electron microscope (SEM) analysis was done with an International Scientific Instruments Model DS130 on surfaces of cryogenically fractured samples that had been sputtered with Au–Pt.

Results

Dynamic Mechanical Properties. The dynamic mechanical properties, storage modulus (E'), loss modulus (E''), and tan δ, of the blends and of the pure polymers were measured at 11 Hz over the temperature range of −150 to +160 °C. The data for PMMA are shown in Figure 1. In this figure, as in all the mechanical spectra figures, the results are shown as smooth curves. These curves are based on tan δ data that was taken at 1–2 °C temperature intervals. The E' and E'' data were calculated at 5 °C intervals with the exception of moduli that were calculated in the area of a transition. In these areas the smallest temperature increments, usually 1 °C, were used. The spectra for pure PMMA are typical and have been described by numerous authors, albeit usually at a different frequency. The α or glass transition appears at 130 °C, as determined by tan δ, and at 107 °C in the E'' spectrum. The β transition, which is generally accepted (*16*) to arise from rotation of the methyl ester side group, is a broad shoulder in the tan δ spectrum and a broad shoulder in the E'' spectrum centering around 25–50 °C. The spectra for bisphenol A PC are shown in Figure 2. The glass transition is located at 150 °C by tan δ and at 143 °C by E''. The β transition is seen as a shoulder in the region of 75–100 °C and has been attributed to packing defects in the glassy state (*17*). The γ transition

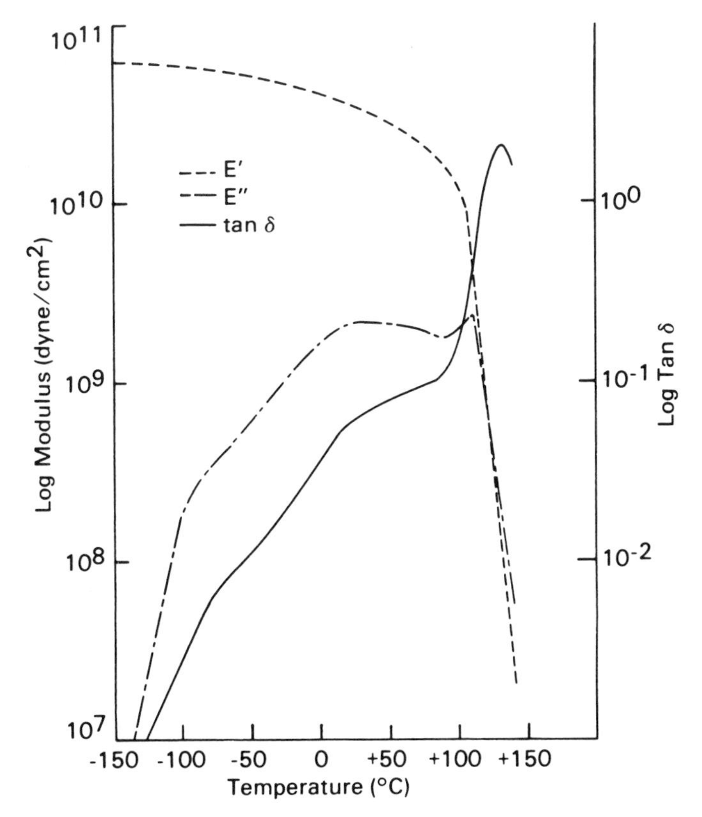

Figure 1. Storage modulus, loss modulus, and tan δ for PMMA at 11 Hz.

results in a broad peak in both E'' and tan δ between -100 °C and -75 °C. The γ peak is considered to be a result of motion of the monomer unit (*17*).

The dynamic spectra for the blends are shown in Figures 3–12. In this chapter, the following nomenclature will be used: For example, 50 PC/50 PMMA indicates a blend containing 50 wt % bisphenol A PC and 50 wt % PMMA. The addition of 2 wt % PC to PMMA to form blend 2 PC/98 PMMA (Figure 3) yielded a 5 °C shift of the α transition as determined by tan δ and a 1 °C shift based on the E''. The data for 15 PC/85 PMMA (Figure 4) are very similar to 2 PC/98 PMMA with only a slight change in the α transition. The indication of a shoulder around -90 °C can be attributed to the γ transition of PC. Blend 25 PC/75 PMMA (Figure 5) also shows small changes in the location of α transition with both tan δ and E''. In the previous blends, E' showed little effects from added PC. In this blend, however, the E' begins to sharply drop off at a higher temperature. The β transition for PMMA and the γ transition for PC now are represented by distinct maxima. The addition of more PC in blend 35 PC/65 PMMA (Fig-

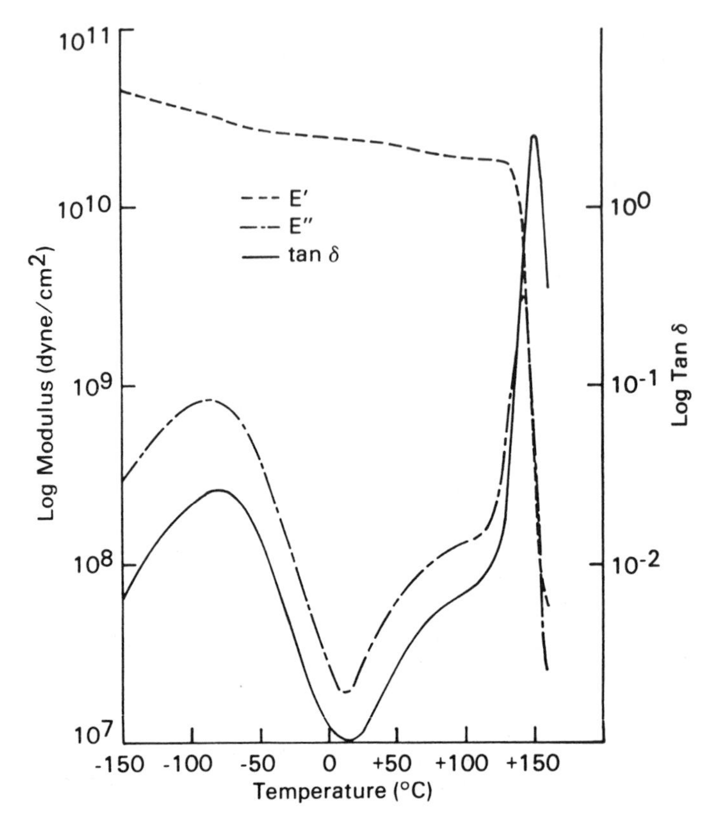

Figure 2. Storage modulus, loss modulus, and tan δ for PC at 11 Hz.

ure 6) causes a slight change in E'—again, the drop off in modulus moves to a slightly higher temperature. The shift in T_g is also small.

Major changes in the dynamic mechanical spectra are noted with the 50/50 blend composition (Figure 7). The α transition for PC is observed as a maximum at 145 °C in the tan δ results and as a shoulder at about 135 °C in the E''. This observation is the first evidence of the α transition for PC in these blends. The α transition for PMMA is now a shoulder in tan δ and a maximum at 115 °C in E''. Both the β transition of PMMA and the γ transition of PC appear as distinct maxima. The influence of the added PC is also seen in the decrease of the slope of E' and the marked increase in values of E' at higher temperatures.

With increased PC content in the blends, both tan δ and E'' show distinct maxima for the two α transitions (Figures 8–12). As the PC content is increased, the magnitude of α transition as determined by E'' increases proportionately and also shows a shift to higher temperatures, reaching a maximum for E'' at 143 °C (Figure 12). The size of the PMMA α transition

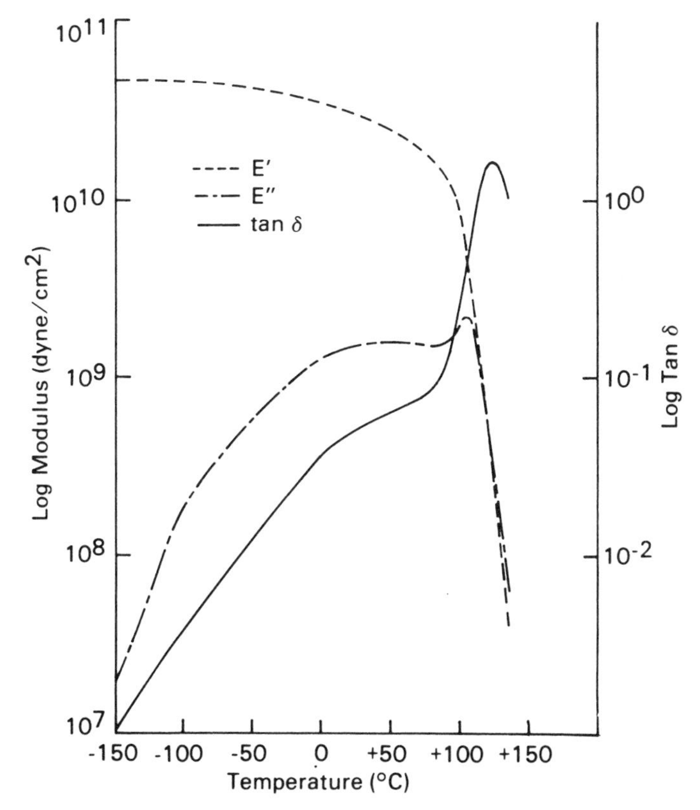

Figure 3. Storage modulus, loss modulus, and tan δ for 2 PC/98 PMMA at 11 Hz.

in the E'' data decreases in this blend series but shows a positive temperature shift from 113 °C for 60 PC/40 PMMA (Figure 8) to 118 °C for 95 PC/5 PMMA (Figure 12). The β transition for PMMA decreases in magnitude in the E'' and tan δ data and the maximum flattens out at high PC contents (Figure 12). The γ transition of PC shows little change in either the size of the maximum or location in temperature.

In the E' data for 50 PC/50 PMMA a double step drop in modulus occurs at about 100 and 125 °C (Figure 7). This double step is also evident in Figure 8 for the 60 PC/40 PMMA blend. This double drop in E' for blends 75 PC/25 PMMA (Figure 9) and 85 PC/15 PMMA (Figure 10) occurs at increasingly higher moduli and at somewhat higher temperatures. Further increases in PC content do not markedly increase E' as seen for 90 PC/10 PMMA (Figure 11) and 95 PC/5 PMMA (Figure 12). The double drop in E' is difficult to discern for these last two blends. The storage modulus is close to that of PC homopolymer.

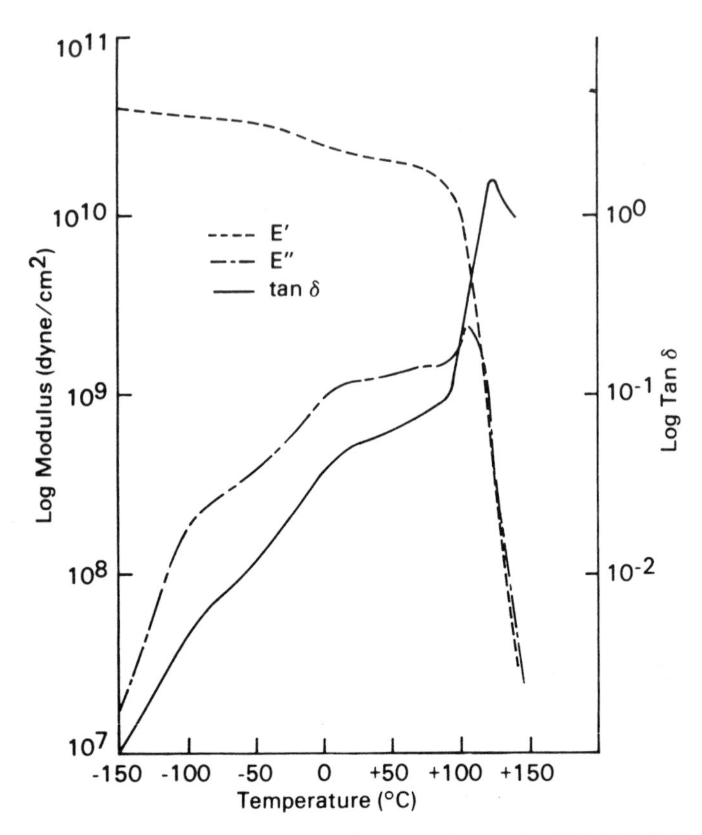

Figure 4. Storage modulus, loss modulus, and tan δ for 15 PC/85 PMMA at 11 Hz.

Blend Morphology. The morphologies of the blends were determined by SEM analysis of cryoscopic fracture surfaces. In Figure 13a, the SEM micrograph of 2 PC/98 PMMA, PMMA is the continuous phase and PC appears as spheres. The few spheres that can be seen have diameters less than 0.5 μm and emerge from the matrix. In the 15 PC/85 PMMA blend (Figure 13b) the size of the PC spheres is unchanged, but the concentration of spheres has increased. Holes are observed in the continuous PMMA phase where PC spheres have been removed during fracture. Adhesion between the spheres and the continuous phase is observed. There are roughly as many holes as there are spheres and the spheres are embedded in the matrix. In Figure 13c, the fracture surface of 25 PC/75 PMMA, the PC spheres exist in a variety of sizes, emerging from the continuous PMMA phase. The beginnings of coalescence of the PC spheres are also observed. For example, in the upper left corner of the micrograph, long (2–3 μm) ridges of coalesced PC are seen. A series of six or seven spheres is seen in the

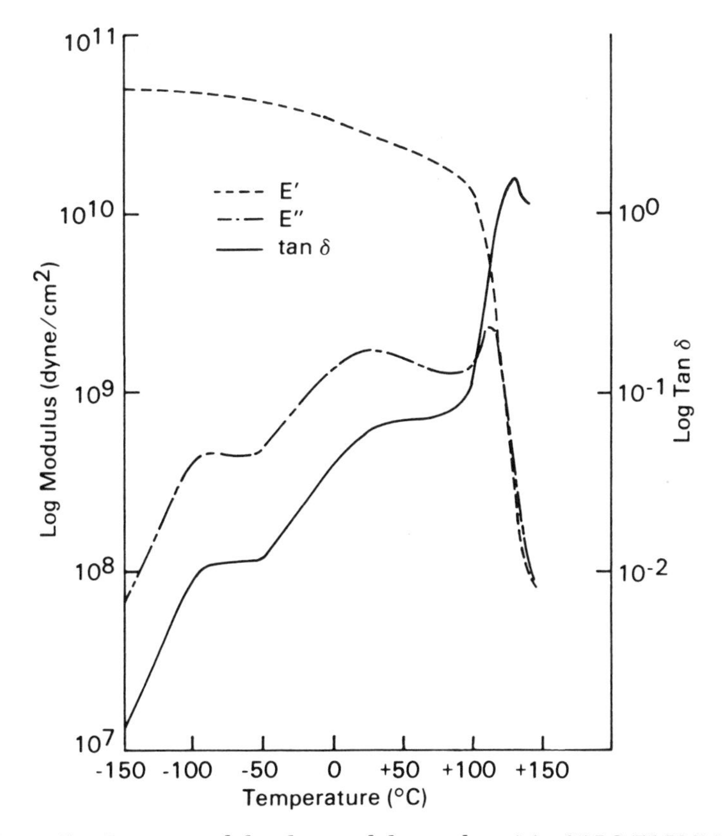

Figure 5. Storage modulus, loss modulus, and tan δ for 25 PC/75 PMMA at 11 Hz.

lower left corner of the same micrograph. Blend 35 PC/65 PMMA (Figure 13d) exhibits further coalescence of the PC spheres. No individual PC spheres are found; rather irregular rows of connected spheres exist. The change in morphology between 35 PC/65 PMMA (Figure 13d) and 50 PC/50 PMMA (Figure 14a) is dramatic. Only lamellae exist in the 50 PC/50 PMMA blend as both phases are now continuous. The continuity of the phases is also seen in 60 PC/40 PMMA (Figure 14b) and 75 PC/25 PMMA (Figure 14c). The holes or voids are formed during fracture by removal of sections of continuous phase. The morphologies of 85 PC/15 PMMA (Figure 14d), 90 PC/10 PMMA (Figure 15a), and 95 PC/5 PMMA (Figure 15b) are less distinctive. Phase separation does not occur and PMMA entities are not formed even at these high PC contents. In fact, with the exception of a decreasing irregularity of the surface, the micrographs are remarkably similar to that of unblended PC.

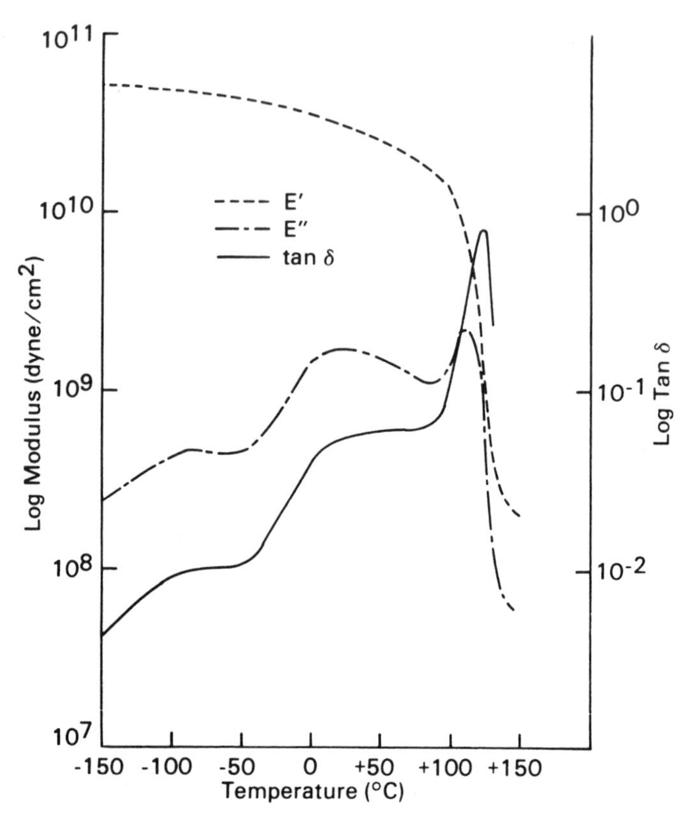

Figure 6. Storage modulus, loss modulus, and tan δ for 35 PC/65 PMMA at 11 Hz.

Discussion

Visual observation shows that the blends are opaque over the entire composition range, a fact that is generally taken as an indication of polymer immiscibility. Also, the two T_g values, as determined by DSC (*15*), of the PMMA and PC are observed over the composition range of the blends. However, even a cursory look at the effect of composition on transition temperature shows a change in temperature with composition. The various transition temperatures are listed in Table I. To avoid a superfluity of subscripts and superscripts, the following nomenclature has been developed for the observed transitions: T_1 is the glass transition (T_α) of the PC component, and T_2 is the glass transition (T_α) of the PMMA. Because the T_β for polycarbonate is obscured in the blends, it will not be listed. T_3 is the T_β of PMMA while T_4 is the T_γ of PC. All transition temperatures were determined as being where the E'' was a maximum. In general, E'' values were calculated with temperature increments of 5 °C, except in the area of a

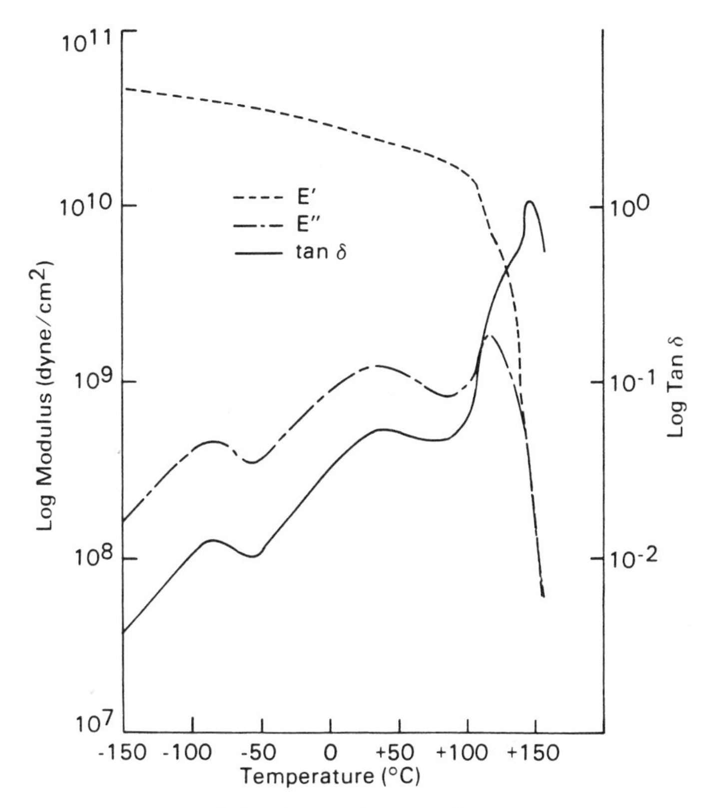

Figure 7. Storage modulus, loss modulus, and tan δ for 50 PC/50 PMMA at 11 Hz.

transition where calculations were made at 1 °C intervals. The glass transition temperatures (T_1 and T_2), as determined by the line intersection method *(18)* from DSC are included for comparison, but the transition temperatures as determined by DMA will be used for most considerations of compositional effects. The lower temperature transitions (T_3 and T_4) as determined by DMA, are very broad, and a temperature is recorded from the E'' calculation as a midpoint temperature. The T_4, the T_γ for PC, does not change with changes in blend composition. This finding is consistent with the assignment of this transition to motion of monomer segments in the polymer backbone. The motion of these segments should be unaffected by the polymer's environment. T_3, the T_β for PMMA (which is very broad), can only be recorded as a temperature range, but it does tend to higher temperatures as the concentration of PC increases. Since T_3 is a result of rotational motion of the ester group of the PMMA, T_3 should be affected by the environment. As the ease with which the ester group can rotate is changed, so should the transition temperature change, as seen by the experimental results. The T_1 for PC does not appear until the PC concentration is

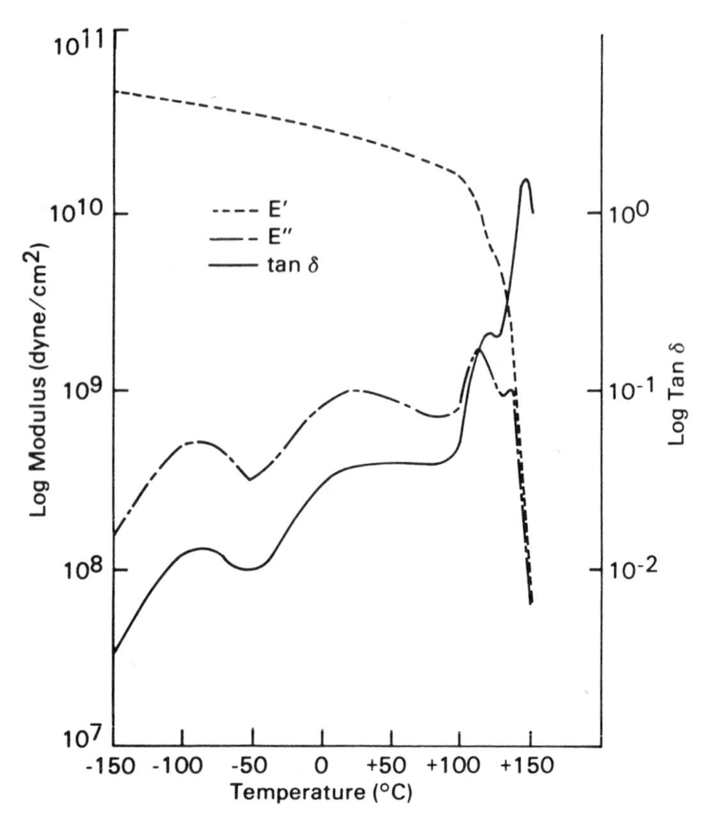

Figure 8. Storage modulus, loss modulus, and tan δ for 60 PC/40 PMMA at 11 Hz.

0.50 weight fraction and then T_1 gradually increases with increased PC content in the blend. The PMMA T_2 is observable over the entire composition range, also increasing with increased PC content. As the blend composition changes from pure PMMA to 0.35 weight fraction PMMA the total change in temperature is 2–3 °C. A large jump in temperature occurs between the 35 PC/65 PMMA blend and the 50/50 blend, and a larger change in transition temperature occurs over the composition range of 0.35–0.95 weight fraction PMMA. Also, for PC contents greater than 0.35 weight fraction, two glass transitions are observed.

Plots of T_1 and T_2 versus composition are shown along with a plot of the Fox equation (*19, 20, 21*) in Figure 16. The dotted lines in the figure indicate the expected T_g values for PC and PMMA if the polymers were completely immiscible. The slightly curved solid line (Fox equation) between the T_g values of the pure polymers predicts the single glass transition for a blend of two miscible polymers. Clearly, for the PC/PMMA blends, intermediate behavior is the case; that is, the polymers are partly soluble in

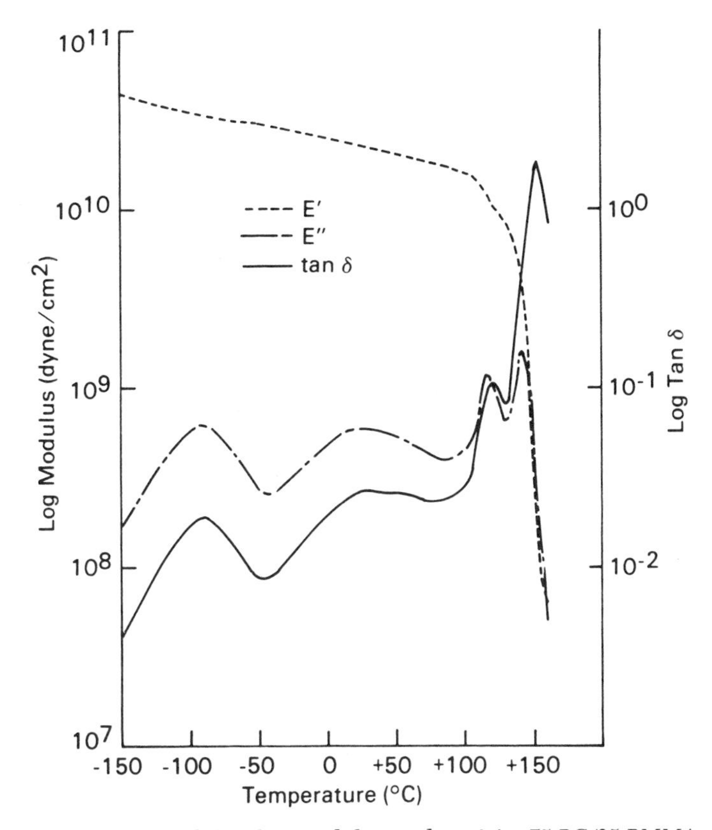

Figure 9. Storage modulus, loss modulus, and tan δ for 75 PC/25 PMMA at 11 Hz.

one another. Extrapolation of the experimental results back to the predicted temperatures allows calculation of the extent of miscibility (i.e., approximately how much the polymers intermix). By using Equation 1

$$W_1 = \frac{T_g - T_{g_2}}{T_{g_1} - T_{g_2}} \tag{1}$$

as derived by Stoelting et al. (22), where T_g is the experimentally determined glass transition temperature of the blend, and T_{g_1} and T_{g_2} are the experimentally determined T_g values of the pure blend components PC and PMMA, respectively, the weight fraction (W_1) of the blend component 1 can be calculated. The results of the calculations based on the glass transition temperatures listed in Table I are shown in Table II. The blend composition listed in Table II are based on the original amounts of PC and PMMA used to prepare the blends. If W_1 is calculated for the 50 PC/50 PMMA, 60 PC/40 PMMA, and 75 PC/25 PMMA blends, the miscibility of

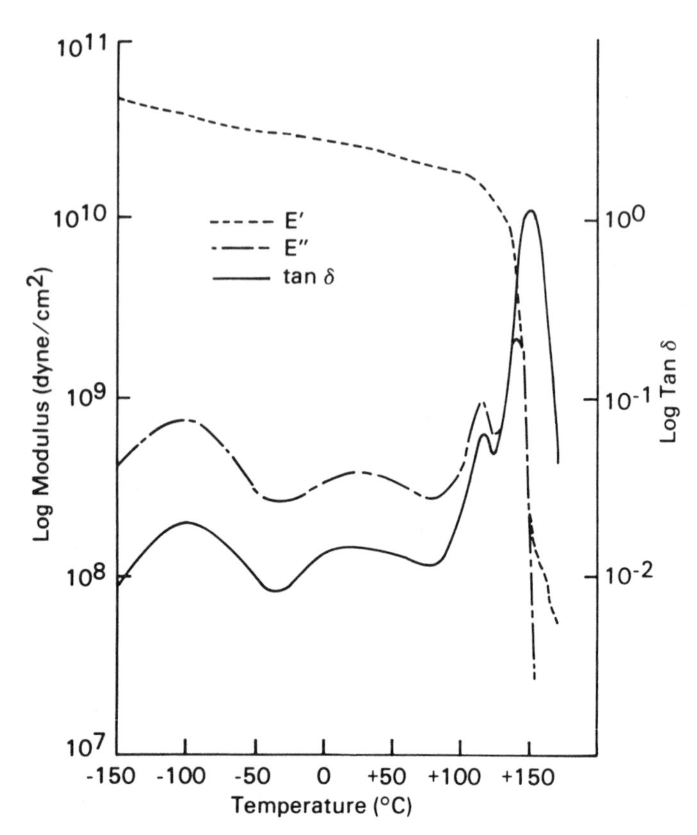

Figure 10. Storage modulus, loss modulus, and tan δ for 85 PC/15 PMMA at 11 Hz.

PC with PMMA and conversely the miscibility of PMMA with PC are approximately equal. For example, in the 50 PC/50 PMMA blend the apparent weight fraction of PMMA in the PC rich phase is 1.00–0.78, or 0.22. This equal miscibility is not observed at PC concentrations greater than 75 wt %, possibly because this empirical analysis is not sufficiently sensitive at very high concentrations of one of the components. The results from calculations based on the DSC measurements are in good agreement with those obtained from DMA measurements. A recalculation of blend compositions from the apparent weight fractions in Table II indicates small changes in actual phase concentrations but does not change the overall observations.

The morphologies observed by SEM agree very well with the results from DMA. The partial miscibility of the two polymers is exemplified by Figure 13c where the PC spheres emerge from the continuous PMMA phase. In cases of nonadhesion of phases where the two polymers are completely immiscible, as with bisphenol A PC/high density PE (*3*), the spheres of the discontinuous phase literally appear to sit in craters formed by the continuous phase.

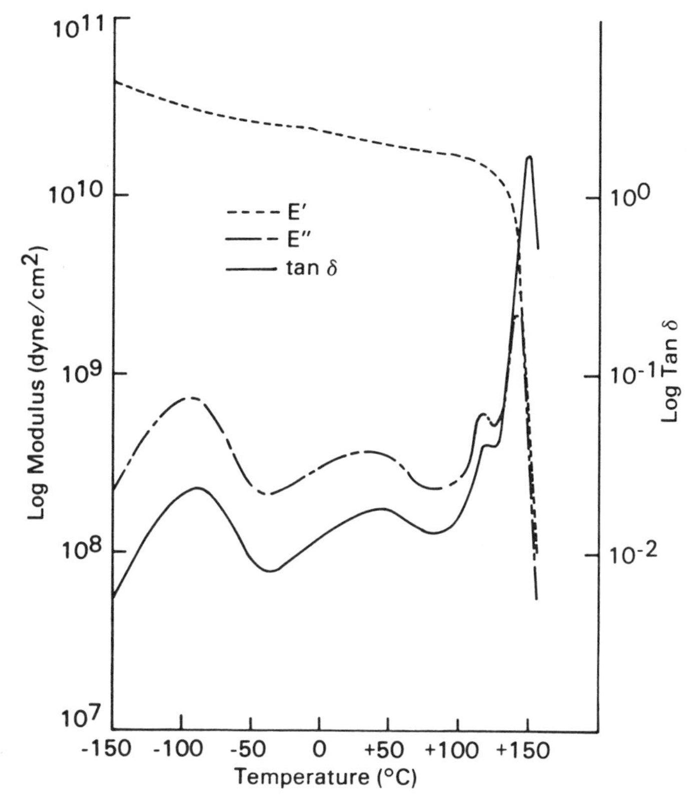

Figure 11. Storage modulus, loss modulus, and tan δ for 90 PC/10 PMMA at 11 Hz.

The double step drop in E', as seen in Figure 7, has also been observed in other partially miscible systems, such as PVC–chlorinated ethylene vinyl acetate copolymer (23). The double step has been described for immiscible polymer systems and correlated with the composition at which the polymers form continuous phases (3). In this PC–PMMA system the SEMs show a change from spheres in a continuous matrix (35 PC/65 PMMA) to the lamellae of two continuous phases (50 PC/50 PMMA) in the same blends that first show the double step in E'.

A plot of the E'' versus weight fraction of PC (Figure 17) at a temperature (i.e., 130 °C) midway between the T_g values of the two polymers dramatically exhibits the effect of blend composition. A sharp jump in modulus occurs between the 35 PC/65 PMMA and 50 PC/50 PMMA blends. This point is precisely where the morphology changes from PC spheres in a continuous PMMA matrix to a lamellar morphology. At the lower weight fractions of PC, the PC is not contributing to the modulus as efficiently as it does when it becomes a continuous phase along with the PMMA at higher

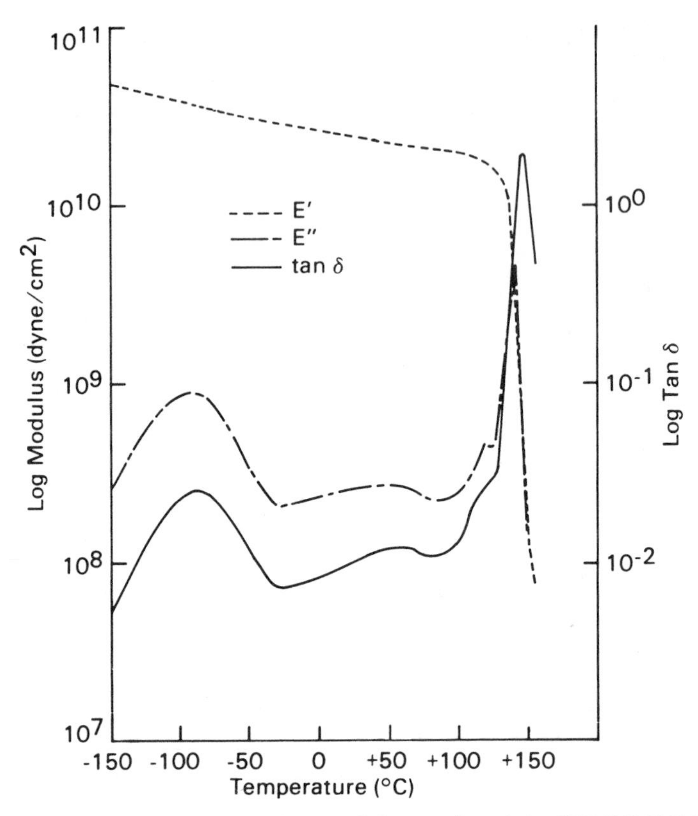

Figure 12. Storage modulus, loss modulus, and tan δ for 95 PC/5 PMMA at 11 Hz.

PC weight fractions. At weight fractions of 0.50 PC and above, the two polymers contribute proportionately to the overall blend modulus.

The question remains as to why there is a partial miscibility between PC and PMMA. This miscibility can be considered to be a result of n–π complex formation between free electrons of the ester groups in the PMMA and the π electrons of the aromatic rings of the PC. Possibly the n–π complex formation contributes to the miscibility of ester–benzene mixtures (*24*). In the case of linear polyesters, for example poly(ethylene succinate), poly(ethylene adipate) (PEA), and poly(ϵ-caprolactone) (PCL), enhanced n–π interactions with bisphenol A PC can be visualized. Blends of these polymers have been shown to be miscible (*13*). Conversely, PS–PCL blends have been found to be immiscible (*25*). This result may be due to the fact that the benzene rings are perpendicular to the PS backbone and the aromatic π electrons are shielded so that there is no interaction with the n electrons of the ester groups. The PMMA–PC blend is an intermediary case. The ester groups of the PMMA are oriented perpendicular to the main

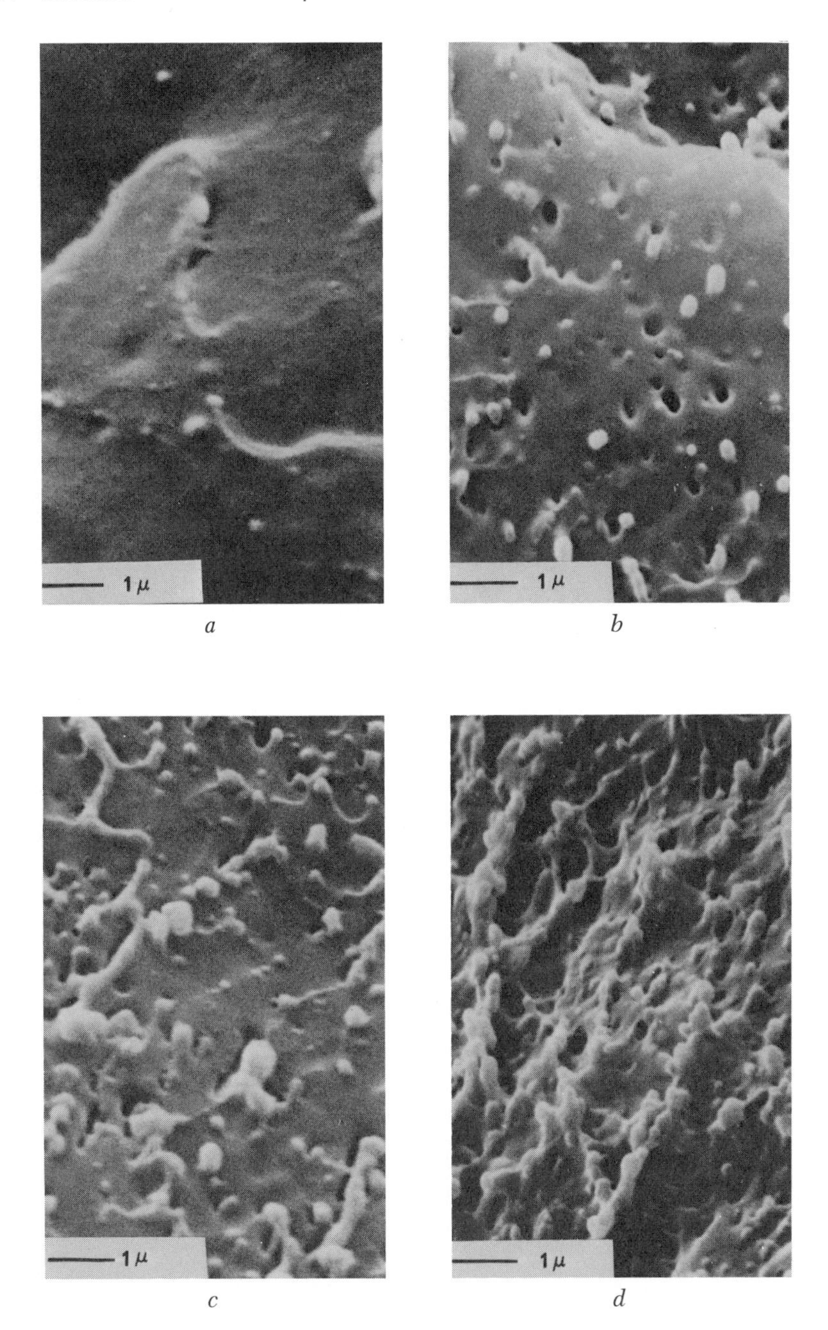

Figure 13. Scanning electron micrograph of fracture surfaces at 10^4 ×. Key: a, 2 PC/98 PMMA; b, 15 PC/85 PMMA; c, 25 PC/75 PMMA; and d, 35 PC/65 PMMA.

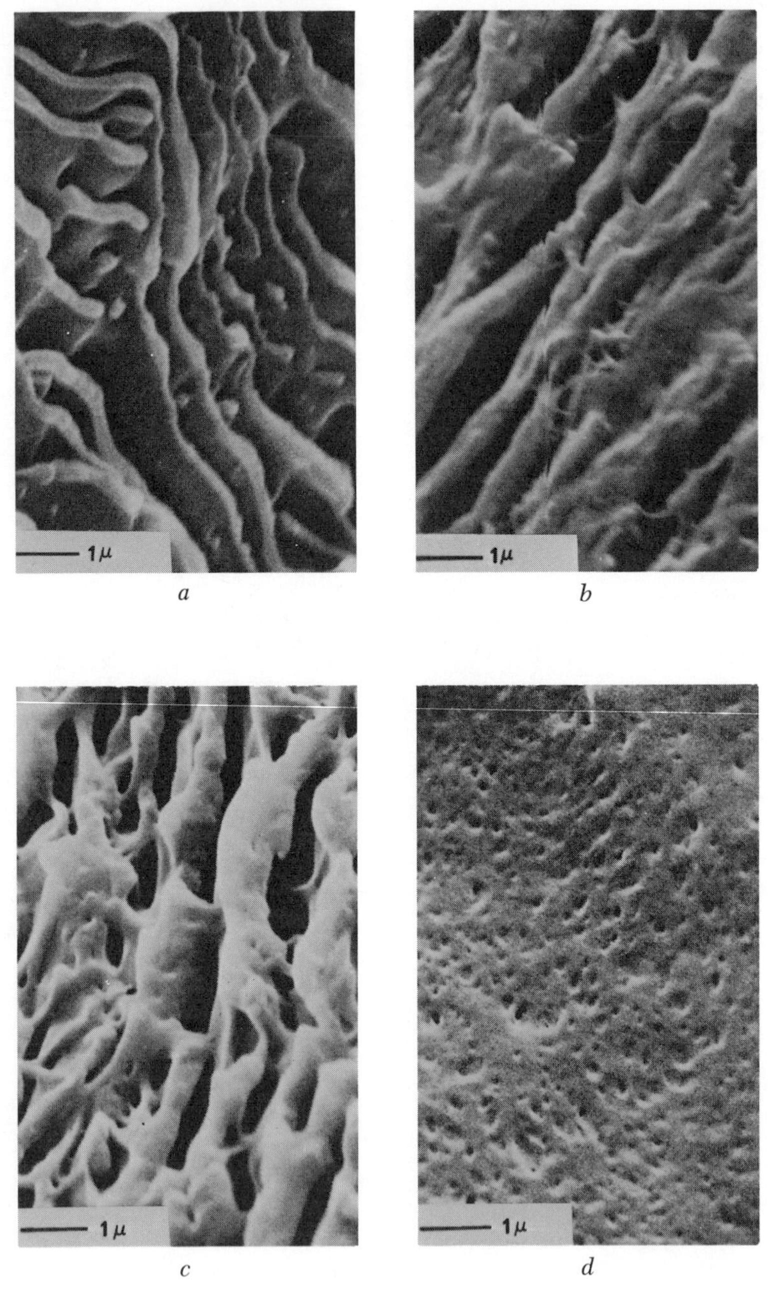

Figure 14. Scanning electron micrograph of fracture surfaces at 10^4 ×. Key: a, 50 PC/50 PMMA; b, 60 PC/40 PMMA; c, 75 PC/25 PMMA; and d, 85 PC/15 PMMA.

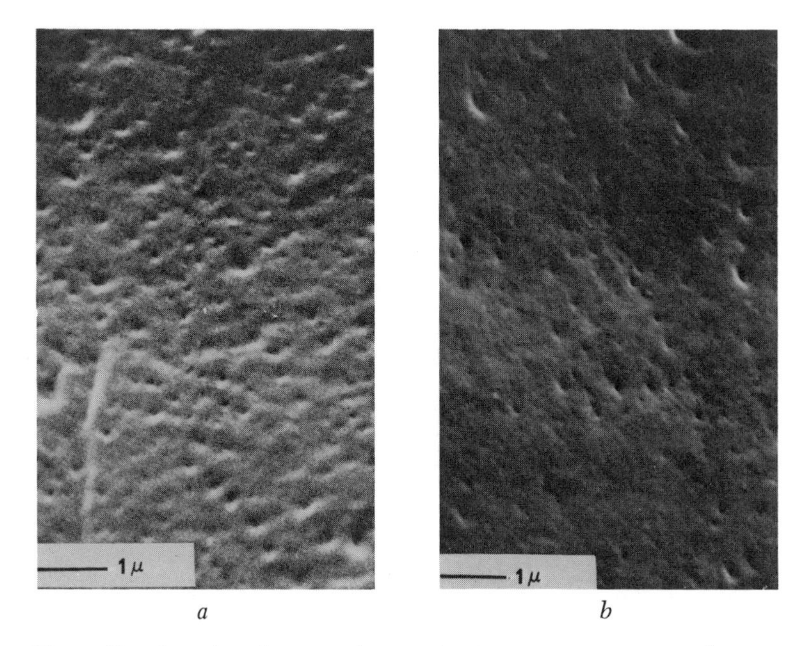

Figure 15. Scanning electron micrograph of fracture surfaces at 10^4 ×. Key: a, 90 PC/10 PMMA; and b, 95 PC/5 PMMA.

Table I. Transition Temperatures of Blends as Determined by DMA and DSC

Polycarbonate	*DMA – Loss Modulus Maxima*				*DSC*[a]	
(wt %)	T_4	T_3	T_2	T_1	T_2	T_1
100	−90	—	—	143	—	150
95	−90	—	118	140	(118)	148
90	−90	(35–45)	117	141	(117)	144
85	−90	(35–40)	116	140	117	144
75	−90	30–35	114	139	115	142
60	−90	20–25	113	136	112	142
50	−90	20–30	114	(135)	111	142
35	(−90)	20–25	108	—	111	140
25	(−90)	25–30	110	—	111	—
15	—	25–30	107	—	112	—
2	—	20–25	106	—	110	—
0	—	25–35	107	—	110	—

NOTE: All transition temperatures are in degrees Centigrade.
T_1 = polycarbonate T_α (glass transition temperature)
T_2 = poly(methyl methacrylate) T_α (glass transition temperature)
T_3 = poly(methyl methacrylate) T_β
T_4 = polycarbonate T_γ
[a]From Gardlund (*15*).

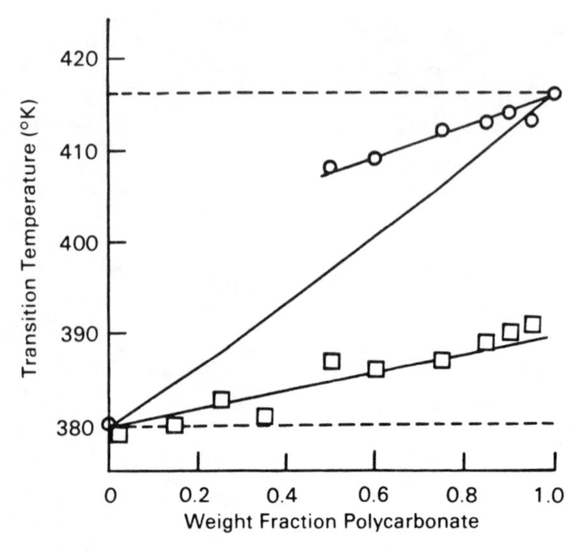

Figure 16. Glass transition temperatures of polycarbonate/poly(methyl methacrylate) blends, as a function of polycarbonate weight fraction. Key: ——, predicted glass transition temperature for miscible polymers; ---, glass transition temperatures for completely immiscible polymers; ○, T_α (polycarbonate) from loss maxima; and □, T_α [poly(methyl methacrylate)] from loss maxima.

Table II. Apparent Weight Fraction (W_1) of PC in the PMMA-Rich and PC-Rich Phases from DMA and DSC Measurements

	W_1 *from Loss Modulus*		W_1 *from DSC*	
Blend	*PMMA-Rich*	*PC-Rich*	*PMMA-Rich*	*PC-Rich*
95 PC/ 5 PMMA	0.31	0.92	0.20	0.95
90 PC/10 PMMA	0.28	0.94	0.18	0.85
85 PC/15 PMMA	0.25	0.92	0.18	0.85
75 PC/25 PMMA	0.19	0.89	0.13	0.80
60 PC/40 PMMA	0.17	0.81	0.05	0.80
50 PC/50 PMMA	0.19	0.78	0.03	0.80
35 PC/65 PMMA	0.03	(0.72)[a]	0.03	0.75
25 PC/75 PMMA	0.08	(0.67)[a]	0.03	—
15 PC/85 PMMA	—		—	—
2 PC/98 PMMA	—		—	—

[a]By extrapolation.

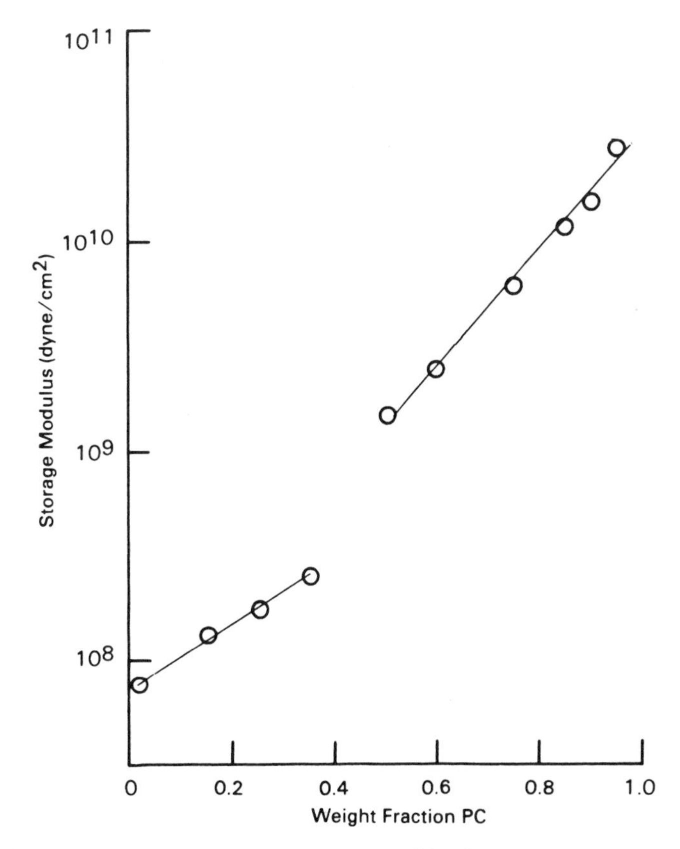

Figure 17. Storage modulus of PC/PMMA blends as a function of composition at 130 °C.

polymer chain but the n electrons can still interact with the π electrons of the benzene rings in the PC chain. Unlike the PS–PCL blends, n–π complexation does occur and PMMA–bisphenol A PC blends exhibit partial miscibility.

Summary

When blends of bisphenol A PC and PMMA are prepared by melt techniques, a partial miscibility of the polymers is discovered. The phase change between 0.35 and 0.50 weight fraction bisphenol A PC has a definite effect on the modulus of the blend system. The partial miscibility is discussed in terms of n–π complex formation between the ester groups of the PMMA and the benzene rings of the bisphenol A PC.

Acknowledgments

The very able technical assistance of M. A. Bator is gratefully acknowledged. The author thanks W. A. Lange for technical expertise in preparing the SEM micrographs of the blend fracture surfaces. Helpful and encouraging discussions with D. R. Paul, G. D. Cheever, and H. B. Hopfenberg were appreciated.

Literature Cited

1. Goldberg, E. P. *J. Polym. Sci.* **1963**, *C4*, 707.
2. Shaw, M. T. *J. Appl. Polym. Sci.* **1974**, *18*, 449.
3. Kunori, T.; Geil, P. H. *J. Macromol. Sci. Phys.* **1980**, *18 (1)*, 93, 135.
4. Stefan, D.; Williams, H. L. *J. Appl. Polym. Sci.* **1974**, *18*, 1451.
5. Dobkowski, Z.; Kohman, Z.; Krajewski, B. In "Polymer Blends"; Kryszewski, M.; Martuscelli, E.; Palumbo, R., Eds.; Plenum: New York, 1980; p. 363.
6. Myers, F. S.; Brittain, J. O., *J. Appl. Polym. Sci.* **1973**, *17*, 2715.
7. Peterson, R. J.; Corneliussen, R. D.; Rozelle, L. T. *Polym. Prepr. Am. Chem. Soc. Div. Polym. Chem.* **1969**, *10*, 385.
8. Majnusz, J. In "Polymer Blends"; Kryszewski, M.; Martuscelli, E.; Palumbo, R., Eds.; Plenum: New York, 1980; p. 349.
9. Varnell, D. F.; Runt, J. P.; Coleman, M. M. *Macromolecules* **1981**, *14*, 1350.
10. Barlow, J. W.; Paul, D. R. *Poly. Eng. Sci.* **1981**, *21*, 985 and references cited therein.
11. Wahrmund, D. C.; Paul, D. R.; Barlow, J. W. *J. Appl. Polym. Sci.* **1978**, *22*, 2155.
12. Nassar, T. R.; Paul, D. R.; Barlow, J. W. *J. Appl. Polym. Sci.* **1979**, *23*, 85.
13. Cruz, C. A.; Barlow, J. W.; Paul, D. R. *J. Appl. Polym. Sci.* **1980**, *25*, 1549.
14. Smith, W. A.; Barlow, J. W.; Paul, D. R. *J. Appl. Polym. Sci.* **1981**, *26*, 4233.
15. Gardlund, Z. G. *Polym. Prepr. Am. Chem. Soc. Div. Polym. Chem.* **1982**, *23*, 258.
16. McCrum, N. G.; Read, B. E.; Williams, G. "Anelastic and Dielectric Effects in Polymer Solids"; Wiley: New York, 1967.
17. Yee, A. F.; Smith, S. A. *Macromolecules* **1981**, *14 (1)*, 54.
18. ASTM D3418
19. Fox, T. G. *Bull. Am. Phys. Soc.* **1956**, *2 (2)* 123.
20. Wood, L. A. *J. Polym. Sci.* **1958**, *28*, 319.
21. Nielsen, L. E. "Mechanical Properties of Polymers and Composites"; Marcel Dekker: New York, 1974; Vol. 1, p. 25.
22. Stoelting, J.; Karasz, F. E.; MacKnight, W. J. *Polym. Eng. Sci.* **1970**, *10 (3)*, 133.
23. Marcincin, K.; Romanov, A.; Pollak, V. *J. Appl. Polym. Sci.* **1972**, *16*, 2239.
24. Grollier, J. P. E.; Ballet, D.; Viallard, A. *J. Chem. Thermodyn.* **1974**, *6*, 895.
25. Kern, R. J. *J. Polym. Sci.* **1956**, *21*, 19.

RECEIVED for review January 20, 1983. ACCEPTED August 15, 1983.

10

Compatibility of a Random Copolymer of Varying Composition with Each Homopolymer

K. FUJIOKA, N. NOETHIGER, and C. L. BEATTY

Department of Materials Science and Engineering, University of Florida, Gainesville, FL 32611

Y. BABA and A. KAGEMOTO

Department of Chemistry, Osaka Institute of Technology, Asahi-ku, Osaka 535 Japan

*The compatibility between polymers, especially poly(styrene) poly(*n-*butyl methacrylate), and poly(styrene-*co-n-*butyl methacrylate) was studied by using visible spectroscopy, differential scanning calorimetry, inverse gas chromatography, and dielectric relaxation spectroscopy. The results showed that the compatibility between these polymers depends not only upon the composition of the copolymer and homopolymer with which it is blended, but also upon the percentage of copolymer chains in the blend. Each experimental technique provides slightly different results concerning the degree and the range of compatibility. These results suggest that a variety of experimental techniques aid in the elucidation of the extent and nature of the compatibility of mixed polymers. The dielectric relaxation data indicate that the compositional regions deemed compatible by other techniques and high frequency dielectric measurements appear to be phase separated when examined at low frequencies. Thus, the compatibility of polymers may not be limited to strictly compatible or incompatible behavior, but may also include intermediate behavior such as partially mixed regions that coexist with phase-separated domains.*

THE QUESTION OF COMPATIBILITY BETWEEN POLYMERS has been well studied for a number of years by using bulk parameters (such as the solubility parameter) and the free energy interaction parameter as guides to possible

0065-2393/84/0206-0149$06.00/0

compatibility (*1*). The lack of success of these approaches for some systems has led to the concept that compatibility may be related to interaction between the specific moieties on the different polymer chains. All of these approaches have had their limitations and have not been universally proven.

However, the length of the chain plays a key role. Many oligomers are compatible with polymers of greatly dissimilar chemical composition, whereas the high molecular weight polymeric analogs are not compatible (*1, 2*). Therefore, an ideal system for studying compatibility would be one in which the degree of compatibility could be varied in a continuous fashion. The degree of compatibility would be held independent of the length of the polymer chain by keeping the molecular weight of the polymer chains used to make the blends constant.

An approach that may satisfy these criteria may be the use of a random copolymer composed of mers of basically incompatible polymers. Thus, polymer chains of a controlled, but variable, chemical composition and of a selected molecular weight could be used in blends with the homopolymer formed from each mer or with random copolymers of different mer ratios. Such a random copolymer would be expected to be compatible with the homopolymer of each of its mers (*3*) over some range of composition of the blend. That is, data on a limited range of compositional compatibility could be obtained by blending a pure homopolymer chain (A) made up of U mers, (A = (U) = UUU. . .) with a random copolymer (B) composed of U and V mers, (B = (V_{1-x}, U_x) = UUVUUV. . .) where x is a number fraction of U units. In addition, this partial range of compositional compatibility could be obtained between random copolymers by making the difference between their x values sufficiently small. Also, it may be possible to determine if partial compatibility states exist instead of strictly compatible or incompatible states, in blended polymer systems.

We used poly(styrene) (PS) and poly(*n*-butyl methacrylate) [P(nBMA)] as the homopolymers (A and A′) and poly(styrene-*co*-*n*-butyl methacrylate) [P(S/nBMA)] of variable chemical composition as B. We studied the compatibility between these polymers [i.e., (A)–(B), (A′)–(B) or (B)–(B)] by using visible spectroscopy (VS), differential scanning calorimetry (DSC), inverse gas chromatography (IGC), and dielectric relaxation spectroscopy (DRS).

Experimental

Materials. PS (obtained from Foster Grant) had a number average molecular weight of approximately 100,000 and a molecular weight distribution ($\bar{M}_w/\bar{M}_n$ of 2.5. P(nBMA) (Polysciences, Inc.) and random copolymers P(S/nBMA) (Scientific Polymer Products, Inc.) also had number average molecular weights of approximately 100,000. The chemical composition (styrene/nBMA) of random copolymers is indicated as 99/1 for 99 wt% styrene and 1 wt% nMBA, and so on. The molecular weight distribution of all of these polymers was basically unimodal with little skewing.

Visible Spectroscopy (VS). Blending was accomplished by dissolving the polymers in chloroform, a good solvent for PS, P(nBMA), and the random copolymer. Solubility parameters, calculated by the group contribution method (*4, 5*) and the heat of vaporization method (*6*), are 9.7, 9.0, and 9.2 $cal^{1/2}cm^{-3/2}$ 25 °C for PS, P(nBMA), and chloroform, respectively. Films of PS, P(nBMA), P(S/nBMA), and blends between polymers were cast from 5 wt% solutions in chloroform and were evaporated slowly with protection from contamination by dust. The resulting films were dried under vacuum at room temperature for approximately 4 days. A Model 552 Perkin-Elmer spectrophotometer was used to measure the visible light transmittance of the cast films. For these measurements, the film thickness was about 0.05 mm for all the polymer films examined.

Differential Scanning Calorimetry (DSC). Sample preparation was carried out in the same manner used for VS. A Perkin-Elmer DSC 2 was used for measuring the glass transition temperature (T_g) of these films. Temperature calibration was achieved by using indium, zinc, and cyclohexane. Approximately 10-mg (20-mg) portions of the cast films were placed into the aluminum pans and sealed (values in parentheses are for the P(nBMA)–P(S/nBMA) system). The heating rate and sensitivity range of the DSC were 10 °C/min and 0.5 mcal/s (2 mcal/s) for each measurement, respectively. Samples were heated from 30 °C (−73 °C) to 160 °C (107 °C) and then cooled back to 30 °C (−73 °C) at 160 °C/min. After holding for 30 min at 30 °C (−73°C) to achieve equilibrium, the samples were heated to 160 °C (107 °C) at 10 °C/min and immediately cooled back to 30 °C (−73 °C) at 160 °C/min. The reported T_g values were determined by this repeated temperature sequence. The T_g was defined as the temperature that corresponded to the midpoint of the observed heat capacity step-change for the glass transition.

Inverse Gas Chromatography (IGC). A Hitachi (Model 163) gas chromatograph equipped with a thermal conductivity detector was used for the inverse gas chromatography studies. The column temperature was detected by a copper–constant thermocouple. The average error in column temperature was ±0.5 °C. The flow rate of helium carrier gas was about 2.5–3.0 mL/min and determined at a precision of ±0.2 mL/min by a soap bubble flowmeter. The column pressure was measured differentially against the atmospheric outlet pressure with a U tube manometer filled with mercury.

The columns were prepared as reported elsewhere (*7, 8*). The polymers were coated from a benzene (solubility parameter 9.2 $cal^{1/2}cm^{-3/2}$ at 25 °C) (*5*) solution onto Chromosorb G (AW-DMCS treated, 70–80 mesh) purchased from Gasukuro Kogyo Co., Ltd. After drying under vacuum for 5 days at room temperature, the coated support was packed into 4-mm I.D. stainless steel columns. The total percent loading of polymer (ca 12%) on the support was determined by calcination. The relative concentration of polymer in the blends was assumed to be identical to that in the original solution. The columns were conditioned under helium for 5 h at a temperature above the T_g of the polymers.

Dielectric Relaxation Spectroscopy (DRS). The dielectric relaxation cell was purchased from Balsbaugh Company, Inc. A Hewlett Packard (HP) Model 4272A multifrequency digital meter (frequency range 10^2–10^5 Hz) was used as the frequency source. This meter has automatic bridge balancing capabilities and was interfaced with an HP 9915A computer for data collection and processing. Plotting of the data was achieved with an HP 9872S plotter.

DRS measurements were made by scanning the temperature at a constant rate of 0.35 °C/min in a Delta Design temperature chamber (Model No. 5100). Temperature rate was controlled by the Delta Design rate programming accessory. The temperature of the sample was measured with an accuracy of ± 0.1 °C by a Fluke 2190A digital thermometer interfaced with the HP 9915A computer. The samples used in this measurement were prepared in the same manner as described for VS.

Results and Discussion

Visible Spectroscopy (VS). A common test of compatibility is opacity. Opacity indicates that scattering of light occurs from domains of differing refractive index, and the lack of light scattering indicates that the domains are approximately less than the wavelength of the radiation used. In this study, visible light of λ 400 and 600 nm was used to measure the transmittance of the cast films.

The transmittance of the cast films is shown in Figures 1 and 2 for the PS–P(S/nBMA), P(nBMA)–P(S/nBMA), and P(S/nBMA)–P(S/nBMA) systems. As seen in Figure 1 for the PS–P(S/nBMA) system, the 99, 95, 90, and 85 wt% styrene P(S/nBMA) copolymers are compatible when blended with PS over the entire composition range. In contrast, the 80, 60, 40, and 20 wt% styrene P(S/nBMA) copolymers when blended with PS seem to be compatible only in a definite range that becomes smaller as the styrene content in the P(S/nBMA) random copolymer decreases.

As seen in Figure 2 for P(nBMA)–P(S/nBMA) blends, the 5 and 10 wt% styrene copolymers are compatible with P(nBMA) over a portion of

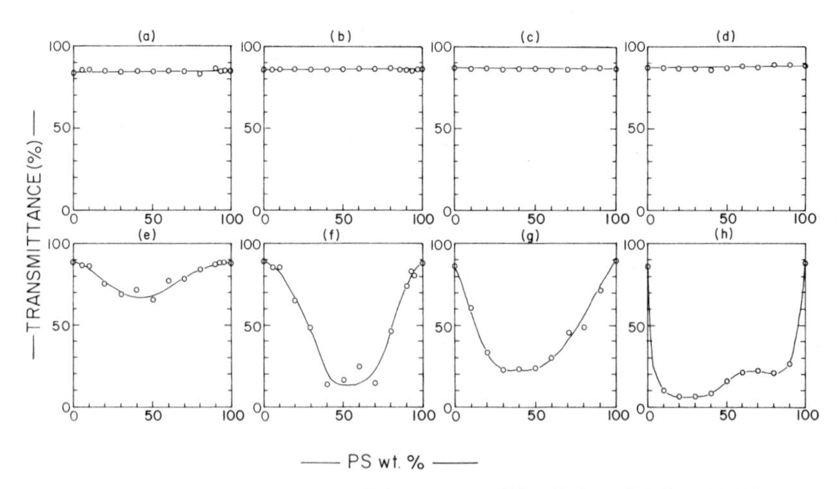

Figure 1. Transmittance at λ 400 nm for PS blended with P(S/nBMA) random copolymers of various chemical compositions: (a) 99/1, (b) 95/5, (c) 90/10, (d) 85/15, (e) 80/20, (f) 60/40, (g) 40/60, and (h) 20/80.

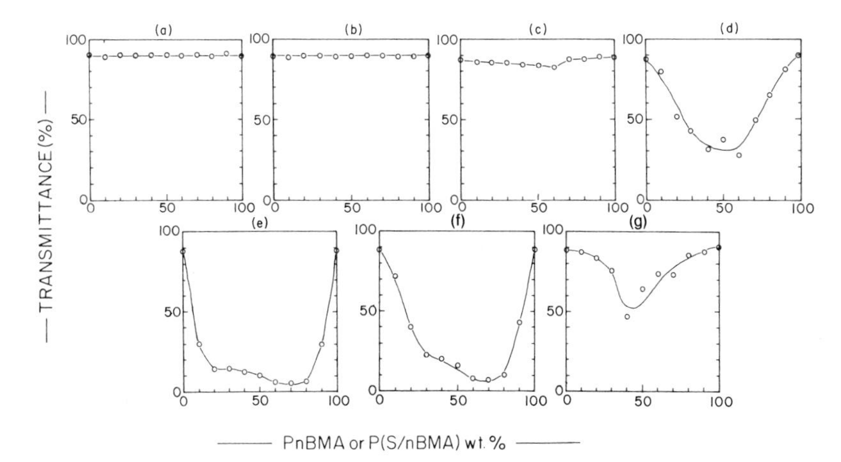

Figure 2. Transmittance at λ 400 nm for P(nBMA) blended with P(S/nBMA) random copolymers of various chemical compositions: (a) 5/95, (b) 10/90, (c) 20/80, (d) 40/60, and (e) 60/40. Transmittance at λ 400 nm for P(S/nBMA)–P(S/nBMA) random copolymers (f) 20/80–80/20 and (g) 40/60–60/40.

the composition range. However, the 20, 40, and 60 wt% styrene P(S/nBMA) random copolymers seem to be compatible with P(nBMA) over a portion of the composition range that becomes smaller as the styrene weight percent of the P(S/nBMA) random copolymer increases. The results for a copolymer/copolymer blend system are also shown in Figure 2. The range of composition that exhibits compatibility for blends formed from 40 wt% styrene P(S/nBMA) and 60 wt% styrene P(S/nBMA) (i.e., the region of high transparency) is greater than that observed for blends of 20 wt% styrene P(S/nBMA) and 80 wt% styrene P(S/nBMA) random copolymers. This result is consistent with the hypothesis that as compositional difference between polymers becomes larger the region of compatibility becomes smaller. To quantify the compatibility for PS–P(S/nBMA), P(nBMA)–P(S/nBMA), and P(S/nBMA)–P(S/nBMA) systems, the following quantity (defined by Equation 1) was formulated to express the extent or degree of compatibility, C, as determined by VS measurements. The derived parameter, C, ranges from 0 to 1; values of 1 are achieved when the blend is optically opaque, and a value of 0 represents an optically compatible blend system.

$$C = 1 - T_r = 1 - \frac{T_m}{T_c} = 1 - \frac{2T_m}{T_a + T_b} \qquad \text{where} \quad T_c = \frac{T_a + T_b}{2} \qquad (1)$$

where T_a, T_b, and T_m are the transmittances at 400 nm of one polymer, the second polymer, and the measured transmittance of the 50/50 wt% of the blend of the two polymers, respectively. The calculated transmittance at

50 wt% for the blends is T_c, and T_r is the ratio of the measured transmittance to the calculated transmittance at 50 wt%. An equation of this form was selected to aid in the comparison of the visible spectroscopy and the Flory–Huggins interaction parameter results. The C values obtained are then plotted against the styrene weight percent in the P(S/nBMA) random copolymer as illustrated in Figure 3. Compatibility between polymers could be obtained by making the composition difference between the blended polymers small. Clearly the value of C for a polymer mixed with itself is zero, as is indicated for the copolymer–copolymer blend system (Figure 3C at 50 wt% PS). In addition, the composition range for compatibility for the P(nBMA)–P(S/nBMA) system is wider than that for the PS–P(S/nBMA) system. The wider composition range for compatibility is thought to be related to the more active or stronger interactions that are possible between *n*-butyl methylacrylate segments than are possible between styrene repeating units. These interactions are probably of the hydrogen bonding type although that question has not been investigated in this work.

Obviously, increased interaction between segments in a polymer chain should affect the interaction parameter between chains. The interaction parameter, χ_{AB} is related to the solubility parameters by Equation 2.

$$\chi_{AB} = \frac{V_r}{RT}(\delta_A - \delta_B)^2 \quad (2)$$

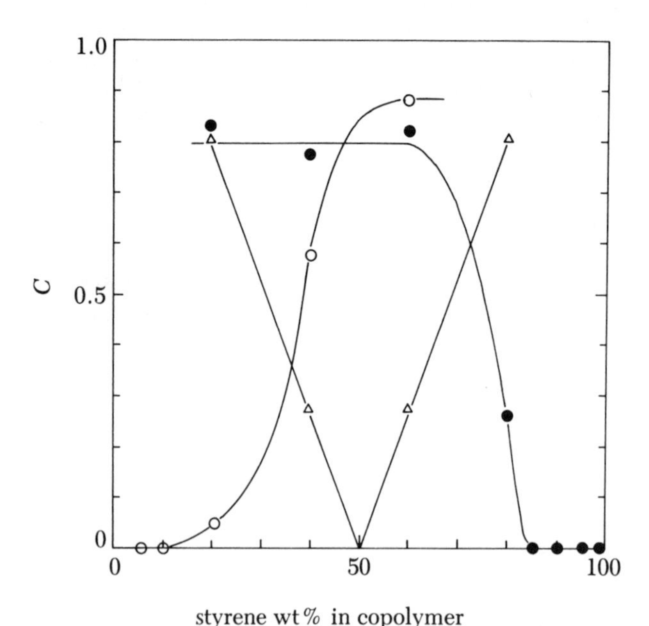

Figure 3. Dependence of extent or degree of compatibility, C, on PS wt% for three blend systems. Key: ●, *PS–P(S/nBMA) system;* ○, *P(nBMA)–P(S/nBMA) system; and* Δ, *P(S/nBMA)–P(S/nBMA) system.*

where V_r is the molar volume of repeating unit, δ is the solubility parameter, R is the molar gas constant, and T is the absolute temperature. The solubility parameters of PS and PnBMA were calculated from the group contribution method (*4*, *5*) by using Equations 3a and 3b:

$$\delta = \rho \Sigma F_i / M \tag{3a}$$

$$F_i = (E_i \nu_0)^{1/2} \tag{3b}$$

where ρ is the density of polymer, M is the molecular weight of the repeating unit, F_i is related to cohesive energy (E_i of each functional group by Equation 3b and ν_0 is the molar volume of the group. Equation 4 was used for the calculation of the solubility parameters, δ_c, of the random copolymers, P(S/nBMA).

$$\delta_c = \rho_1 (\Sigma F_{i,1}/M_1) \phi_1 + \rho_2 (\Sigma F_{i,2}/M_2)\, \phi_2 \tag{4}$$

where ϕ_1 and ϕ_2 are the volume fractions of each component in the random copolymers. The densities of PS and P(nBMA) required for the calculation were obtained from the literature (*9*, *10*) and the density of P(S/nBMA) was estimated by assuming that the intermediate mixtures were linear (the rule of linear mixtures). The values of V_r and T used for the calculation were 100 cm^3/mol and 298 K, respectively.

The calculated values of χ_{AB} are plotted against styrene weight percent in the P(S/nBMA) used in the three blends system described in Figure 4. The value of $(\chi_{AB})_{cr}$ (also shown in Figure 4) was estimated at $X_A = X_B = 1000$ by using Equation 5

$$(\chi_{AB})_{cr} = \frac{1}{2}\left[\frac{1}{X_A{}^{1/2}} + \frac{1}{X_B{}^{1/2}}\right]^2 \tag{5}$$

In general, if $\chi_{AB} > (\chi_{AB})_{cr}$, then the two polymers should be incompatible. The greater the difference between χ_{AB} and $(\chi_{AB})_{cr}$, the more incompatible the polymers should be. If $\chi_{AB} \leq (\chi_{AB})_{cr}$, then the two polymers should be compatible at that composition. Comparing Figures 3 and 4 shows that the values of C obtained from VS are in fair agreement with the χ_{AB} prediction of compatibility for values of χ_{AB} below $(\chi_{AB})_{cr}$. The similarity of the observed and predicted trends extends across the composition ranges for copolymers blended with P(nBMA) as well as PS, and to the compatibility of the copolymer/copolymer blends as a function of composition. Thus, VS may be a convenient and simple method to measure not only the compatibility between polymers but also the degree of incompatibility that is achieved.

The composition of the blend for which χ_{AB} becomes less than the critical value, $\leq (\chi_{AB})_{cr}$, (as determined by the group contribution calculation route) agrees well with the composition for which the experimentally de-

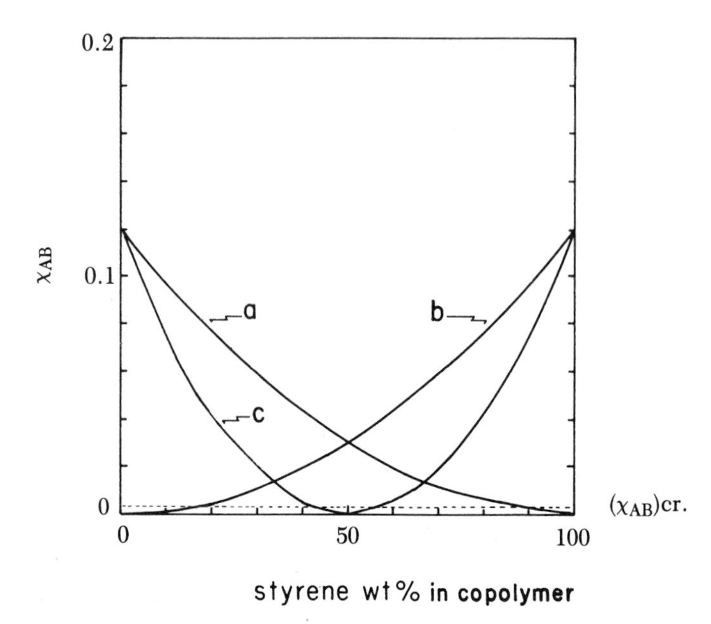

Figure 4. χ_{AB} *calculated from the group functional method for styrene wt% in PS–P(S/nBMA) blends. Key: a, PS–P(S/nBMA); b, P(nBMA)–P(S/nBMA); and c, P(S/nBMA)–P(S/nBMA).*

termined (inverse phase gas chromatography) interaction parameter becomes negative. Thus, the group contribution route for estimating χ_{AB} and $(\chi_{AB})_{cr}$ appears to offer a simple, short route for approximating the compositional range of compatibility of polymer/copolymer and polymer/copolymer blend systems.

Glass Transition Temperature (T_g) by DSC. To confirm the results of VS, the glass transition temperatures of the blends were measured with DSC. The T_g values obtained are shown in Figures 5 and 6. In the case of 60, 80, and 85 wt% styrene P(S/nBMA)–PS blends (Figure 5), two T_g values were observed (indicative of phase separation). For blends of 90, 95, and 99 wt% styrene P(S/nBMA) copolymers with PS, only one T_g was observed. This result suggests that the range of compatibility is dependent upon the chemical composition of the copolymer as measured by the range of existence of one or two T_g values. The 60, 80, and 85 wt% styrene P(S/nBMA) appear to be compatible with PS in the range of 10–20, 10–15, and 50–60 wt% PS in the blend, respectively. On the other hand, 90, 95, and 99 wt% styrene P(S/nBMA) appear to be totally compatible with PS as determined by DSC measurements.

The T_g observed for the P(nBMA)–20/80 P(S/nBMA) system (Figure 6) indicates that these polymers are compatible. Thus, the composition range of compatibility for the P(nBMA)–P(S/nBMA) system is wider than that for

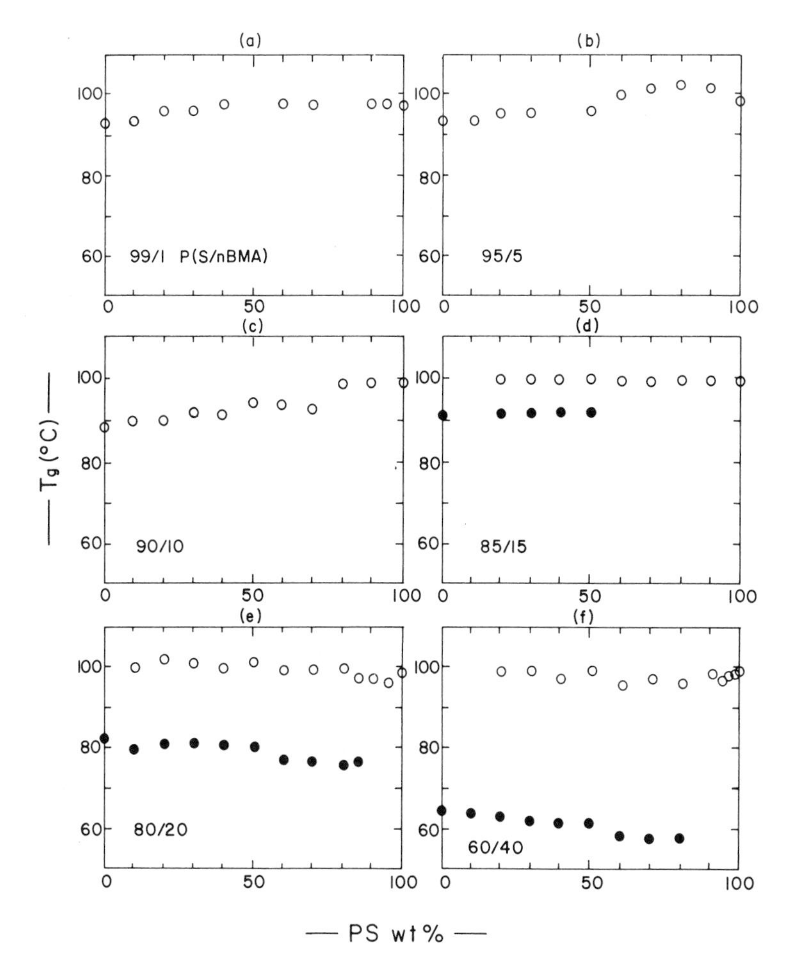

Figure 5. Glass transition temperature of PS blended with P(S/nBMA) random copolymers of various chemical composition. (a) 99/1, (b) 95/5, (c) 90/10, (d) 85/15, (e) 80/20, and (f) 60/40.

the PS–P(S/nBMA) system. This finding is analogous to the results found by using VS.

Inverse Gas Chromatography (IGC). The existence of either a compatible blending (one T_g) or an incompatible blending (two T_g values) that is dependent upon the chemical composition of the random copolymer used in the blend suggests that the interaction parameter between the random copolymer and the homopolymer is a function of the chemical composition of the random copolymer. Thus, IGC has been used to determine the interaction parameter, χ'_{23}, of the random copolymer and the homopolymer in the PS–80/20 P(S/nBMA) and PS–60/40 P(S/nBMA) blend systems.

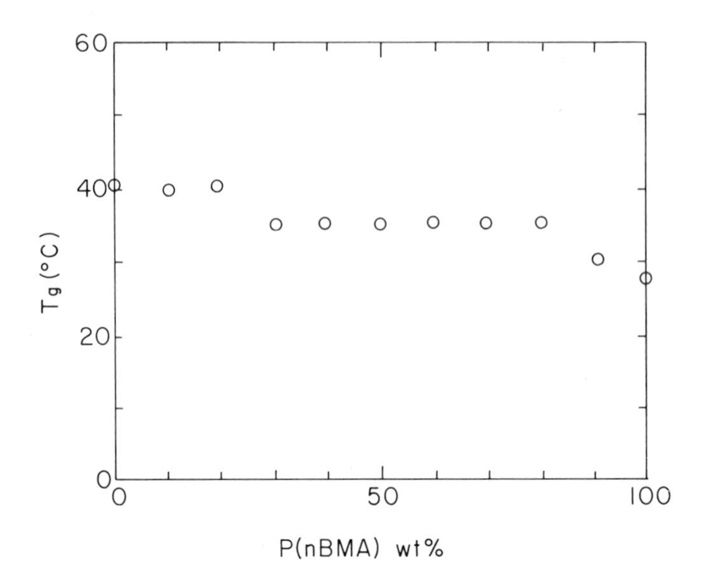

Figure 6. Glass transition temperature of P(nBMA) blended with 20/80 P(S/nBMA) as detected by DSC.

The specific retention volume, V_g° (cubic centimeters per gram) was computed by using the method in Reference 11. The Flory–Huggins interaction parameters (*12*), χ, at infinite dilution of the probe were determined (*13*) from the measured specific retention volume for the benzene probe molecule in each of the pure phases (χ_{12} and χ_{13}) (*13*) and in the mixed stationary phase ($\chi_{1(23)}$) (*14*) in Equations 6a, 6b, and 7.

$$\chi_{12} = \ln\left[\frac{RT\nu_2}{V_g^\circ V_1 P_1^\circ}\right] - \left(1 - \frac{V_1}{\bar{M}_2 + \bar{M}_3}\right) - \frac{P_1^\circ}{\mathrm{RT}}(B_{11} - V_1) \qquad (6a)$$

$$\chi_{13} = \ln\left[\frac{RT\nu_3}{V_g^\circ V_1 P_1^\circ}\right] - \left(1 - \frac{V_1}{\bar{M}_2 + \bar{M}_3}\right) - \frac{P_1^\circ}{RT}(B_{11} - V_1) \qquad (6b)$$

$$\chi_{1_{(23)}} = \chi_{12}\phi_2 + \chi_{13}\phi_3 - \chi_{23}'\phi_2\phi_3 = \ln\left[\frac{RT(\omega_2\nu_2 + \omega_3\nu_3)}{P_1^\circ V_g^\circ V_1}\right]$$

$$- \left(1 - \frac{V_1}{\bar{M}_2 + \bar{M}_3}\phi_2 \frac{V_1}{\bar{M}_2 + \bar{M}_3}\right)\phi_3 - \frac{P_1^\circ}{RT}(B_{11} - V_1) \qquad (7)$$

where R is the gas constant; T is the temperature (Kelvin); V_g° is the specific retention volume; ν_2 and ν_3 are the specific volume of polymer 2 and polymer 3, respectively; P_1° is the vapor pressure of pure solvent; V_1 is the molar volume of solvent; B_{11} is the second virial coefficient of solvent; $\bar{M}_2$ and $\bar{M}_3$ are the number average molecular weights of the polymers; and ω_2 and ω_3 are the weight fractions of polymers in the mixed stationary phase.

The χ'_{23} values for the PS–80/20 P(S/nBMA) and the 60/40 P(S/nBMA) blend systems are shown in Figure 7. In the PS–80/20 P(S/nBMA) system, the χ'_{23} values for blend compositions less than 90 wt% PS weight fraction are positive or zero (incompatibility should be found). Whereas, in the range of more than 90 wt% PS, the χ'_{23} values are negative (these systems should be compatible). In the PS–60/40 P(S/nBMA) system, the value of χ'_{23} versus PS wt% depends on the temperature of measurement. The results obtained at 393 and 413 K indicate that compatibility can be achieved at greater than 85 wt% PS in the PS–60/40 P(S/nBMA) blend. However, the result at 445 K shows that these two polymers are incompatible over the entire composition range. We suspect that at 445 K the lower critical solution temperature (LCST) has been exceeded. These results suggest that IGC can be used to study critical solution temperatures (CST) (*15, 16*).

Symmetric IGC peaks were obtained for both the incompatible and compatible blends reported here. The retention volume was repeatable to better than ± 3.0%.

Dielectric Relaxation Spectroscopy (DRS). DRS measurements were made on the PS–80/20 P(S/nBMA) blend system as a function of blend composition to determine if the other techniques used (VS, DSC, and IGC) were sufficiently sensitive to interactions at the segmental and molecular level. The effect of frequency on the dielectric tan δ versus temperature plots is presented for this blend system in Figures 8a and 8b, which show the results for the PS homopolymer and the 80/20 P(S/nBMA) random copolymer, respectively. Figures 8c and 8d show the DRS results of 80 wt%

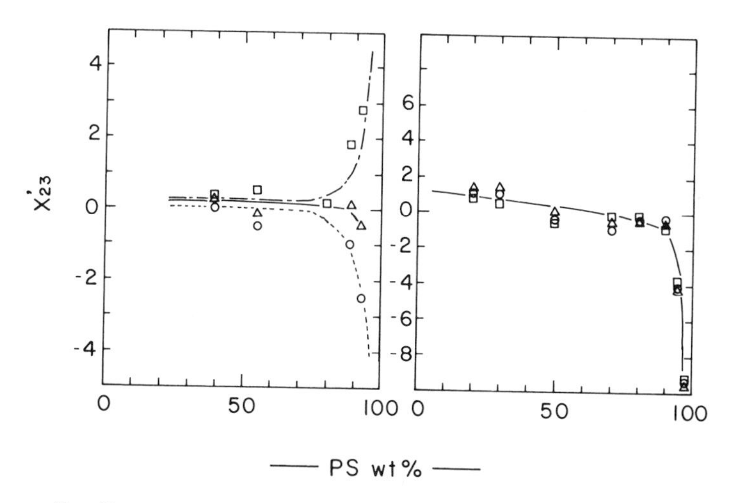

Figure 7. Free energy interaction parameter χ_{23} between polymers for (left) PS–60/40 P(S/nBMA) and (right) PS–80/20 P(S/nBMA) blend systems. Key: ○, 393 K; △, 413 K; and □, 443 K.

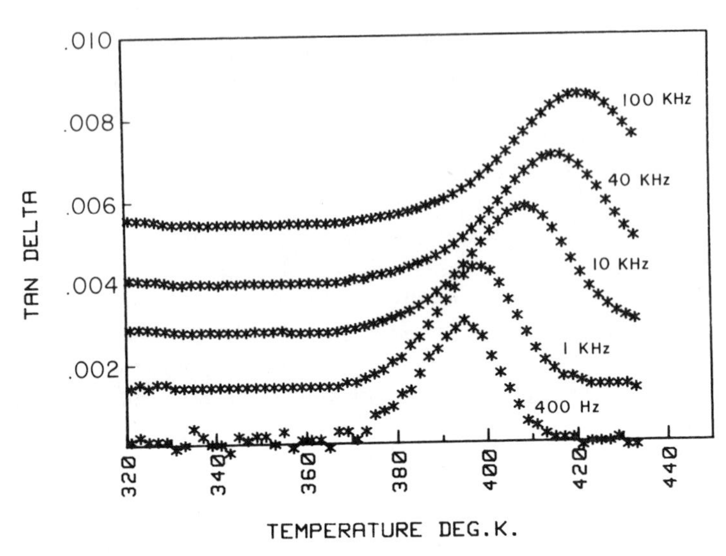

Figure 8a. Dielectric tan δ of PS vs. temperature as a function of measurement frequency.

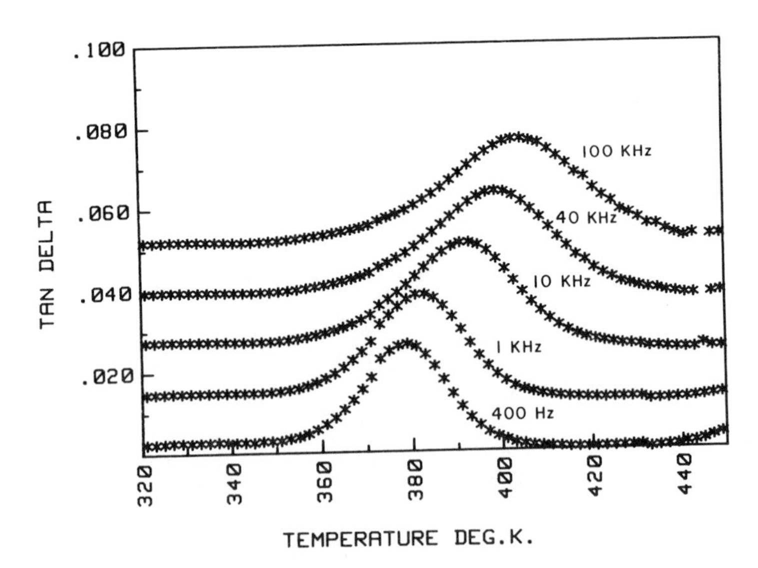

Figure 8b. Dielectric tan δ of 80/20 P(S/nBMA) vs. temperature as a function of measurement frequency.

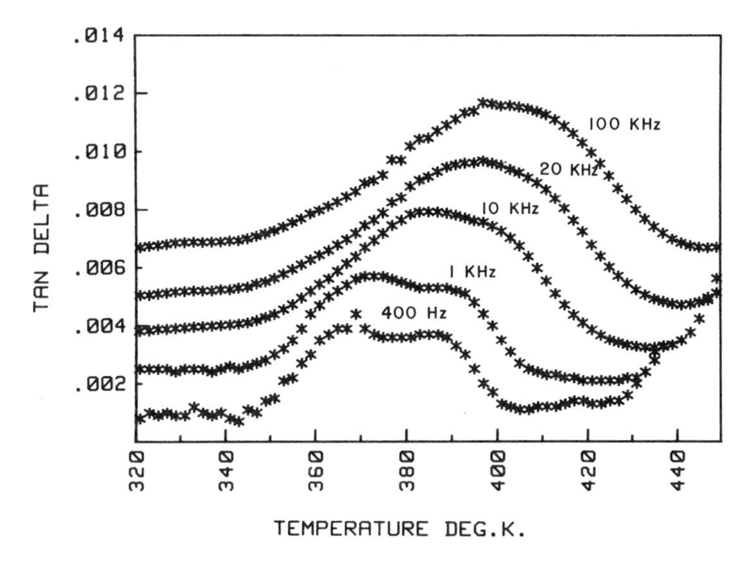

Figure 8c. Tan δ of PS blended with 80/20 P(S/nBMA) [80% PS, 80/20 20% P(S/nBMA)] vs. temperature as a function of measurement frequency.

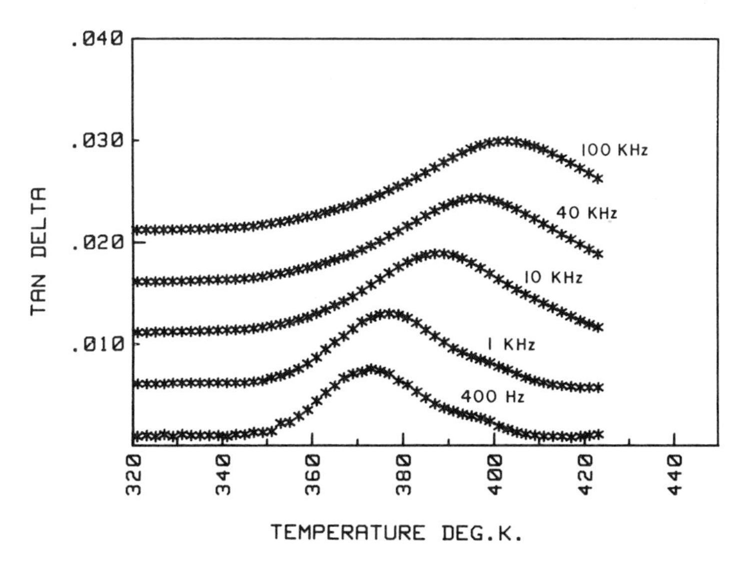

Figure 8d. Tan δ of PS blended with 80/20 P(S/nBMA) [60% PS, 80/20 40% P(S/nBMA)] vs. temperature as a function of measurement frequency.

PS–80/20 P(S/nBMA) and 60 wt% PS–80/20 P(S/nBMA) blends, respectively. As seen in Figure 8a, the T_g shifts to higher temperatures as the measurement frequency is increased. This shift was expected for all the homopolymers and random copolymers. In all of the blends that have the blend composition of 40–70 wt% PS or greater, only one apparent T_g occurs at high frequencies (100 kHz). However, in most cases (≤ 90 wt% PS) the blends exhibit two peaks at low frequencies (≤ 400 Hz). Clearly the low frequency DRS data suggest that phase separation occurs at the molecular level and that the two glass transition peaks merge at the higher measurement frequencies. These results also clearly indicate that the detection of phase separation depends upon the sensitivity and volume of material that is probed by the particular technique used. Thus, of the techniques used, the DRS data at high frequencies agree better with the DSC and VS data in terms of compatible versus incompatible behavior. The low frequency DRS data allow detection of phase separation at the segmental or molecular level, even in the blend composition ranges that appear to be compatible via the other two techniques.

To understand better the effect of measurement frequency on the detection of one or two phases, i.e. T_g values, transition maps were constructed for the series of blends of the homopolymer PS blended with the 80 wt% styrene/20 wt% *n*-butyl methacrylate random copolymer. The data in Figure 9a indicate that both PS and the 80/20 P(S/nBMA) random copolymer exhibit typical Williams–Landel–Ferry type behavior for their glass transition relaxation process. These data are shown on the transition maps for the blends by dashed lines to be used as reference lines for comparison of the blend data with the homopolymer behavior. The 90 wt% PS homopolymer–10 wt% P(S/nBMA) random copolymer blend transition map in Figure 9b illustrates that the high frequency single peak almost perfectly coincides with the PS homopolymer data and that the high frequency copolymer peak is not observable. However, the low frequency data do not coincide well with the previously obtained data for the unblended polymers. The data suggest that the PS-rich phase, or the PS-like relaxation process, may be antiplasticized (T_g is increased); whereas the random copolymer-rich phase, or the random copolymer-like relaxation process, may be plasticized (T_g is lowered). The transition map for the 80 wt% PS/20%–(80/20) P(S/nBMA) random copolymer blend (Figure 9c) illustrates different behavior: The highest frequency T_g peak is depressed to values below the 80/20 P(S/nBMA) random copolymer T_g. This effect suggests that plasticization may be evident even in the high frequency measurement regime. In addition, two relaxation processes are observed at frequencies below 100 kHz. Both occur at lower temperatures than the relaxation processes that are observed for the unblended polymers. Figures 9d and 9e show that, when data are taken at closer frequency intervals, a transition from a double process to a single merged apparent relaxation process can be observed.

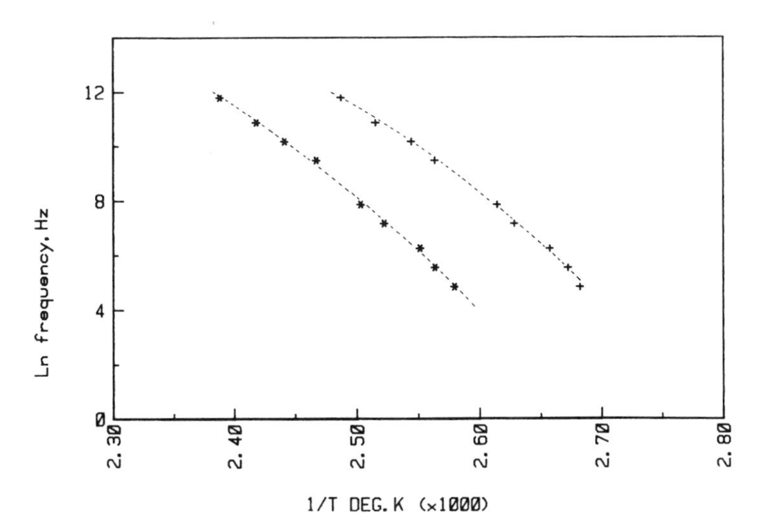

*Figure 9a. Glass transition temperature transition map of PS and 80/20 P(S/nBMA) unblended polymers. Key: *, PS; and +, P(S/nBMA).*

In these data as well as in Figure 9f, the blends formed are incompatible as measured by the DSC and VS techniques.

The transition map data are summarized in Figures 10a and 10b. The relaxation process that occurs at high temperature in the low frequency regime for all of these blend compositions (the upper set of curves in Figure 10b) is nearly identical with the PS homopolymer data (100% PS). These results suggest that little mixing occurs at the phase boundary and that the

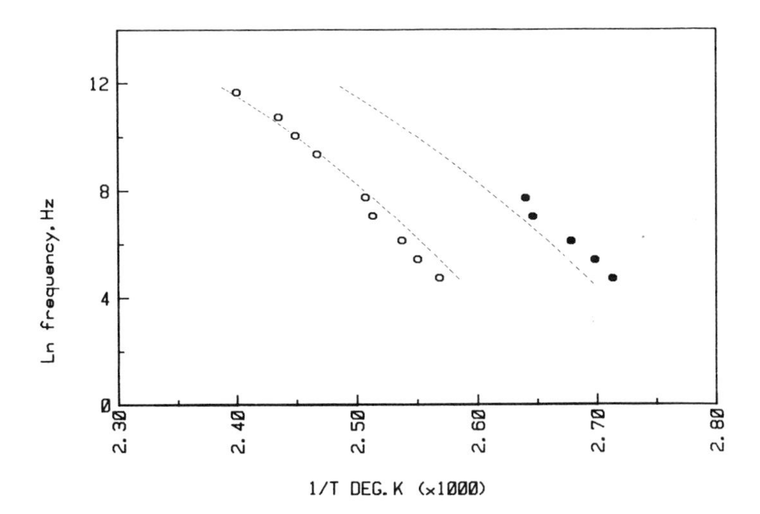

Figure 9b. Glass transition temperature transition map of the 90% PS–80/20 10% P(S/nBMA) blend. Key: ○, 90% PS; ●, 10% 80/20 P(S/nBMA); and ---, unblended polymer reference lines.

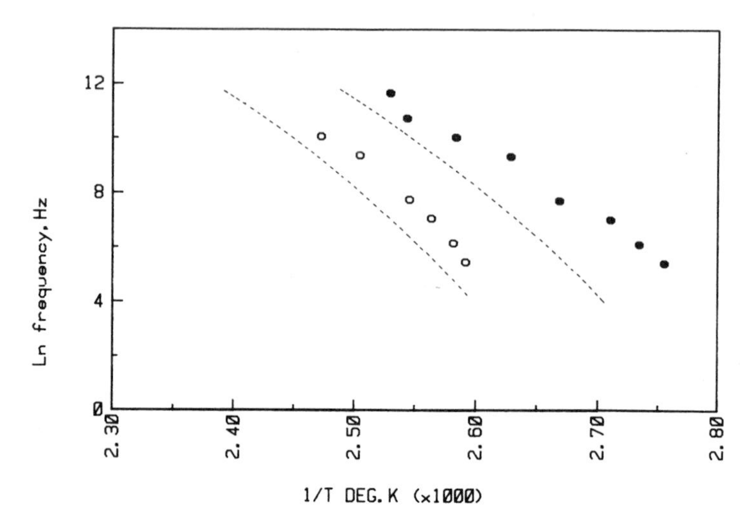

Figure 9c. Transition map of the 80% PS–80/20 20% P(S/nBMA) blend. Key: ○, *80% PS;* ●, *80/20 20% P(S/nBMA); and ---, unblended polymer reference lines.*

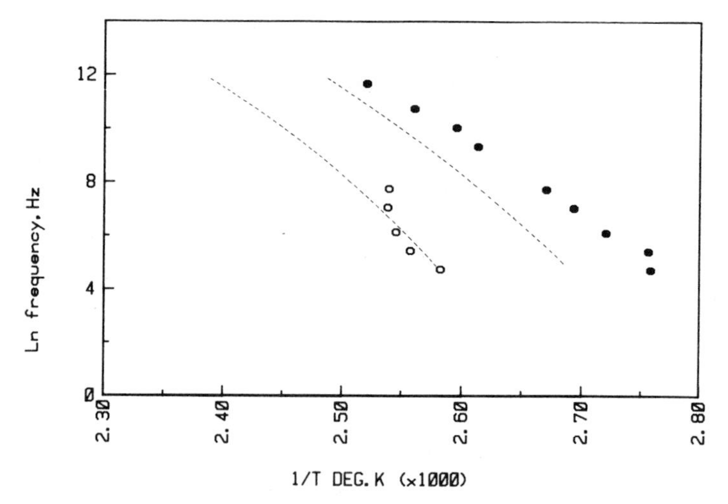

Figure 9d. Transition map of the 70% PS–80/20 30% P(S/nBMA) blend. Key: ○, *70% PS;* ●, *80/20 P(S/nBMA); and ---, unblended polymer reference lines.*

PS-rich phase is quite incompatible. In contrast, the low frequency data for the random copolymer-rich phase occur at a lower temperature for all the blend compositions at 200 Hz. These results suggest, as mentioned previously, that plasticization may be occurring. However, plasticization by a higher glass transition polymer such as PS seems unlikely. A more likely plasticizing species would be water (PnBMA is hydroscopic), but all of the

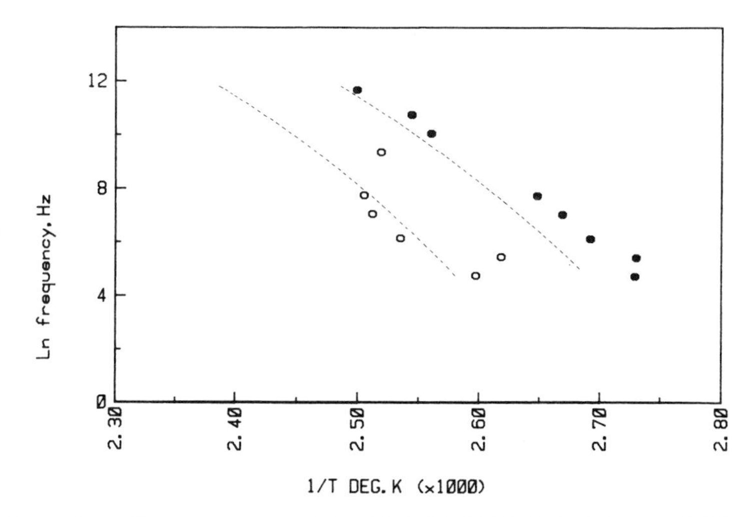

Figure 9e. Transition map of the 60% PS–80/20 40% P(S/nBMA) blend. Key: ○, 60% PS; ●, 40% P(S/nBMA); and ---, unblended polymer reference lines.

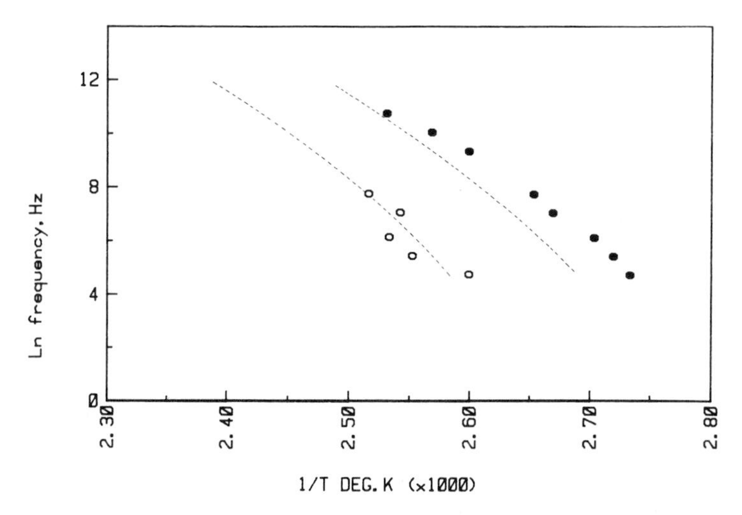

Figure 9f. Transition map of the 40% PS–80/20 60% P(S/nBMA) blend. Key: ○, 40% PS; ●, 80/20 60% P(S/nBMA); and ---, unblended polymer reference lines.

blended materials, as well as the PS and random copolymer species, that were blended were carefully vacuum-dried and handled identically. Thus, the mechanism for a reduction of the temperature of this blend relaxation process is unresolved.

The high frequency (single relaxation process) T_g appears to gradually decrease in the high weight percent PS (≥ 90 wt%) blend region. This result

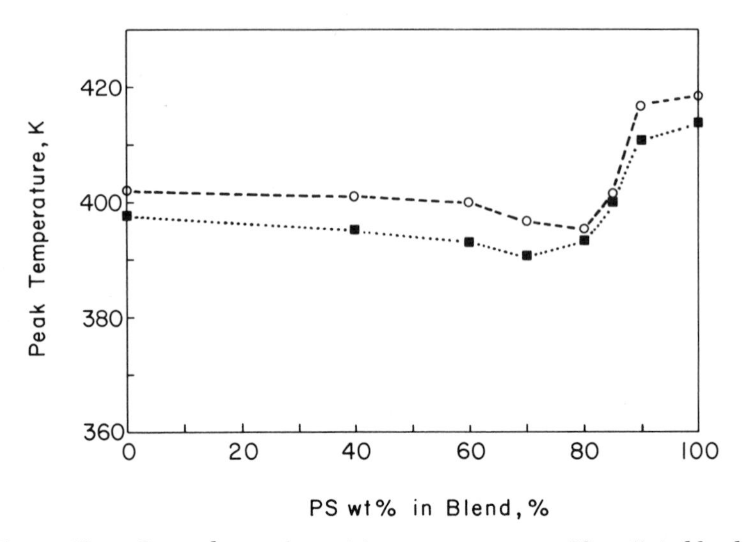

Figure 10a. Dependence of transition temperature on PS wt % in blend at high dielectric relaxation measurement frequencies. Key: ○, *100 kHz; and* ■, *40 kHz.*

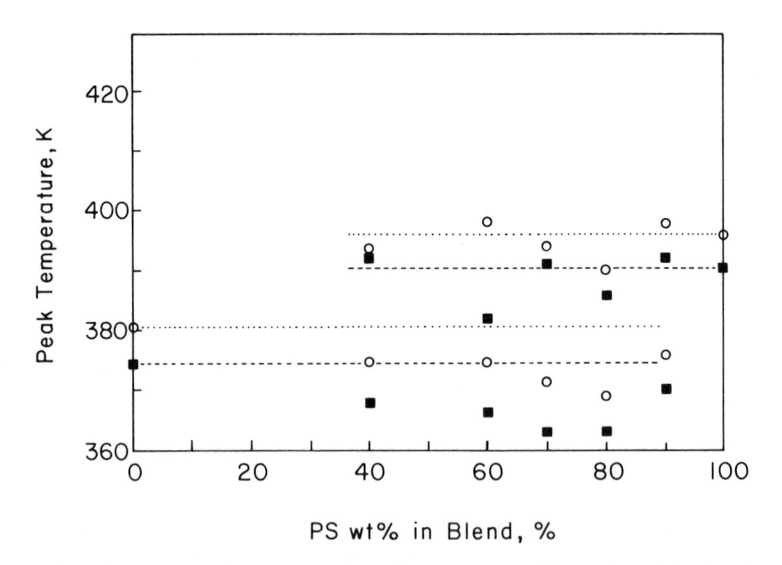

Figure 10b. Dependence of peak temperatures on PS wt % in blend at low measurement frequencies. Key: ■, *200 Hz; and* ○, *1 kHz. The dashed and dotted lines represent the transition temperatures observed at the specified measurement frequency for the PS and P(nBMA) polymers.*

would be expected for a compatible blend system. The composition range over which this behavior occurs corresponds rather well with the range of compatibility of these two polymers as determined by both DSC and VS. In the composition range where incompatibility is observed by DSC and VS, the high frequency data indicate that the combined relaxation process occurs at a lower temperature than the PS glass transition temperature. The temperature of this combined relaxation process appears to pass through a minimum in the 60–80 wt% PS for this blend pair and to increase gradually to the copolymer T_g as the PS content is decreased in the blend system. This minimum in the high frequency data (40 kHz) corresponds roughly to the copolymer rich phase in the low frequency (200 Hz) data. These minima indicate that although the two polymers appear to phase separate there may be significant mixing at the phase boundary near the 70 wt% PS blend composition. The results on this blend system to date suggest that similar additional research on polymer blends by a variety of investigative methods is warranted.

Summary. VS allows the determination of the concentration range for which a random copolymer remains transparent when mixed with a homopolymer of one mer from the copolymer. DSC and IGC suggest that, although the concentration range of compatibility may be smaller than that determined by VS, a range of compatibility does exist. DRS data indicate that the range of compatibility is even smaller, if it exists at all, than the range of compatibility detected by VS and DSC. The detectability of two phases (i.e., two T_g values) depends on the frequency of measurement. At concentrations less than 10 wt% of the 80/20 P(S/nBMA) random copolymer blended with PS, the high frequency T_g approaches the value for pure PS. At higher concentrations of the P(S/nBMA) copolymer, the T_g of the blend is lower than that of the pure P(S/nBMA) copolymer at all frequencies of measurement.

Acknowledgments

This work was partially supported by the Microstructure Center of Excellence and the Air Force Office of Scientific Research Contract F49620–80–C–0047.

Literature Cited

1. Paul, D. R.; Newman, S. "Polymer Blends"; Academic Press; New York, 1978; p. 115. Olabisi, O.; Robeson, L. M.; and Shaw, M. T."Polymer–Polymer Miscibility"; Academic Press: New York, 1979; pp. 19, 117.
2. DiPaola-Baranyi, G. *Polym. Prepr., Am. Chem. Soc., Div. Polym. Chem.* **1980**, *21*, No. 2 214.
3. Gennes, P. de "Scaling Concepts in Polymer Physics", Cornell Univ. Press: Ithaca and London **1979**, p. 100.

4. Small, P. A. *J. Appl. Chem.* **1953**, *3*, 71.
5. Van Krevelen, D. W. "Properties of Polymers", 2nd Ed., Elsevier: Amsterdam, 1976; p. 134.
6. Brandrup, J.; Immergut, E. H. "Polymer Handbook", 2nd Ed. IV 337, 1975.
7. DiPaola-Baranyi, G. *Macromolecules* **1981**, *14*, 683.
8. DiPaola-Baranyi, G. *Macromolecules* **1981**, *14*, 1456.
9. Olabisi, O.; Simha, R. *Macromolecules* **1975**, *8*, 206.
10. Hocker, H.; Blake, G. J.; Flory, P. J. *Trans Faraday Soc.* **1971**, *67*, 2251.
11. Littlewood, A. B.; Phillips, C. S. A.; Price, D. T. *J. Chem. Soc.* **1955**, 1480.
12. Flory, P. J. "Principles of Polymer Chemistry"; Cornell University Press: Ithaca, New York, 1953.
13. Patterson, D.; Tewari, Y. B.; Schreiber, H. P.; Guillet, J. E. *Macromolecules* **1971**, *4*, 356.
14. Deshpande, D. D.; Patterson, D.; Schreiber, H. P.; Su, C. S. *Macromolecules* **1974**, *7*, 530.
15. Delmas, G.; Patterson, D. *J. Paint Technol.* **1962**, *34*, 677.
16. Kwei, T. K.; Nishi, T.; Robert, R. F. *Macromolecules* **1974**, *7*, 667.

RECEIVED for review April 11, 1983. ACCEPTED July 13, 1983.

RHEOLOGY, PROCESSING, AND PROPERTIES OF HETEROGENEOUS POLYMER BLENDS

11

Rheological Behavior of Blends of Nylon with a Chemically Modified Polyolefin

HSIAO-KEN CHUANG and CHANG DAE HAN

Department of Chemical Engineering, Polytechnic Institute of New York, Brooklyn, NY 11201

An experimental study was conducted to investigate the rheological behavior of a heterogeneous polymer blend system consisting of Nylon 6 and a chemically modified polyolefin. The polyolefin is believed to be an ethylene-based multifunctional polymer, although its exact molecular structure has not been disclosed. For the study, phase contrast microscopy was employed for identifying the individual components in the blend; differential scanning calorimetry (DSC) was used for investigating the melting behavior; and a dynamic mechanical analyzer (DMA) was used for investigating the transition behavior of the blends. In the polyolefin-rich blends, the Nylon 6 is the dispersed phase, whereas in the Nylon-rich blends, the polyolefin forms the dispersed phase. The rheological measurements taken do not follow the additivity rule, and show evidence of chemical interaction between the Nylon 6 and polyolefin phases.

TWO TYPES OF POLYMER BLENDS can be prepared: heterogeneous (i.e., immiscible or incompatible) and homogeneous (i.e., miscible or compatible). At a given temperature, homogeneous blends give rise to a single phase in which individual components are mutually soluble in one another. The properties of these blends usually obey the rule of mixtures; although sometimes, physical–mechanical properties superior to those of the individual components have been observed. Various experimental techniques, such as electron microscopy, small-angle X-ray diffraction, light scattering, thermal analysis, and dynamic mechanical analysis, have been used to investigate the question of miscibility (*1*, *2*).

The majority of heterogeneous polymer blends contain a matrix phase of rigid resin and a dispersed phase of flexible resin. Naturally, the charac-

0065–2393/84/0206–0171$06.00/0

teristics of the blend will be directly influenced by the chemical structure, molecular weight (MW) and molecular weight distribution (MWD) of each component, and by the ratio of each component in the blend. However, knowledge of these variables, although necessary, is not sufficient for the prediction of the ultimate mechanical–physical behavior of polymer blends. They are strongly influenced by many other factors, such as the size, shape, distribution and relative deformability of the flexible dispersed phase, and the nature and extent of adhesion between the phases (3). These factors affect the blend morphology, which is also strongly influenced by the processing conditions (i.e., the deformation and thermal histories). Interfacial adhesion in heterogeneous polymer blends has a profound influence on the mechanical properties of such systems.

Basically, the processing of heterogeneous polymer blends involves the following considerations: (1) the rheological properties of the individual constituents; (2) the effectiveness of mixing; (3) control of the microstructure in the solid state; and (4) control of the mechanical–physical properties of the blends.

Currently, no rigorous theory can predict which of the two components in a blend will form the discrete phase dispersed in the other component. The state of dispersion (or the mode of dispersion) depends on the rheological properties of the individual polymers, which in turn are influenced by their molecular weights (3–6). It would be worth investigating how the phase interactions, if any, might be affected as one polymer or both, present either as the continuous phase or discrete phase, crystallize as the molten polymer blend solidifies during processing.

In heterogeneous polymer blends, many interrelated variables affect the mechanical–physical properties of the finished product. For instance, the method of blend preparation (i.e., the method of mixing the polymers and the intensity of mixing) controls the morphology of the blend (i.e, the state of dispersion, domain size, and its distribution), which, in turn, controls the rheological properties of the blend. Currently, no comprehensive theory can predict the mechanical–physical properties of a heterogeneous polymer blend in terms of its processing variables. Therefore, an investigation is needed into the processing–morphology–property relationships of heterogeneous polymer blends.

As part of our continuing effort for establishing these relationships, we have recently conducted an experimental investigation with blends of Nylon 6 and an ethylene-based multifunctional polymer (Du Pont CXA 3095). The choice of these blends was based on our earlier experimental evidence (7) that, when the two resins were coextruded forming two stratified layers, strong adhesion occurred between the layers. This observation has prompted us to speculate that blends of these two resins (Nylon 6 and CXA 3095) might give rise to some interesting properties. In this chapter, we shall present the highlights of our experimental results.

Experimental

Materials. The Nylon 6 was a fiber-spinning grade, supplied by American Enka Company. The CXA 3095 (Du Pont Co.) is believed to be an ethylene-based multifunctional polymer; however, neither the chemical composition nor the molecular structure was disclosed to us by the resin manufacturer.

We prepared four blends, with the following blend ratios (by weight): Nylon/CXA = 80/20; Nylon/CXA = 60/40; Nylon/CXA = 40/60; and Nylon/CXA = 20/80. A twin-screw compounding machine (Werner & Pfleiderer) was used.

Sample Preparation. The compounded pellets were compression molded in the form of disks for the rheological measurements. The compression-molded films were also used for dynamic mechanical analysis (DMA), as well as for taking photomicrographs to investigate the state of dispersion in the blend.

Thermal and Thermomechanical Analyses. To determine the melting behavior of the materials, we used a Du Pont 910 differential scanning calorimeter (DSC); for measuring the thermomechanical behavior of the materials, we used a Du Pont 981 dynamic mechanical analyzer (DMA) in conjunction with a 1090 thermal analyzer. Before measurements were taken, the samples were dried in a vacuum oven at 65 °C for 48 h, and measurements were carried out in a nitrogen atmosphere at a heating rate of 10 °C/min.

Rheological Measurement. A cone-and-plate rheometer (a Weissenberg Model R-16 rheogoniometer) was used to measure steady shearing flow properties at various temperatures. Because Nylon 6 has a melting point of approximately 220 °C and CXA 3095 begins to decompose at approximately 250 °C, the permissible temperature range for rheological measurements (and hence for the melt processing operation) was 220–250 °C.

Melt Drawability (Stretchability) Test. To test the drawability of the blends, we conducted a simple extrusion experiment by using a melt indexer equipped with a cooling chamber and a takeup device. The following experimental procedure was employed: The capillary and reservoir sections of the melt indexer were charged with the pellets and heated up to a desired temperature. A known weight was placed on top of the piston rod, and the flow rate was measured. This measurement allowed us to calculate the linear velocity of the melt leaving the die exit. The extrudate, upon exiting from the die, was passed through a cooling chamber into which nitrogen at room temperature was blown gently, and was pulled by a takeup device controlled by a speed controller. The maximum takeup speed was determined by pulling the thread until the thread broke. From this data we determined the maximum draw-down ratio. In the experiment, we used two different lengths of capillary, which gave L/D ratios of 4 and 6.3 (D = 2.10 mm).

Results and Discussion

Morphology of the Nylon/CXA Blends. The compression-molded specimens were first put into microtom capsules filled with a liquid, uncured epoxy resin. After the epoxy resin was solidified with the aid of a curing agent, the embedded specimen was microtomed into thin films of about 5 μm thick. The films were then examined under a phase-contrast microscope. Figure 1 shows photomicrographs of four blend samples in which the *dark* area represents the Nylon phase and the *white* area represents the CXA phase.

Figure 1 shows that, in the CXA-rich blends, the Nylon forms the discrete phase, dispersed in the continuous CXA phase, and in the Nylon-rich blends, the CXA forms the discrete phase dispersed in the continuous Nylon phase. The blend ratio appears to have played a predominant role in determining which of the two components forms the discrete and which forms the continuous phase.

Melting Behavior of the Nylon/CXA Blends. Figure 2 displays thermograms of the blends investigated. Nylon 6 has a melting point of 222 °C, and CXA has two melting points, 126 and 98 °C. The Nylon/CXA = 20/80 and Nylon/CXA = 40/60 blends have three melting points (222, 126, and 98 °C), whereas the Nylon/CXA = 80/20 and Nylon/CXA = 60/40 blends have two melting points (222 and 98 °C). No depression of the melting point occurred in either the Nylon-rich or CXA-rich blends, except that in the Nylon-rich blends (i.e., Nylon/CXA = 80/20 and Nylon/CXA = 60/40)

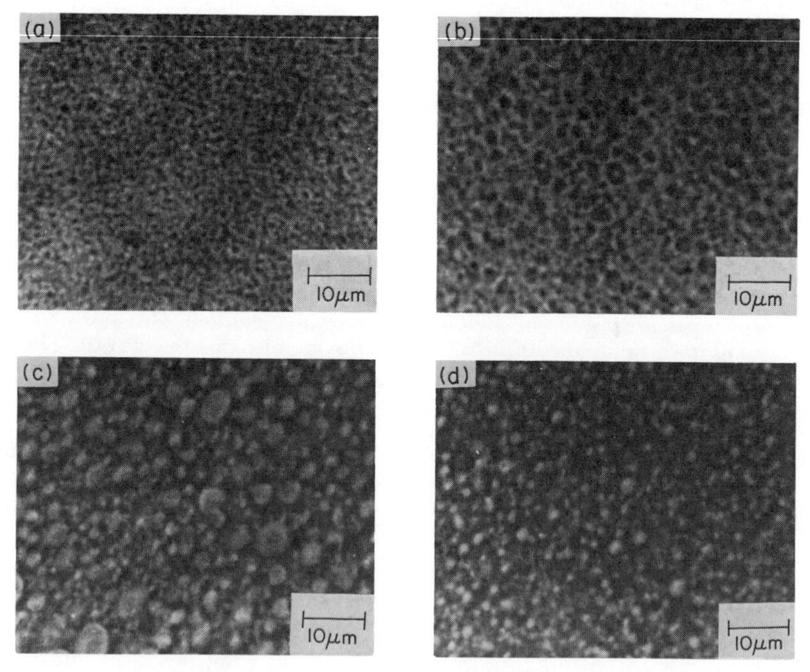

Figure 1. Photomicrographs of the compression-molded specimens of the Nylon/CXA blend system: (a) Nylon/CXA = 20/80, (b) Nylon/CXA = 40/60, (c) Nylon/CXA = 60/40, and (d) Nylon/CXA = 80/20.

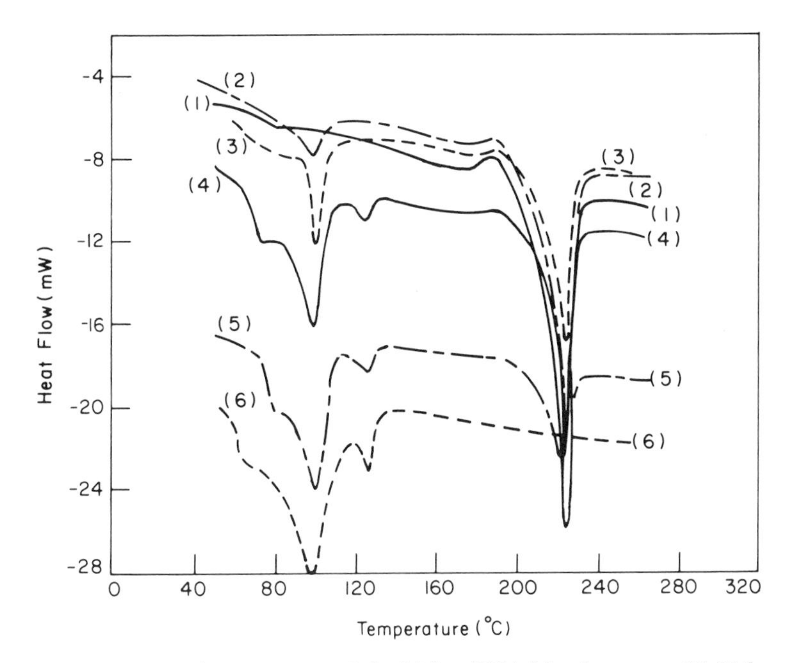

Figure 2. DSC thermograms of the Nylon/CXA blend system: (1) Nylon, (2) Nylon/CXA = 80/20, (3) Nylon/CXA = 60/40, (4) Nylon/CXA = 40/60, (5) Nylon/CXA = 20/80, and (6) CXA 3095.

the secondary melt point (126 °C) of the CXA disappears completely in the thermogram.

Because, at present, we do not know the exact molecular structure of the CXA, we cannot relate the individual melting point with a particular structure. Nevertheless we know that the CXA is a blend of two polymers, one of which is a copolymer. Therefore, two distinct structures in the CXA must be responsible for the existence of two melting points. Of great interest is the disappearance of the second melting point (126 °C) in the thermograms of the Nylon-rich blends. On the basis of the photomicrographs shown in Figure 1, the disappearance of the secondary melting point (126°C) in the thermogram occurs only when the CXA forms the discrete phase. The reason for this seemingly peculiar melting behavior is worth investigating in the future.

Dynamic Mechanical Behavior of the Nylon/CXA Blends. Figure 3 presents plots of the loss modulus (E'') of the blends investigated. Pure Nylon has three peaks: α peak at 64 °C, β peak at -50 °C, and γ peak at -120 °C, although the γ peak is not shown in Figure 3. The α peak is believed to be associated with the glass transition, the β peak is attributed to the existence of the polar group forming the hydrogen bonds in the polyamide, and the γ peak is thought to be associated with the crankshaft rotation of the $-(CH_2)_n$ group in the main chain of the polyamide (8).

Figure 3 shows one peak at -25 °C for CXA 3095. Because the molecular structure of the CXA 3095 is not known, we cannot comment on which type of functional group in the CXA 3095 is responsible for the peak. Also, in the Nylon-rich blends, the α transition of Nylon is shifted slightly toward lower temperatures in the Nylon/CXA = 80/20 and Nylon/CXA = 60/40 blends, and the β transition of Nylon is shifted somewhat toward higher temperatures (-38 and -32 °C in the Nylon/CXA = 80/20 and Nylon/CXA = 60/40 blends, respectively). The rather noticeable shift of the β transition of Nylon in the Nylon-rich blends seems to indicate some kind of chemical interaction between the CXA 3095 and the Nylon 6.

In the CXA-rich blends, as shown in Figure 3, the peak (-25 °C) of the CXA 3095 is shifted slightly toward lower temperatures in the Nylon/CXA = 20/80 and Nylon/CXA = 40/60 blends, and no peak representing the Nylon 6 is observed. Therefore, the Nylon 6 has little influence on the transition behavior of the CXA-rich blends.

Figure 4 displays plots of the elastic (Young's) modulus (E') of the blends investigated. The E' decreases as the Nylon content in the blend is decreased, and at temperatures of 50 °C and higher, the Nylon 6 in the CXA-rich blends contributes little to the modulus of the CXA 3095.

Rheological Behavior of the Nylon/CXA Blends. Figure 5 gives plots of viscosity η versus shear rate $\dot{\gamma}$, and Figure 6 shows plots of normal stress

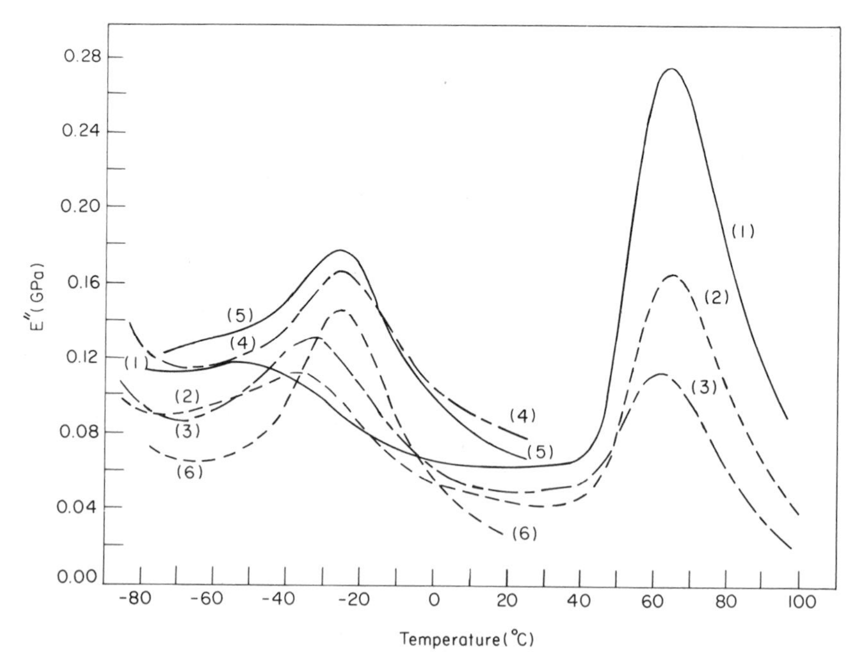

Figure 3. Loss modulus vs. temperature for the Nylon/CXA blend system: (1) Nylon 6, (2) Nylon/CXA = 80/20, (3) Nylon/CXA = 60/40, (4) Nylon/CXA = 40/60, (5) Nylon/CXA = 20/80, and (6) CXA 3095.

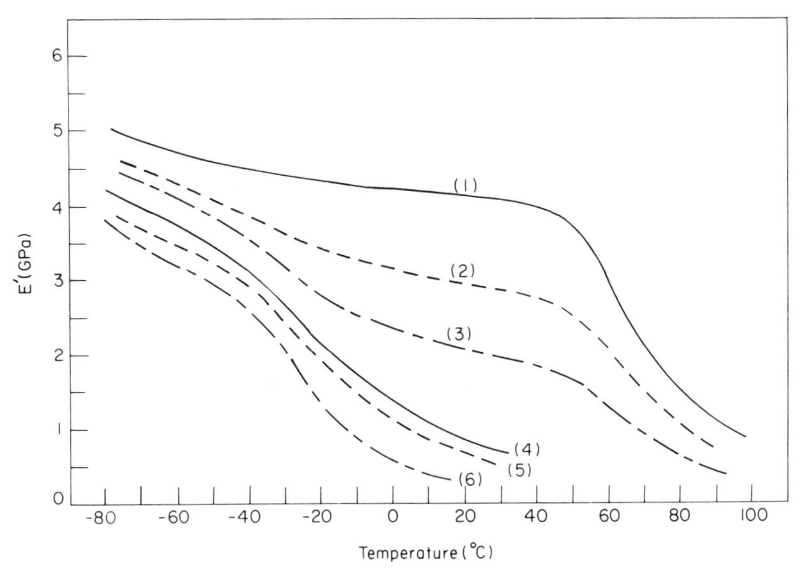

Figure 4. Elastic modulus vs. temperature for the Nylon/CXA blend system: (1) Nylon 6, (2) Nylon/CXA = 80/20, (3) Nylon/CXA = 60/40, (4) Nylon/CXA = 40/60, (5) Nylon/CXA = 20/80, and (6) CXA 3095.

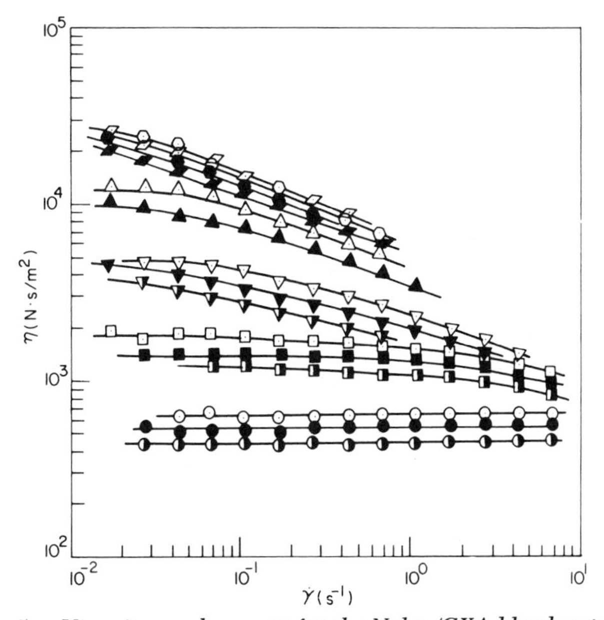

Figure 5. Viscosity vs. shear rate for the Nylon/CXA blend system at various temperatures (°C): (1) Nylon—⊙, 230; ●, 240; ◑, 250; (2) Nylon/CXA = 80/20—⊡, 230; ■, 240; ◨, 250; (3) Nylon/CXA = 60/40—▽, 230; ▼, 240; ▼, 250; (4) Nylon/CXA = 40/60—◐, 220; ⬡, 230; ⬢, 240; (5) Nylon/CXA = 20/80—▰, 220; ▱, 230; ▰, 240; and (6) CXA 3095—▲, 220; △, 230; ▲, 240.

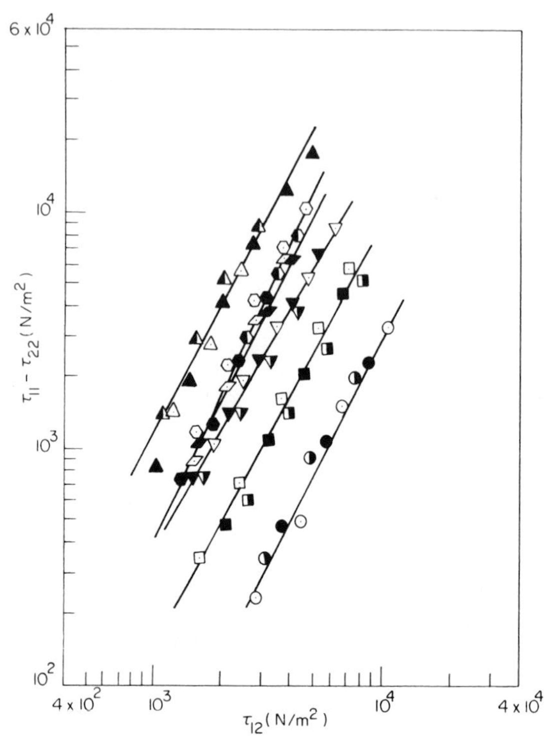

Figure 6. Normal stress difference vs. shear stress for the Nylon/CXA blend system at various temperatures. Key: see Figure 5.

difference $\tau_{11} - \tau_{22}$ versus shear stress τ_{12}, for the blends investigated, at various melt temperatures. Figure 5 shows that η decreases as the melt temperature increases and that the η of the CXA-rich blends is greater than that of the CXA 3095 alone. Figure 6 shows that the $\tau_{11} - \tau_{22}$ of the blends lies between those of Nylon 6 and CXA 3095, although the $\tau_{11} - \tau_{22}$ of the Nylon/CXA = 40/60 blend appears to be greater than that of the Nylon / CXA = 20/80 blend. When plotted against τ_{12}, $\tau_{11} - \tau_{22}$ becomes independent of the temperature. This effect has been reported (*3*, *5*, *6*, *9*) and a phenomenological explanation, though qualitative, has been presented (*10*). However, plots of η versus τ_{12} still exhibit temperature dependency, as shown in Figure 7. The use of τ_{12} (*3*, *6*, *9*) is appropriate for correlating the rheological properties (e.g., η and $\tau_{11} - \tau_{22}$) of multiphase polymer systems.

Figure 8 displays the variation of η with blend ratio, and Figure 9 shows the variation of $\tau_{11} - \tau_{22}$ with blend ratio, for the Nylon/CXA blends. Neither η nor $\tau_{11} - \tau_{22}$ follow the additivity rule. The shape of the η-blend ratio curve, shown in Figure 8, is quite different from that observed in the

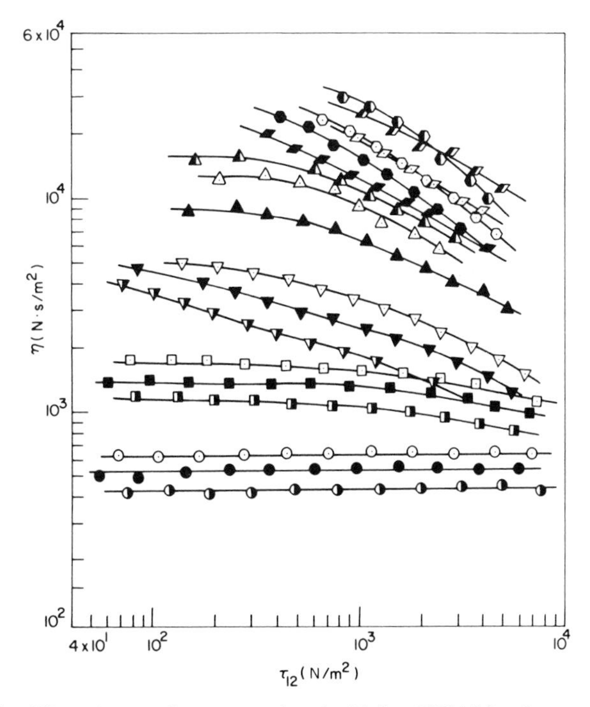

Figure 7. Viscosity vs. shear stress for the Nylon/CXA blend system at various temperatures. Key: see Figure 5.

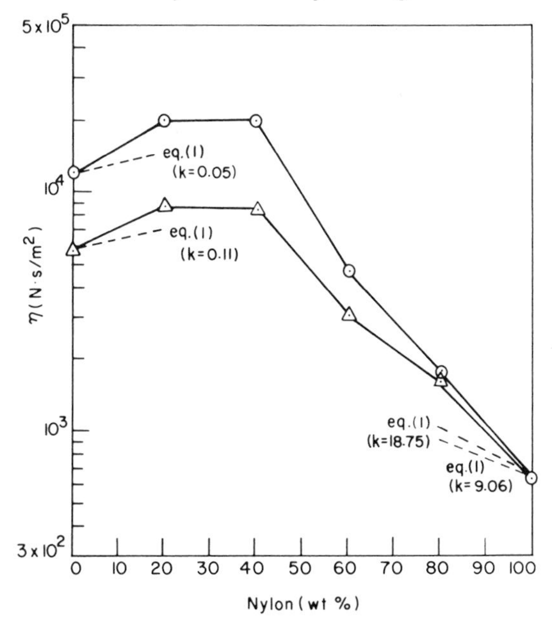

Figure 8. Viscosity vs. blend ratio for the Nylon/CXA blend system (T = 230 °C) at two different shear rates (s^{-1}). Key: ⊙, 0.043; and △, 0.427.

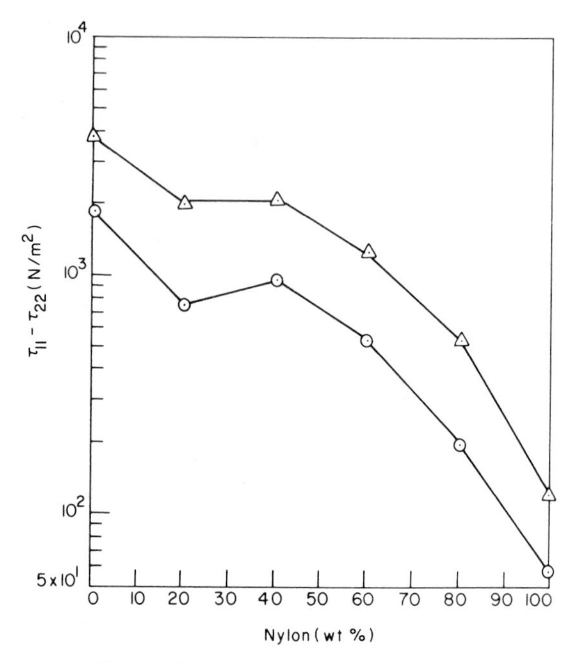

Figure 9. Normal stress difference vs. blend ratio for the Nylon/CXA blend system at two different shear stresses (N/m²). Key: ⊙, 2 × 10³; and △, 3 × 10³.

polypropylene (PP)/polystyrene (PS) blends (*11*), shown in Figure 10. The PP/PS blends form a heterogeneous phase, and the PP and PS have *no* chemical interaction at the interface and do not form an interphase.

When dealing with heterogeneous two-component blends having little chemical interaction between phases, a viscosity minimum or maximum is observed (*3–6, 9*). In those situations, the viscosity of one component is decreased by adding a small amount of the other component (for instance, *see* Figure 10). The viscosity–blend ratio curve shows a *negative* slope in the vicinity of the viscosities of the individual components. However, the viscosity–blend ratio curve of the Nylon/CXA blends (Figure 8) shows a *positive* slope in the vicinity of the viscosities of either components, Nylon and CXA. In view of the fact that the deformation rates for the curves shown in Figure 9 are very small, both the Nylon and CXA phases probably follow Newtonian behavior.

The well-known classical theory by Taylor (*12*) predicts the viscosity η of an emulsion by

$$\eta = \eta_0 \left[1 + \frac{(5k + 2)}{(2k + 2)} \phi \right] \tag{1}$$

where η_0 is the viscosity of the suspending medium, k is the ratio of the dispersed phase viscosity to the continuous medium viscosity, and ϕ is the volume fraction of the dispersed phase. Equation 1 assumes that no chemical interaction occurs between the dispersed and continuous phases, that the dispersed phase forms spherical droplets of uniform size, and that the droplets retain their spherical shape in the shear flow field.

We used the rheological data in Figure 5 to determine how applicable Equation 1 may be for predicting the experimentally observed viscosities. The results are shown in Figure 8 for comparison purposes. The experimental results show *greater* values of η than the theoretical predictions. We fully recognize that Equation 1 does not include the effect of the droplet size on η, whereas in reality the droplet size greatly influences the η of an emulsion. As may be seen in Figure 1, the dispersed phase in the Nylon/CXA blends has varying sizes of discrete particles, which is *not* consonant with one of the several assumptions made in the theoretical development of Equation 1. Nevertheless, we believe that the higher experimentally observed viscosities of the Nylon/CXA blends may be attributable to some kind of chemical interaction at the interface between the Nylon and CXA phases. If no chemical interaction occurs between the phases, we would *not* expect the experimentally observed viscosities to be greater than the theoretically predicted ones. This speculation must be proven (or disproven) by an independent experimental (or theoretical) investigation in the future.

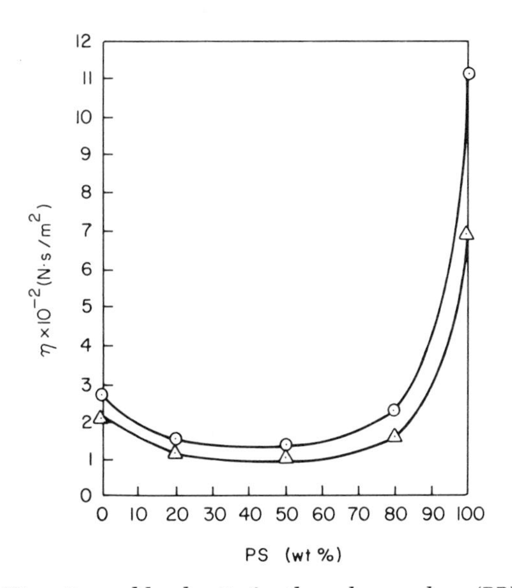

Figure 10. Viscosity vs. blend ratio for the polypropylene (PP)/polystyrene (PS) blend system ($T = 200$ °C) at two different shear stresses (N/m^2) (11). Key: ⊙, 4.1×10^4; and △, 4.8×10^4.

Melt Drawability of the Nylon/CXA Blends. Figure 11 displays the observed melt drawability (or stretchability) of the Nylon/CXA blends in a simple extrusion experiment. The maximum draw-down ratio $(V_L/V_o)_{max}$ used in Figure 11 may be considered to be a simple, effective test for determining how much more stretchable one material is than others, if the tests are conducted under the same extrusion conditions. Figure 11 demonstrates two important points. First, the $(V_L/V_o)_{max}$ of the CXA-rich blends depends on deformation history (i.e., L/D ratio), whereas the $(V_L/V_o)_{max}$ of the Nylon-rich blends does not. This result is understandable because the elasticity of the CXA-rich blends is greater than that of the Nylon-rich blends, as may be seen in Figure 6. Second, except for the Nylon/CXA = 40/60 blend, the blends give rise to a melt drawability better than that expected from the additivity rule. Why the Nylon/CXA = 40/60 blend gives rise to such a poor melt drawability is inexplicable from the experimental evidence.

Concluding Remarks

From the observations on the E' of the blends, given in Figure 4, the Nylon-rich blends (e.g., Nylon/CXA = 80/20 blend) seem to be more attrac-

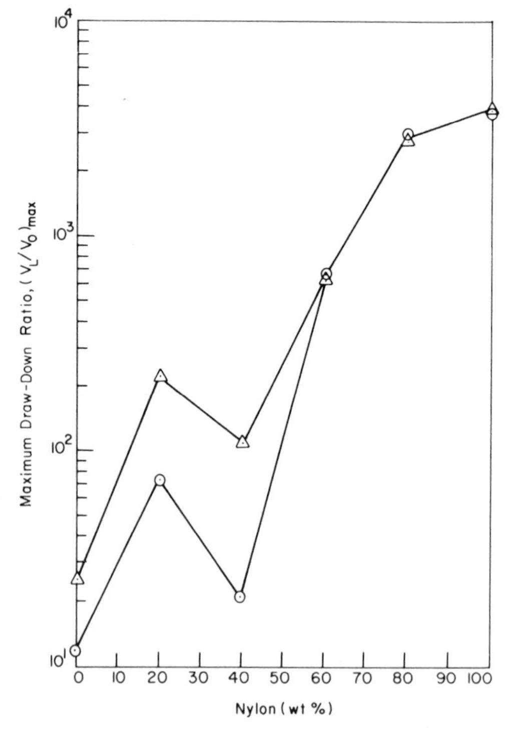

Figure 11. Maximum drawn-down ratio vs. blend ratio for the Nylon/CXA blend system (T = 255 °C), extruded through two different capillaries. Key: ⊙, L/D = 4.0; and △, L/D = 6.3.

tive than the CXA-rich blends. The CXA has a higher viscosity than the Nylon (*see* Figure 6) and therefore, when subjected to a flow field, the CXA will not deform as easily as the Nylon. The Nylon-rich blends retain essentially the same melting and transition behaviors as those of Nylon (*see* Figures 2 and 3). We speculate that, during melt processing, some chemical interactions occurred at the interface between the CXA and Nylon phases and gave rise to unique properties (e.g., melt drawability) in the Nylon-rich blends. In the future, we shall put our efforts into unraveling the mechanism(s) of chemical interaction that may exist in Nylon/CXA blends.

Acknowledgment

We acknowledge, with gratitude, that American Enka Company supplied us with the large quantity of the Nylon 6 resin, and that the Du Pont Company supplied us with the CXA 3095.

Literature Cited

1. "Polymer Blends"; Paul, D. R.; Newman, S. Eds.; Academic: New York, 1978.
2. Olabisi, O; Robeson, L. M.; Shaw, M. T. In "Polymer–Polymer Miscibility"; Academic: New York, 1979.
3. Han, C. D. "Multiphase Flow in Polymer Processing"; Academic: New York, 1981.
4. VanOene, H. *J. Colloid Interface Sci.* **1972**, *40*, 40.
5. Han, C. D.; Yu, T. C. *J. Appl. Polym. Sci.* **1971**, *15*, 1163.
6. Han, C. D.; Kim, Y. W. *Trans. Soc. Rheol.* **1975**, *19*, 245.
7. Kim, Y. J.; Han, C. D., Unpublished research, 1981.
8. Kawaguchi, T. *J. Appl. Polym. Sci.* **1959**, *2*, 56.
9. Han, C. D. In "Rheology in Polymer Processing"; Academic: New York, 1976.
10. Han, C. D.; Lem, K. W. *Polym. Eng. Reviews* **1982**, *2*, 135.
11. Han, C. D.; Kim, Y. W.; Chen, S. J. *J. Appl. Polym. Sci.* **1975**, *19*, 2831.
12. Taylor, G. I. *Proc. R. Soc. London, Ser. A* **1932**, *138*, 41.

RECEIVED for review January 20, 1983. ACCEPTED August 29, 1983.

12

Controlled Ingredient-Distribution Mixing: Effect on Some Properties of Elastomer Blend Compounds

BIING-LIN LEE

Research and Development Center, The BFGoodrich Company, Brecksville, OH 44141

*The crack growth and heat buildup of carbon black loaded rubber blends were studied as a function of the distribution of carbon black in the individual rubber phases. The blends are styrene-butadiene rubber/polybutadiene (SBR/PBD), natural rubber/*cis*-polybutadiene (NR/PBD), and styrene-butadiene copolymer/butadiene-acrylonitrile copolymer (SBR/NBR). Regardless of the difference in chemical structure of these rubber blends, the results show that a better crack-growth resistance compound contains proportionately more carbon black in the major rubber phase. A simple analogy of rubber blends to the rubber-modified thermoplastics is proposed to interpret these findings. The heat buildup of the blends is also affected by the distribution of carbon black. However, the extent of the carbon black distribution effect is influenced by the total amount of carbon black in the blends.*

ANY ONE TYPE OF POLYMER MAY NOT POSSESS all the physical–mechanical properties desired in a finished product, so two or more polymers are mixed together with nonpolymer solid fillers to meet the required properties. Hundreds of homopolymers are commercially available. When two or more homopolymers are to be mixed, or a homopolymer is to be mixed with other nonpolymeric materials, a number of fundamental and practical questions must be answered: (1) Will the mixture of two or more polymers be compatible? (2) What is the morphology of the final mixed blend? Which polymer is the continuous phase? What is the size of the dispersed polymer phase? (3) What are the processing characteristics of the blends? (4) How can one be sure of obtaining "good" dispersion of fillers? Some of these questions on polymer blends have already been discussed (*1–15*). When a solid particle phase is added to polymer blends, an additional ques-

0065-2393/84/0206-0185$07.25/0

tion is how the particles distribute in each polymer phase and how this distribution affects the compound's properties (*16–21*). In other words, when dealing with multiphase polymer systems, one of the most important aims is to optimize the physical–mechanical properties desired in the finished product via appropriate processing. This optimization is especially needed in the blends of two or more rubbers that are used for tires, conveyor belts, hoses, or other rubber goods.

Rubber products that are flexed in service frequently fail because of the appearance and growth of cracks (*22, 23*). Therefore, improving the crack-growth resistance of rubber products is industrially important. Various theories for some model experimental results on crack growth of rubbers have been advanced (*23–33*). In reality, however, the mixing of carbon black in a rubber blend compound could result in a nonuniform distribution of carbon black and, consequently, affect the crack growth (*17*). Despite its industrial importance, the effect of heterogeneity in rubber blends, which results from nonuniform distribution of carbon black in the individual rubber phases, is seldom systematically studied.

We have investigated the reinforcement of uncured and cured rubbers (*34, 35*) and the morphology of carbon black loaded styrene–butadiene rubber/polybutadiene[1] (SBR/PBD) blends (*20, 21*). We reported that the distribution of carbon black in SBR and PBD phases is governed by the method of mixing. In the mechanical cross-mixing of carbon black master batches, carbon black always stays in the original rubber phase. That is, no significant amount of carbon black migration from one phase to the other phase occurs during Brabender-mixed master batching with a second rubber (*17, 20, 21*).

In this chapter, we report some new data and summarize some of our comparative experimental studies on how the location of carbon black in several rubber blends influences the crack-growth resistance (*36, 37*). The mixing process to control the carbon black distribution in rubber phases, essentially, is an extension of our previous mixing studies (*20, 21*). The results show how crack-growth resistance can be improved by the nonhomogeneous distribution of carbon black. We will then discuss some possible important mechanisms to improve crack-growth resistance, stress relief, and low heat buildup. The investigation of such mechanisms is a part of our research into the relationships of processing–morphology–properties of rubber blends (*20, 21, 34, 35, 38, 39*).

Experimental

Materials. The three rubber blend compounds studied are System I, SBR : PBD : carbon black (60 : 40 : 85 and curatives); System II, natural rubber : PBD : carbon black (75 : 25 : 45 and curatives); and System III, SBR : NBR : carbon black (70 : 30 : 82.5 and curatives). The ingredients and weights are as follows:

[1]The ASTM designation of *cis*-polybutadiene is BR.

SBR/PBD blends—System I

oil extended PBD [PBD, 100; oil, 37.5] (55.0)
oil extended SBR [SBR, 100; oil, 37.5] (82.5)
petroleum oil (22.2)
furnace carbon black (85)
sulfur, stearic acid, zinc oxide, accelerators, antioxidants, and antiozonants.

NR/PBD blends—System II

natural rubber (75)
oil extended PBD [PBD, 100; oil, 37.5] (34.6)
furnace carbon black (45)
sulfur, petroleum oil, paraffin wax, stearic acid, zinc oxide, accelerators, antioxidants, and antiozonants.

SBR/NBR Blends—System III

NBR[2] (30.0)
SBR [SBR, 57.4%; HAF carbon black, 35.8%; and oil, 6.8%] (122.46)
HAF carbon black (38.70)
zinc oxide, stearic acid, accelerators, antioxidants, antiozonants, plasticizer, and sulfur.

[2]HYCAR 1043.

SBR and PBD master batches—System I

Component	*PBDMB*	*SBRMB*
Oil extended PBD (PBD, 100; oil, 37.5)	55.0	—
Oil extended SBR (SBR, 100; oil, 37.5)	—	82.5
Petroleum oil	6.2	16.2
Furnace black	a	85 − a

The sulfur, stearic acid, zinc oxide, accelerators, antioxidants, and antiozonants were incorporated in the SBRMB and PBDMB at the ratio of 60/40, according to the rubber weight ratio.

Amounts of carbon black in SBR and PBD master batch pairs that were cross-mixed to form the SBR/PBD compounds

	SBRMB (rubber A-MB)			*PBDMB (rubber B-MB)*		
SBR/PBD Blend CPD	*CPD*	*85 − a (carbon black)*	ϕ_2*	*CPD*	*a (carbon black)*	ϕ_2(%)
356	352	51	20.6	351	34.0	21.3
362	358	42.5	17.7	357	42.5	25.4
368	364	34.0	14.7	363	51.0	29.0
373	370	63.8	24.4	369	21.3	14.5
378	375	76.5	27.9	374	8.5	6.3

*ϕ_2 is defined as the volume fraction of carbon black in the master batch.

NR and PBD master batches—System II

Component	*NRMB*	*PBDMB*
Oil extended PBD (PBD, 100; oil, 37.5)	—	34.6
Natural rubber	75	—
Furnace carbon black	b	45 – b

The sulfur, stearic acid, zinc oxide, petroleum oil, paraffin wax, accelerators, antioxidants, and antiozonants were incorporated in the NRMB and PBDMB at the ratio of 75/25, according to the rubber weight ratio.

Amounts of carbon black in NR and PBD master batch pairs that were cross-mixed to form the NR–PBD compounds

	(rubber A-MB)			*(rubber B-MB)*		
NR–PBD CPD	*CPD*	*b (carbon black)*	ϕ_2(%)	*CPD*	*45 – b (carbon black)*	ϕ_2(%)
384	382	33	16.7	383	12	13.5
389	387	45	21.0	388	0	0
395	392	20	10.9	393	25	24.6

SBRMB and NBRMB master batches—System III

	CPD308		*CPD322*		*CPD329*	
Component	*SBRMB CPD307*	*NBRMB CPD306*	*SBRMB CPD321*	*NBRMB CPD320*	*SBRMB CPD328*	*NBRMB CPD327*
SBR	122.46	—	122.46	—	122.46	—
NBR	—	30	—	30	—	30
HAF carbon black	38.7	—	—	38.70	30.40	8.3

The zinc oxide, stearic acid, accelerators, antioxidants, antiozonants, plasticizer, and sulfur were incorporated in the NBRMB/SBRMB at the ratio of 30/70, according to the rubber weight ratio.

Amounts of carbon black in SBR and NBR master batch pairs that were cross-mixed to form SBR–NBR compounds

	(rubber A-MB) SBRMB		*(rubber B-MB) NBRMB*	
SBR–NBR CPD	*CPD*	*Parts*	*CPD*	*Parts*
308	307	82.5	306	0
322	321	43.8	320	38.7
329	328	74.2	327	8.3

Mixing. A model B Banbury mixer along with a 10-in. mill was used to fabricate the compounds. Carbon black was incorporated into the individual rubber phases in different ways to enhance the modulus difference between the individual master batches. We reported earlier (*21*, *38*) that a three-component system may be reduced to a two-component system by considering the carbon black preloaded master batch as a continuum. The detailed mixing methods are described as follows:

FREE BLACK MIXING. Carbon black was added to a preblend of rubber in the mixer. Curing agents were added on the 10-in. mill. Figure 1 shows the detailed procedure.

CARBON BLACK MASTER BATCH APPROACH. To prepare each rubber blend, the individual rubber phases were premixed with an appropriate amount of the ingredients, including carbon black and curing agents. We call these master batches Rubber A-MB and Rubber B-MB. Appropriate quantities of Rubber A-MB and Rubber B-MB were then finally cross-mixed on the 10-in. mill to generate a compound with the same ingredients and weights as those listed under "Materials." Figure 2 shows the detailed mixing procedure. The compositions of the master batch pairs studied are shown in the boxes. The ranges of the ratio of the amounts of carbon black in the Rubber A-MB and Rubber B-MB pairs provide a wide variety of modulus ratios of Rubber A-MB/Rubber B-MB.

For the mixing of Systems I and II, the rotor speed of the mixer was 77 RPM, and the initial chamber temperature was 93 °C. The running water control valve was open during mixing to remove the heat generated by the shear mixing. The dump temperature was controlled, and did not exceed 149 °C. For System III, the initial chamber temperature was 82 °C, and the dump temperature was about 132 °C and not over 138 °C. The mill temperature was set at 71 °C.

Curing. The compounded rubber cure time was evaluated in the Monsanto rheometer 100 (*40*). Optimum cure time + 10 min at a specific temperature was the cure time selected in all cases (149 °C for System I, 138 °C for System II, and 146 °C for System III). The rheometer test conditions were oscillation disk frequency, 1.66 Hz; and oscillation amplitude, half-cycles of 1 ± 0.002°. The optimum cure time was defined as the time required to achieve 90% of maximum torque in a Monsanto rheometer. This instrument has also been used to study the kinetics of vulcanization (*41*), the reinforcing characteristics of carbon black in rubber (*42*), and dispersive mixing (*34*).

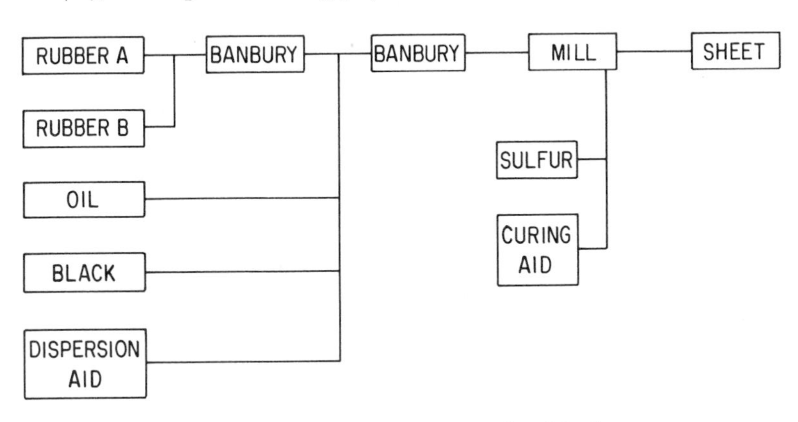

Figure 1. Block diagram of free carbon black mixing.

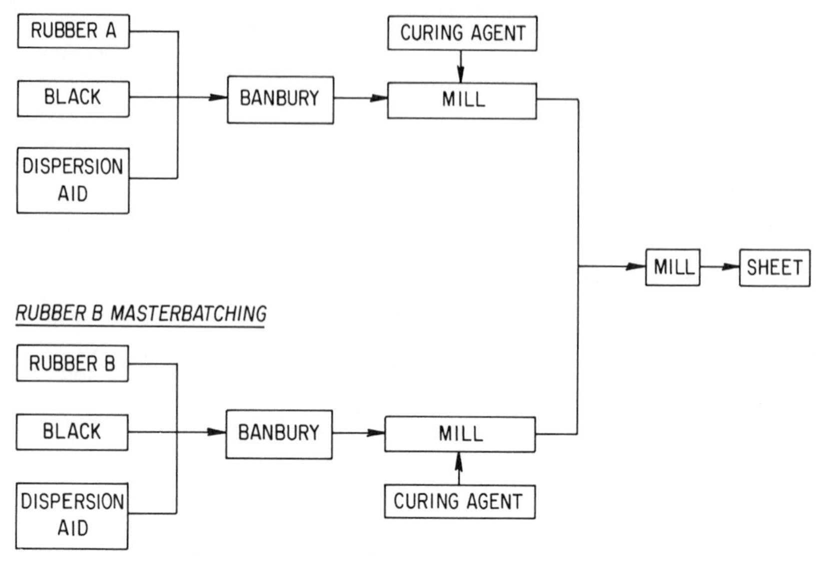

Figure 2. Block diagram of controlled carbon black distribution mixing process.

Crack Growth Test. Crack growth was tested in a rotating ring crack growth tester that was devised by Beatty and Juve (*23*). This apparatus is compact and vibration-free, and utilizes a ring-type specimen. The test conditions are as follows: chamber temperature, 70 °C for Systems I and II, room temperature for System III; load, 1.36 kg (or 3 lb); crack at start, 2.5 mm; and rotation speed, 320 cycles/min.

As pointed out by Beatty and Juve (*23*), groove cracking is a continuous problem in the running of tires. Cracking takes place in two steps: initiation and growth. Initiation is inevitable in service because of ozone cracking, cuts, and nicks. Therefore, crack growth is an important concern.

Heat Buildup Test. For the blends of SBR/PBD and NR/PBD (Systems I and II), the hysterisis heat buildup was also measured. The operation procedures as outlined in ASTM D623 (*43*) were employed. The flexometer was operated at 9.8×10^5 Pa load and 17.5% constant deflection starting at 48 °C. The temperature rise after 25 min of deflection was measured. The relationship of the heat buildup to the dynamic mechanical properties of carbon black loaded rubber vulcanizates has been discussed (*35*).

Stress–Strain Test. The stress–strain data were obtained from an Instron tensile tester. The test conditions were as follows: type of sample, ovals; cross head speed, 25.4 cm/min; and room temperature.

Results and Discussion

Location of Carbon Black. An important feature associated with the mixing of carbon black and multicomponent elastomer systems is the identification of the location of carbon black in the mixed blends. Figure 3 shows some of our attempts in the morphological investigation of these blends. This picture is the electron micrograph of the cross-mixed NR/PBD blend, CPD 389. Distinct, clear PBD domains containing no carbon black are obvious. This blend was prepared so that all of the carbon black was mechanically premixed in the NR phase. The presence of distinct domains suggests that no carbon black was transferred from the NR to the PBD phase during the cross-mixing process. This observation is in agreement with the observations by Hess et al. (*44, 45*) and Marsh et al. (*46*) on low-level carbon black loaded NR/PBD blends. In our earlier study of SBR/PBD blends (*20, 22*), we had concluded that no significant amount of carbon black was transferred from one phase to the other on Brabender-mixed master batching with a second gum rubber. Of course, considerable work on the development of microscopic techniques should be done to be able to unambiguously identify the morphology of highly carbon black loaded rubber blends.

Modulus of the Individual Master Batches. The stress–strain data of the individual master batches loaded with different amounts of carbon black are plotted in Figures 4–7 for Systems I and II. The stress shown in these figures is expressed as:

$$\sigma_t = \sigma_u(1 + \epsilon) \tag{1}$$

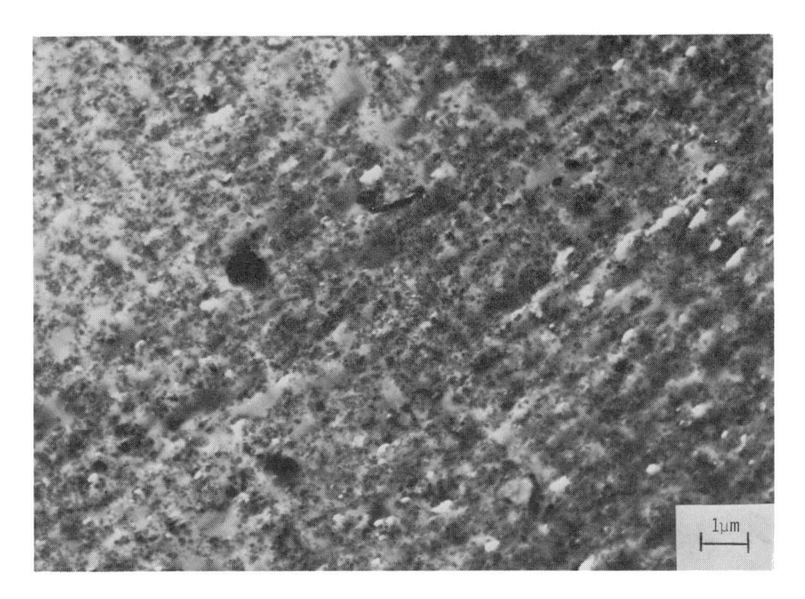

Figure 3. Electron micrograph of CPD 389. All the carbon black was premixed in the NR phase.

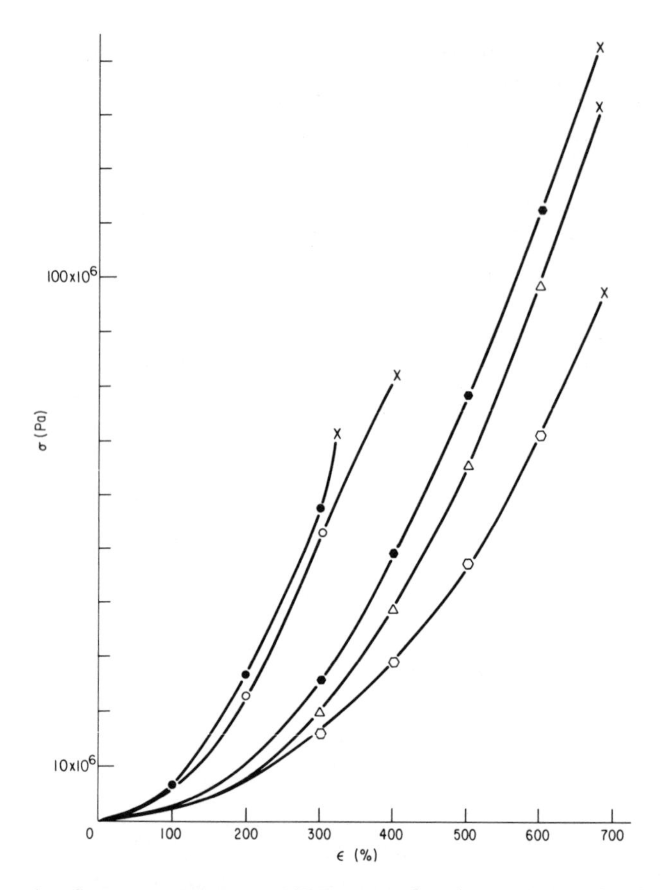

Figure 4. Stress–strain curve of SBR master batches (System I) loaded with different amounts of carbon black. Key: ×*, break point;* ⬢*, CPD 352;* △*, CPD 358;* ⬡*, CPD 364;* ○*, CPD 370; and* ●*, CPD 375. (Reproduced with permission from Ref. 36. Copyright 1982,* J. Appl. Polym. Sci.)

where σ_u is the uncorrected stress that may be defined as F/A_0, A_0 is the initial cross-sectional area of the tested sample, and ϵ is elongation.

The tensile elastic moduli (E) of these master batches are plotted in Figures 8 and 9. As expected, the moduli of the individual master batches can be varied by adjusting the amount of carbon black. The order of the moduli ratio of the master batch pairs that were cross-mixed to form the blend compounds is as follows: For System I:

$$\frac{E_{\mathrm{CPD\ 375}}}{E_{\mathrm{CPD\ 374}}} > \frac{E_{\mathrm{CPD\ 370}}}{E_{\mathrm{CPD\ 369}}} > \frac{E_{\mathrm{CPD\ 352}}}{E_{\mathrm{CPD\ 351}}} > \frac{E_{\mathrm{CPD\ 358}}}{E_{\mathrm{CPD\ 357}}} > \frac{E_{\mathrm{CPD\ 364}}}{E_{\mathrm{CPD\ 363}}} \qquad (2)$$

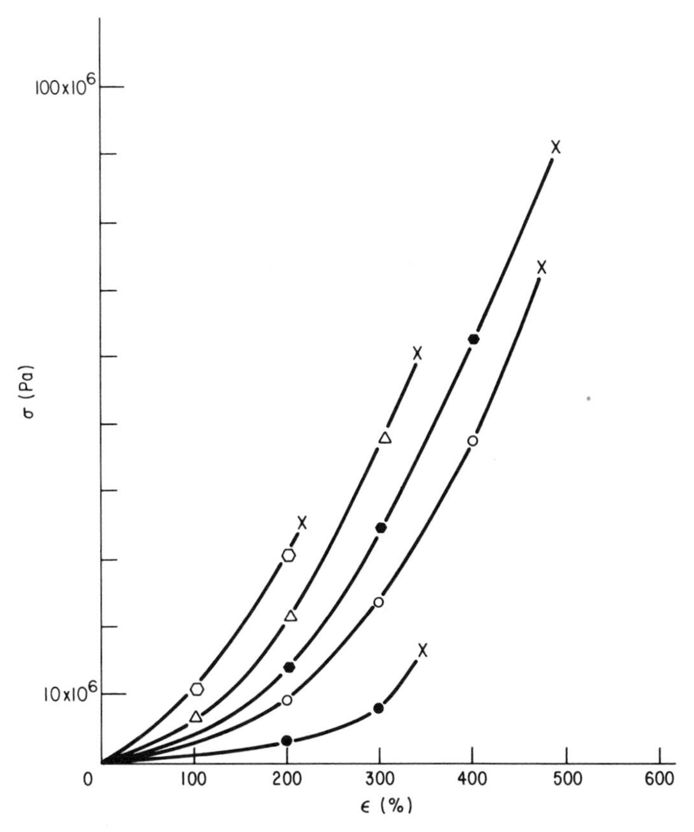

Figure 5. Stress–strain curve of PBD master batches (System I) loaded with different amounts of carbon black. Key: ×, break point; ⬢ CPD, 351; △, CPD 357; ⬡, CPD 363; ○, CPD 369; and ●, CPD 374. (Reproduced with permission from Ref. 36. Copyright 1982, J. Appl. Polym. Sci.)

For System II:

$$\frac{E_{\mathrm{CPD\ 387}}}{E_{\mathrm{CPD\ 388}}} > \frac{E_{\mathrm{CPD\ 382}}}{E_{\mathrm{CPD\ 383}}} > \frac{E_{\mathrm{CPD\ 392}}}{E_{\mathrm{CPD\ 393}}} \tag{3}$$

Later we will discuss how this moduli ratio of the master batch pairs affects crack growth.

Modulus of the Blend Compounds. Table I lists the moduli of the cross-mixed blend compounds with different amounts of carbon black in the master batches. If the slight differences in moduli as measured by the Instron stress–strain tests are significant, they may be a result of the differences of morphology. However, we do not have definite proof.

Crack Growth of Blend Compounds. Figures 10–13 summarize the results of the crack growth of the blended compounds as tested by the ring

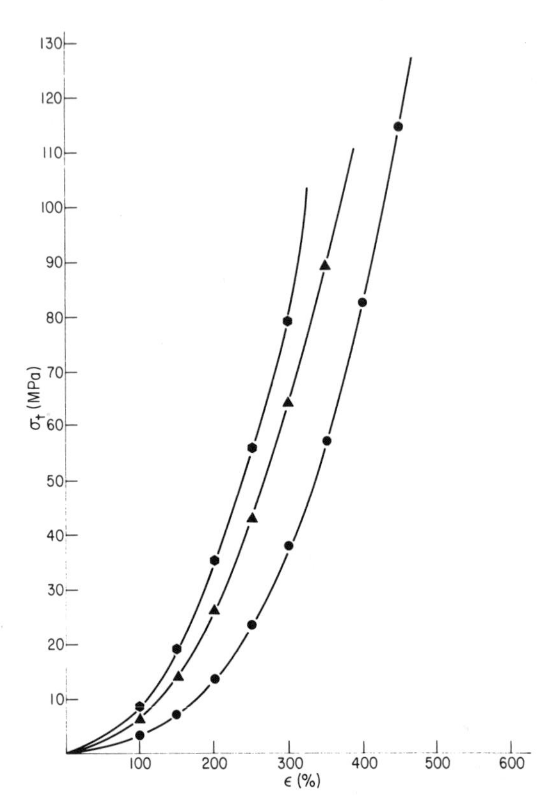

Figure 6. Stress–strain curve of NR master batches (System II) loaded with different amounts of carbon black. Key: ⬢, *CPD 387;* ▲, *CPD 382; and* ●, *CPD 392.*

flex tester. Figure 10 describes the blends of SBR–PBD, Figure 11A describes an all natural rubber compound (i.e., 100 phr of NR and 45 phr of carbon black), and Figure 11B describes the NR–PBD blend compounds. Figures 12 and 13 show the crack growth of the blends of SBR–NBR that were cross-milled at two different levels of watt-hours. Several important points follow:

1. The presence of a second rubber component in the first rubber matrix usually improves the crack-growth resistance of the first rubber compound (see Figures 11A and 11B and Reference 33). More strictly speaking, on the basis of our results as indicated in Figures 10–13, the crack-growth resistance of a carbon black loaded rubber blend is strongly dominated by the distribution of carbon black in the individual rubber phases.
2. The compounds that were prepared by the controlled carbon black distribution process (Figure 2) give a variety of crack-

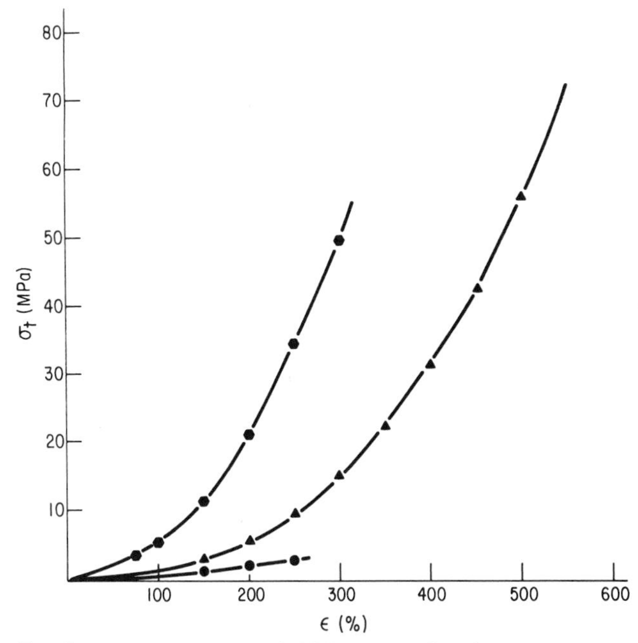

Figure 7. Stress–strain curve of PBD master batches (System II) loaded with different amounts of carbon black. Key: ▲, CPD 383; ●, CPD 388; and ⬢, CPD 393.

growth resistance curves. The better crack-growth resistance compound is the one with more carbon black initially in the major rubber phase.

3. The free carbon black mixing process usually is considered an easy and convenient mixing process. However, as far as the crack-growth resistance is concerned, this mixing process is not the best.
4. The term "cross-mixing" is not a new process; the physical properties of rubber compounds are dependent on the method of mixing. We demonstrate that controlled carbon black distribution in these chemically dissimilar rubber blends can be accomplished by this old cross-mixing process.
5. The crack-growth resistance of a carbon black loaded rubber blend is also affected by the energy of cross-mixing. Longer mixing reduces the crack-growth resistance (*see* Figures 12–13). In other words, the blend morphology also influences the crack-growth resistance.

We will now discuss how the heterogeneous distribution of carbon black introduces some possible mechanisms affecting the crack growth of rubber blends. We will first propose the analogy of rubber blends to rubber-modified thermoplastics, then discuss the possibly favorable factors for improving the crack-growth resistance—stress relief and low heat buildup.

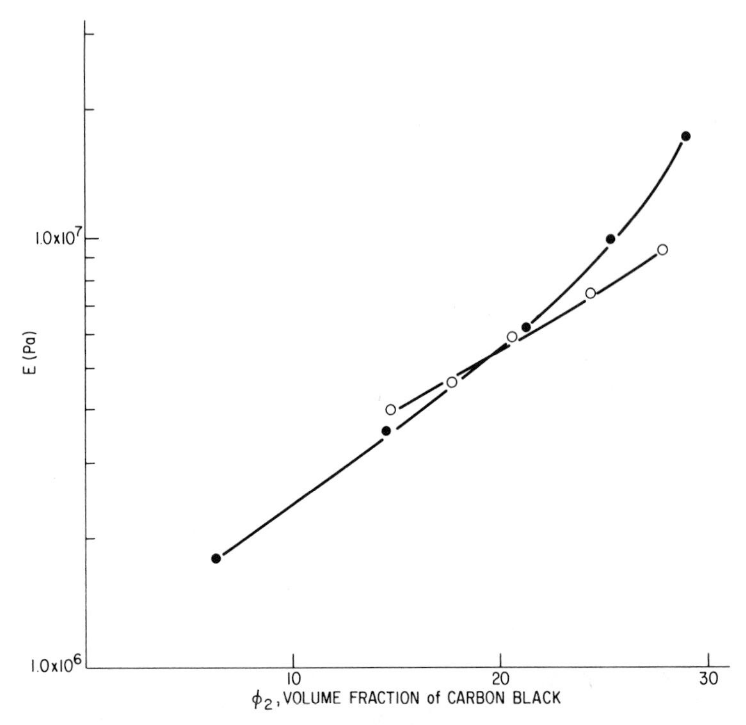

Figure 8. Tensile (elastic) modulus of SBRMB (○) and PBDMB (●) as a function of carbon black loadings (System I). (Reproduced with permission from Ref. 36. Copyright 1982, J. Appl. Polym. Sci.)

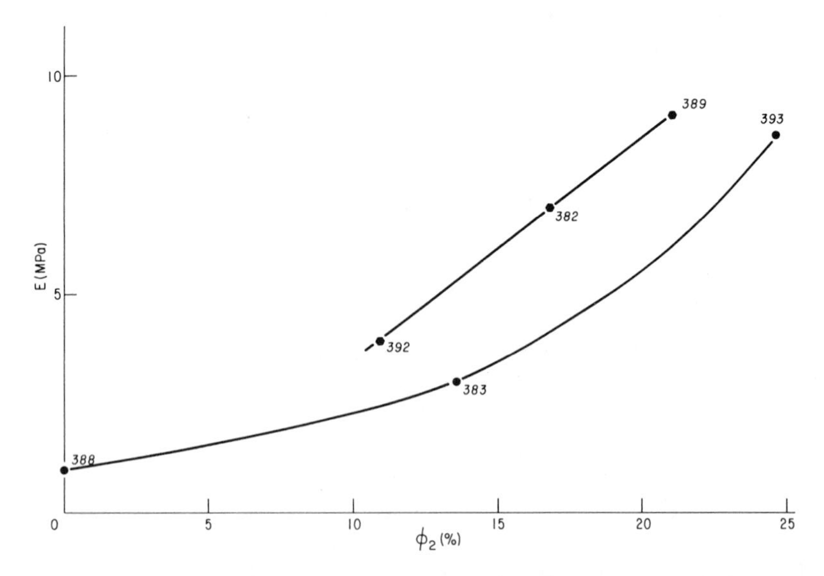

Figure 9. Tensile (elastic) modulus of NRMB (●) and PBDMB (●) as a function of carbon black loading (System II).

Table I. Tensile Properties of the Rubber Blends

System I—Tensile elastic modulus (E) and 300% modulus (σ_{300}) of the SBR/PBD blend compounds prepared by different mixing methods

CPD	E (MPa)	σ_{300} (MPa)
350	6.29	7.82
356	6.25	7.17
362	6.69	7.40
368	6.08	6.43
373	5.88	9.78
378	5.18	9.92

System II—Tensile elastic modulus (E) and 300% stress (σ_{300}) of the NR/PBD blend compounds prepared by different mixing methods

CPD	E (MPa)	σ_{300} (MPa)
381	5.47	11.34
384	5.46	10.67
389	4.55	10.58
395	5.18	10.71
382	6.69	16.01

Note: CPD 382 is a total NR compound, that is, 100 phr NR and 45 phr carbon black.

System III—Tensile properties of SBR–NBR blends

CPD	W_x (W-h)[a]	ϵB (%)	σ_B (MPa)	σ_{300} (MPa)
303	—	417	14.64	11.08
308	100	333	18.75	16.53
309	50	319	17.08	15.90
322	100	340	17.55	15.54
323	50	354	17.88	15.30
329	100	333	15.67	14.24
330	50	374	17.00	13.85

[a] W_x = (watt-hour/1200 gram) for cross-mixing the SBRMB and NBRMB on a 10-in. mill.

Effect of Modulus on Crack-Growth Resistance of Rubber Blends. The crack-growth resistance of rubber compounds is affected by the modulus levels of the compounds. Auer et al. (*47*), for example, determined the crack-growth resistance of SBR as a function of modulus over a wide range of accelerator and sulfur concentrations and proposed the following relation:

$$\log \frac{t}{L} = a - bM \tag{4}$$

where t/L is the crack-growth resistance in thousand cycles per inch, M is the 300% modulus, and a and b are constants. Equation 4 states that the

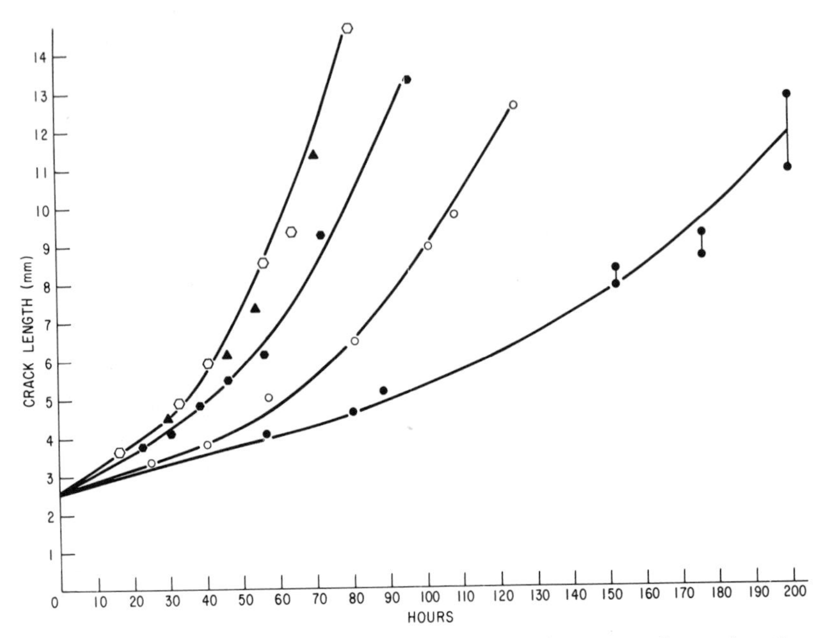

Figure 10. Crack growth of the SBR/PBD blend compounds as a function of method of mixing (System I). Key: ▲, CPD 350; ⬢, CPD 356; ⬡, CPD 368; ○, CPD 373; and ●, CPD 378. (Reproduced with permission from Ref. 36. Copyright J. Appl. Polym. Sci.)

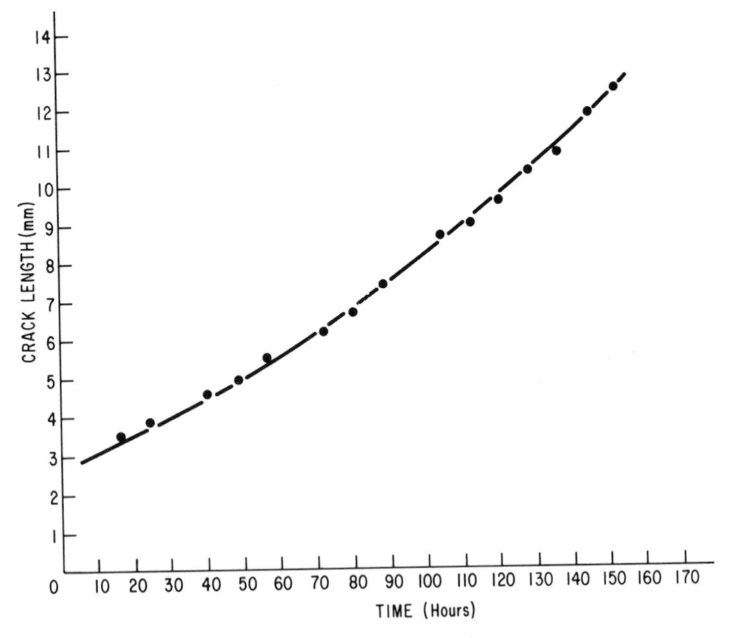

Figure 11A. Crack growth of an all NR compound.

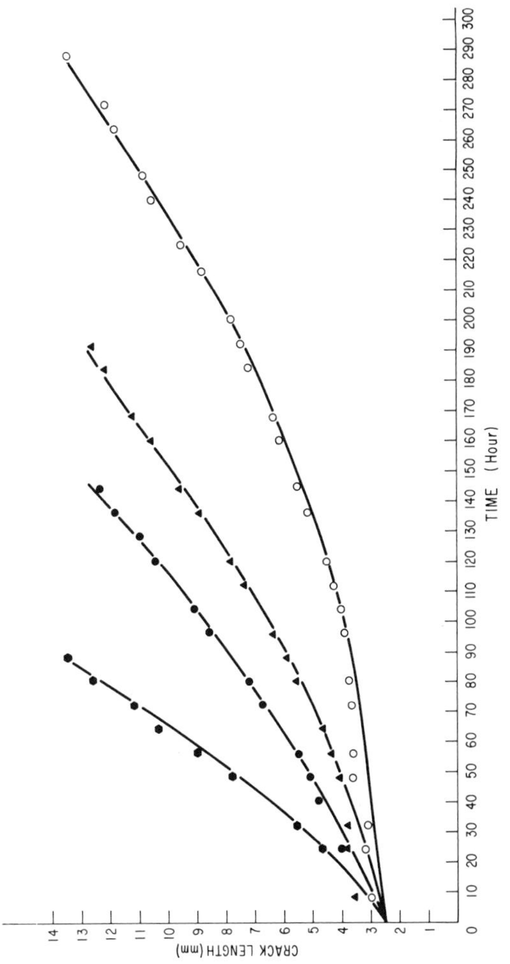

Figure 11B. Crack growth of the NR/PBD blend compounds as a function of method of mixing. Key: ●, *CPD 381 (free carbon black mixing);* ▲, *CPD 384;* ○, *CPD 389; and* ⬢, *CPD 395.*

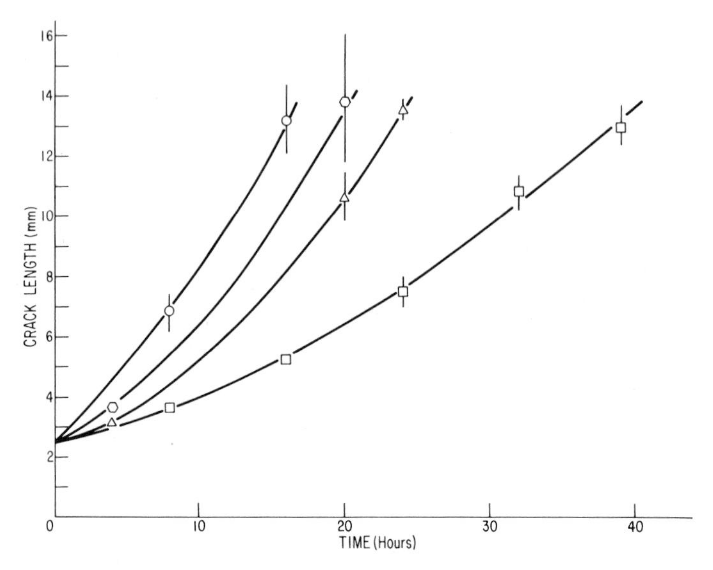

Figure 12. Flex crack growth of SBR/NBR blends (System III), cross-mixing energy = 100 W-h/1200 g. Key: ○, CPD 303; Δ, CPD 308; ⬡, CPD 322; and □, CPD 329. (Reproduced with permission from Ref. 37. Copyright 1982, Poly. Eng. Sci.)

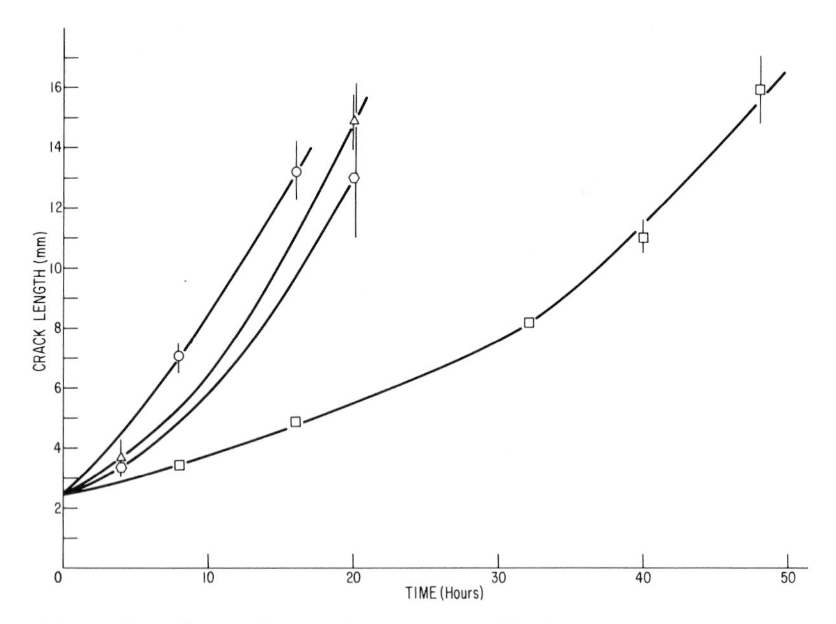

Figure 13. Flex crack growth of SBR/NBR blends (System III), cross-mixing energy = 50 W-h/1200 g. Key: ○, CPD 303; Δ, CPD 309; ⬡, CPD 323; and □, CPD 330. (Reproduced with permission from Ref. 37. Copyright 1982, Poly. Eng. Sci.)

higher the modulus of the compound is, the poorer its crack-growth resistance is.

By carefully comparing the results in Figures 10–13 and Table I, we observe no definite correlation of the crack-growth resistance of rubber blends with 300% modulus as indicated by Equation 4. Thus, for the crack growth of rubber blends in which the ingredients and concentrations are identical, the 300% modulus is not a primary factor affecting the crack-growth resistance.

Effect of Modulus Ratio of the Master Batch Pairs on Crack Growth. To quantify the results, Figures 14 and 15 show the crack-growth resistance as a function of the modulus ratio of the corresponding master batch pairs. These figures illustrate the significant effect of the carbon black distribution in the individual rubber phases on $T_{5\times}$, the time required for crack growth to five times the original crack length. The modulus, as measured by the Instron tensile test, is subject to some inherent errors, but the order of magnitude should not be affected. The $T_{5\times}$ is shifted from 90 to 200 h as E_{SBRMB}/E_{PBDMB} changes from 1 to 5 for the blends of SBR–PBD, and $T_{5\times}$ of the blends of NR–PBD can be tripled as E_{NRMB}/E_{PBDMB} changes from 1 to 10.

The curves in Figures 10–12 can be generalized in the following form:

$$\frac{dL}{dt} = AL^n \tag{5}$$

where A and n are constants.

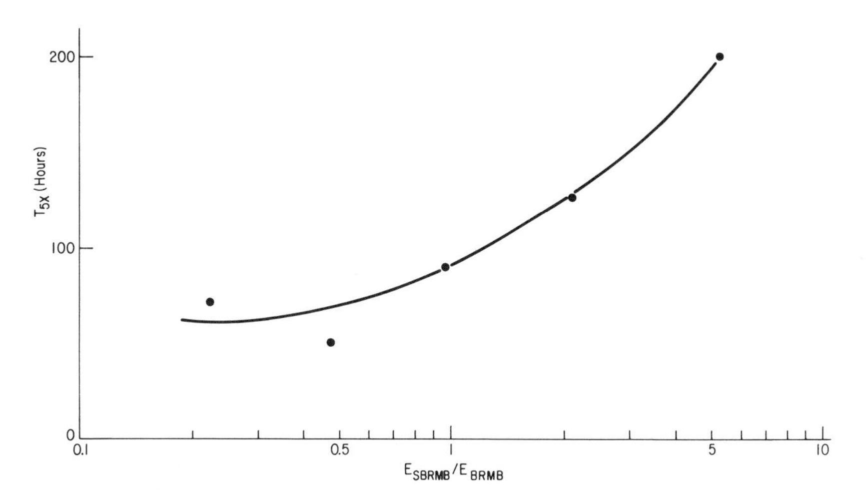

Figure 14. System I $T_{5\times}$ *of various SBR/PBD blend compounds as a function of the modulus ratio of the master batch pairs. (Reproduced with permission from Ref. 36. Copyright 1982,* J. Appl. Polym. Sci.)

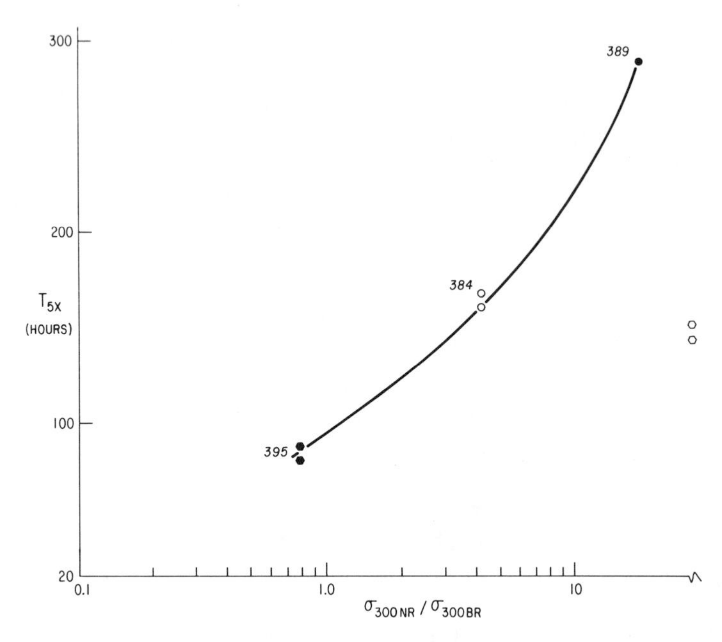

Figure 15. System II $T_{5\times}$ *of various NR/PBD blend compounds as a function of the modulus ratio of the master batch pairs. Key:* ⬡*, CPD 381 (free carbon black mixing);* ○*, CPD 384;* ●*, CPD 389; and* ⬢*, CPD 395.*

Just for the purpose of illustration, a semilogarithmic plot of crack length versus time (or number of cycles) of the SBR–PBD blends is shown in Figure 16, where the value of n is equal to 1. Equation 5, in general, can fit the data, except in the region where the crack is small.

From Figure 16, the order of the constant A is as follows:

$$A_{378} < A_{373} < A_{356} < A_{368} \tag{6}$$

From Equation 5, a small A implies better crack growth resistance. Our results, as indicated in Equations 2 and 6, showed that a smaller value of A was obtained for the compound in which the modulus ratio E_{SBRMB}/E_{PBDMB} was larger.

The form of Equation 5 was probably first proposed by Greensmith (*28*) for the system containing one rubber phase. No single values of A can be obtained in the blends, presumably because of the presence of heterogeneity as reflected by the nonuniform distribution of carbon black in the rubber blend. As will be discussed later, the presence of heterogeneity, which gives a stress concentration and serves as an energy dissipator, may be more significant in affecting the crack growth of rubber blends.

Analogy of Rubber Blends to Rubber-Modified Thermoplastics. An interesting feature, as indicated in Figures 10–16 and Equations 2 and 6, is

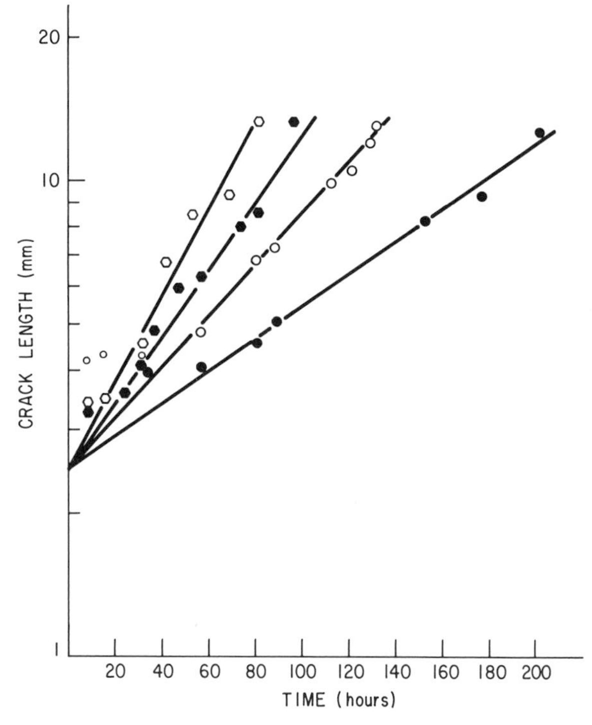

Figure 16. Semilog relation of crack length of SBR/PBD blend compounds vs. time of cycling (System I). Key: ●, CPD 378; ○, CPD 373; ⬡, CPD 368; and ⬢, CPD 356. (Reproduced with permission from Ref. 36. Copyright 1982, J. Appl. Polym. Sci.)

the importance of the relative moduli of the individual rubber master batches. The crack-growth resistance of the rubber blends may be significantly improved if one rubber phase is of a lower modulus. This result leads us to speculate, as far as the crack growth is concerned, on the analogy of rubber blends to rubber-modified thermoplastics. Brittle polymers can be converted into high impact materials by the addition of a low-modulus rubber phase. The low-modulus rubber phase can dissipate large amounts of energy. A general review on this subject is available (*48–51*).

If carbon black loaded rubber blends, to some extent, can be considered analogous to the rubber-modified thermoplastics, then the following criteria should be followed during mixing of rubber blends to improve the crack-growth resistance of the blend:

1. One phase is high modulus, and the other phase is low modulus. In the practical and useful rubber blend products, the higher modulus phase should be the major rubber phase (*see* Figure 17a).
2. A desirable blend morphology to improve the crack-growth

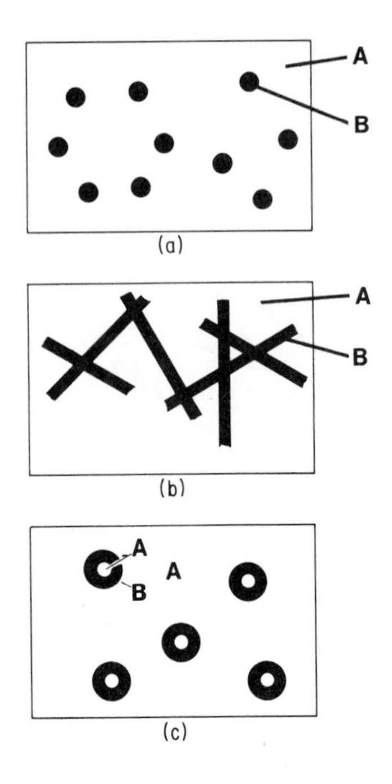

Figure 17. Some desirable blend morphology to improve crack-growth resistance of rubber blend compounds Key: A, major component, high modulus phase; and B, minor component, low modulus phase.

resistance is that both of the high and low modulus master batch phases are continuous (*see* Figure 17b).

3. If a discreted, dispersed morphology is inevitable, the crack-growth resistance of rubber blends can also be further improved via occlusion dispersion, as illustrated in Figure 17c.
4. There should be good adhesion between the high and low modulus phases.

This "crude" analogy implies that, besides the relative distribution of carbon black in different rubber phases, the crack-growth resistance of rubber blends must be considered with respect to the factors operative in other phenomena in the multicomponent polymer systems. These factors include the type, shape, and aggregation of filler particles (*52, 53*); the state of dispersion of filler particles (*34*); the type, shape, and orientation of the minor rubber phase; the properties and the thickness of the interfaces between rubber–rubber and rubber–filler (*20*); and the tendency to co-cure between the rubber phases (*38, 54–57*). Obviously more work should be done to examine the validity of this analogy.

Effect of Stress Concentrators on Crack Growth. The presence of stress concentrators, resulting from heterogeneity, is another vital factor in determining the crack growth of the materials. For example, the stress at the tip of a crack is concentrated according to the equation (*58*):

$$\sigma_m \doteq \sigma_o \left\{ 1 + 2\left(\frac{a}{R}\right)^2 \right\} \tag{7}$$

where σ_o is the applied stress, σ_m is the maximum stress at the crack tip (radius of curvature R), and a is the length of the crack. If the mixed rubber blends follow the criteria discussed, the crack may be retarded because the radius of curvature of the second, dispersed rubber phase may be greater than that of the crack tip. Thus, the intensity of the stress concentration is decreased. Wang et al. (*59*) studied the fracture of adhesive joints and reported that the stress concentration that developed in the adhesive crack tip is an inverse function of the modulus ratio of the joint modulus to the modulus of the adhesive. This result may be further employed to explain the effect of stress concentrator on crack growth. As the crack tip passes the low-modulus dispersed phase, the stress concentration at the crack tip is significantly relieved. Hence, the presence of the lower modulus dispersed phase should retard the crack propagation and increase the fatigue life of the rubber blends. Thus, the presence of intermediate low-modulus layers (Figure 17c) would reduce the level of stress in the materials, and in turn, would increase crack-growth resistance.

Heat Buildup. Another important property in determining the performance of rubber blend products is heat buildup, which is related to mechanical damping. Figure 18 shows the heat buildup of System I, SBR/PBD blends, and the relationship of heat buildup to the crack-growth resistance $T_{5\times}$. Obviously, the heat buildup of our SBR/PBD blends is strongly influenced by the distribution of carbon black. The compound with low heat buildup also has better crack-growth resistance.

Figure 19 shows the effect of the location of carbon black on ΔT of the NR/PBD blends. The ΔT of the blends, however, is not sensitive to the location of carbon black in the individual rubber phases.

Now, one may ask why the location of carbon black affects ΔT of SBR/PBD blends, but has no effect on NR/PBD blends. This difference in effect could be a result of the loading of the carbon black; the carbon black loading is 85 phr in the SBR/PBD compound and only 45 phr in the NR/PBD compound. The amount of carbon black aggregates in the rubber compounds is a function of the carbon black loading and the wetting characteristics of the rubber to the carbon black surface (*35, 53*). Also, the presence of carbon black aggregates significantly increases the heat buildup (*35*). So, at low carbon black loading (45 phr in NR/PBD), ΔT is low and not sensi-

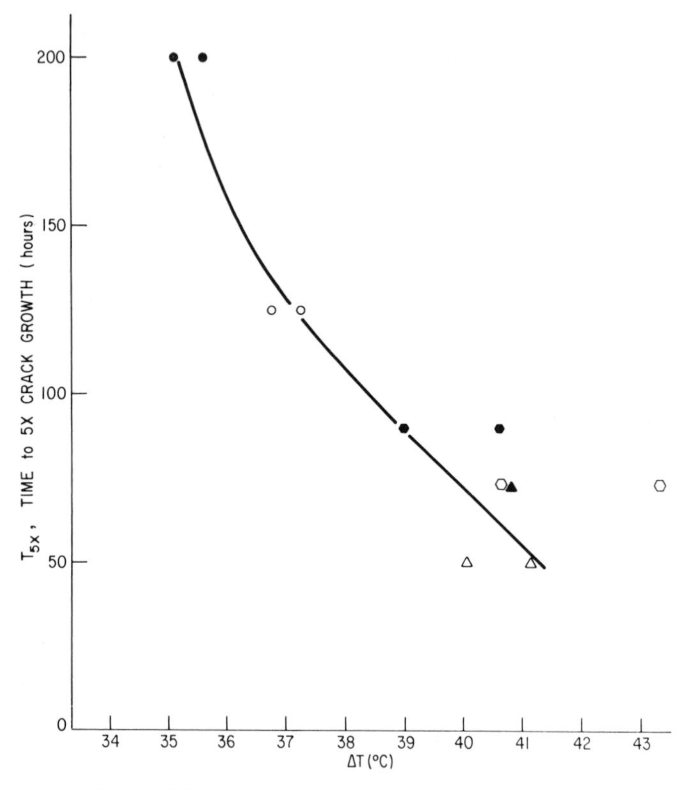

Figure 18. System I $T_{5\times}$ *of the SBR/PBD compounds as a function of heat buildup* (ΔT). *Key:* ▲, *CPD 350;* ⬢, *CPD 356;* △, *CPD 362;* ⬡, *CPD 368;* ○, *CPD 373; and* ●, *CPD 378. (Reproduced with permission from Ref.* 36. *Copyright 1982,* J. Appl. Polym. Sci.)

tive to the location of carbon black in the rubber phases, primarily because of the smaller amount of carbon black particle–particle friction in the mixed compounds. As the carbon black loading increases (85 phr in SBR/PBD blend) the chance of more carbon black particle–particle contacts increases. So, at higher carbon black loading, ΔT is higher and significantly depends on the location (or distribution) of carbon black in the rubber phases.

Conclusions

In carbon black loaded rubber blend compounds, crack growth was found to be sensitive to the heterogeneous distribution of carbon black. A better crack-growth resistant compound can be obtained with the blends containing proportionately more carbon black in the major rubber phase.

Blends that are cross-mixed for shorter periods of time show better crack-growth resistance than blends cross-mixed for longer periods. This

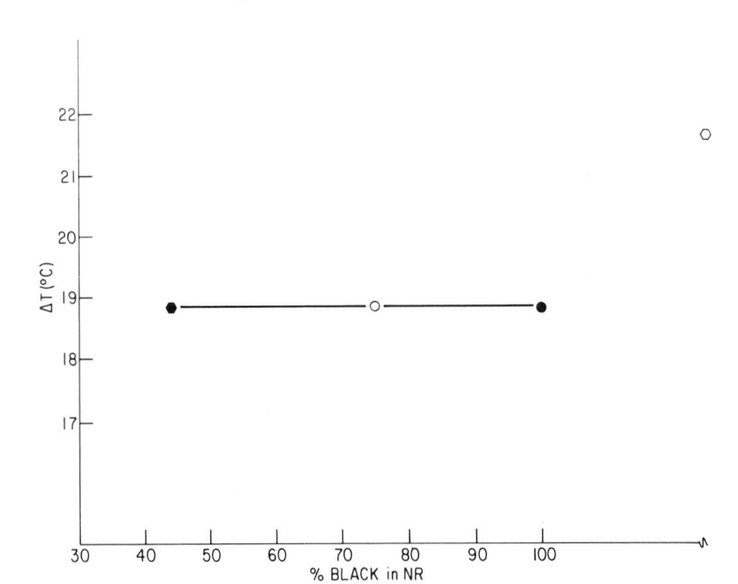

Figure 19. Heat buildup (ΔT) of the NR/PBD blend compounds as a function of the distribution of carbon black (System II). Key: ⬡, CPD 381 (free carbon black mixing); ○, CPD 384; ●, CPD 389; and ⬢, CPD 395.

finding further suggests the importance of the blend morphology on product performance. Furthermore, the orientation of the elongated, dispersed rubber phase should lead to blends with significant anisotropic properties.

Heat buildup of rubber blends may also be affected by heterogeneous carbon black distributions. However, heat buildup is also influenced by the amount of carbon black in the blends.

Acknowledgments

I thank The BFGoodrich Company for support in this research. I acknowledge the Polymer Processing Group of Corporate Technical Support for performing the tedious mixing experiments, and Physical Testing for performing the ring flex test. Appreciation is also extended to Jack Morris and Lance Jameson for the Instron tensile test. Particular thanks are due to Chloe Singleton, Tom Kelly, and Liane Schneeweis for the microscopy work.

Nomenclature

SBR Styrene–butadiene copolymer
NBR Butadiene–acrylonitrile copolymer
PBD *cis*-Polybutadiene
NR Natural rubber

σ_u Uncorrected stress
ϵ Elongation
ϵ_B Elongation at break
E Tensile elastic modulus
L Crack length
M 300% modulus
A,n Constants
σ_m Maximum stress at crack tip
σ_o Applied stress

Literature Cited

1. Keskkula, M. M., ed., "Polymer Modification of Rubbers and Plastics", Appl. Polym. Symp. No. 7, Wiley (Interscience), New York, 1969.
2. Bruins, P. F., ed., "Polyblends and Composites", Appl. Polym. Symp. No. 15, Wiley (Interscience), New York, 1970.
3. "Multicomponent Polymer Systems"; Platzer, N. A. J., ed.: ADVANCES IN CHEMISTRY SERIES No. 99; American Chemical Society: Washington, D.C., 1971.
4. VanOene, H. *J. Colloid Interface Sci.* **1972**, *40*, 448.
5. Starita, J. M. *Trans. Soc. Rheol.* **1972**, *16*, 339.
6. Nielsen, L. E. "Mechanical Properties of Polymers and Composites"; Marcel Dekker: New York, 1974.
7. Lee, B. L.; White, J. L. *Trans. Soc. Rheol.* **1974**, *18*, 469.
8. Lee, B. L.; White, J. L. *Trans. Soc. Rheol.* **1975**, *19*, 481.
9. "Copolymers, Polyblends, and Composites"; Platzer, N. A. J., ed.; ADVANCES IN CHEMISTRY SERIES No. 142; American Chemical Society; Washington, D.C., 1975.
10. Manson, J. A.; Sperling, L. M. "Polymer Blends and Composites"; Plenum: New York, 1976.
11. "Polymer Blends"; Paul, D. R.; Newman, S., eds.; Academic Press: New York, 1978.
12. "Science and Technology of Rubber"; Eirich, F., ed.; Academic Press: New York, 1978.
13. "Multiphase Polymers"; Cooper, S. L.; Ester, G. M., eds.; ADVANCES IN CHEMISTRY SERIES No. 176; American Chemical Society: Washington, D.C., 1979.
14. Olabisi, O.; Robeson, L. M.; Shaw, M. T. "Polymer-Polymer Miscibility"; Academic Press: New York, 1979.
15. Han, C. D. "Multiphase Flow in Polymer Processing"; Academic Press: New York, 1981.
16. Corrish, P. J.; Powell, B. D. W. *Rubber Chem. Technol.* **1974**, *47*, 481.
17. Sircar, A. K.; Lamond, T. G.; Pinter, P. E. *Rubber Chem. Technol.* **1974**, *47*, 48.
18. Gessler, A. M.; Hess, W. M.; Medalia, A. I. *Plastics and Rubber: Processing* **1978**, *3*, 37.
19. McDonel, E. T.; Baranwal, K. C.; Andries, J. C. In "Polymer Blends"; Paul, D. R., Newman, S., Eds: Academic Press, 1978; Ch. 19.
20. Lee, B. L.; Singleton, C. *J. Appl. Polym. Sci.* **1979**, *24*, 2169.
21. Lee, B. L. *Polym. Eng. Sci.* **1981**, *21*, 294.
22. Dillon, J. H. In "Advances in Colloid Science"; Interscience Publishers: New York, 1950; Vol. 3, p. 219.
23. Beatty, J. R.; Juve, A. E. *Rubber Chem. Technol.* **1965**, *38*, 719.
24. Thomas, A. G. *J. Polym. Sci.* **1958**, *31*, 467.

25. Lindley, P. B.; Thomas, A. G. Proc. Fourth Inter. Rubber Conference, London, 1962; p. 268.
26. Gent, A. N.; Lindley, P. B.; Thomas, A. G. *J. Appl. Polym. Sci.* **1964**, *8*, 455.
27. Lake, G. L.; Lindley, P. B. *J. Appl. Polym. Sci.* **1964**, *8*, 707.
28. Greensmith, H. W. *J. Appl. Polym. Sci.* **1964**, *8*, 1113.
29. Lake, G. L. *Rubber Chem Technol.* **1972**, *45*, 309.
30. Eirich, F. R.; Smith, T. L. In "Fracture"; H. Liebowitz, ed.; Academic Press: New York, 1972; Vol VII, Ch. 7.
31. Eirich, F. R. In "Colloques Internationaux du C.N.R.S.", No. 231—Interactions entre les Elastomeres et les Surfaces Solides Ayant une Action Renforcante, September 24–26, 1973.
32. Beatty, J. R. *Rubber Chem. Technol.* **1964**, *37*, 1341.
33. Beatty, J. R. presented at the Int. Rubber Symp., Rubber Division, American Chemistry Society, San Francisco, CA (October 5–8, 1976); also *Rubber Chem. Technol.* **1977**, *50*, 429.
34. Lee, B. L. *Rubber Chem. Technol.* **1979**, *52*, 1019.
35. Lee, B. L. presented at the 39th Society of Plastics Engineers ANTEC Meeting, May 1981, Boston, MA.
36. Lee, B. L. *J. Appl. Polym. Sci.* **1982**, *27*, 3379.
37. Lee, B. L. *Polym. Eng. Sci.* **1982**, *22*, 902.
38. Lee, B. L. "Multiple Polymer Processing: Controlled-Ingredient-Distribution Mixing and its Effect on the Covulcanization of Elastomer Blend Compounds"; presented at the 55th Annual Meeting of the Society of Rheology, Knoxville, TN, October 16–20, 1983; also to appear in *J. Rheology*.
39. Lee, B. L.; Singleton, C. J. *J. Macromol. Sci., Phys.*, in press.
40. Monsanto Rheometer 100, Operation Manual.
41. Redding, R. B.; Smith, D. A. *J. Polym. Sci., Part C* **1970**, *30*, 491.
42. Wolff, S.; Westlinning, M. *Rubber J.* **1968**, *150*, 24.
43. ASTM D623, Method A.
44. Scott, C. E.; Callan, J. E.; Hess, W. M. *J. Rubber Res. Inst. Malays.* **1969**, *22*, 242.
45. Hess, W. H.; Chirico, V. E. *Rubber Chem. Technol.* **1977**, *50*, 301.
46. Marsh, P. A.; Voet, A.; Price, L. O.; Mullens, T. J. *Rubber Chem. Technol.* **1968**, *41*, 344.
47. Auer, E. E.; Doak, E. W.; Schaffner, I. J. *Rubber Chem. Technol.* **1958**, *31*, 185; also *Rubber World* **1957**, *135*, 876.
48. Williams, Mann, J. R. In "Physics of Glassy Polymers"; Haward, R. N. ed.; Wiley: New York, 1973, Ch. 8.
49. Haward, R. N. *Br. Polym. J.* **1970**, *2*, 209.
50. Bragaw, C. G. In "Multicomponent Polymer Systems"; Platzer, N. A. J., ed.; ADVANCES IN CHEMISTRY SERIES No. 99; American Chemical Society: Washington, D.C.; 1971, p. 86.
51. Kambour, R. P. *J. Polym. Sci., Macromol. Rev.* **1973**, *7*, 1.
52. Nielsen, L. E.; Lee, B. L. *J. Compos. Mater.* **1972**, *6*, 136.
53. Lee, B. L.; Nielsen, L. E. *J. Polym. Sci., Polym. Phys. Ed.* **1977**, *15*, 683.
54. Gardiner, J. B. *Rubber Chem. Technol.* **1968**, *41*, 1312.
55. Gardiner, J. B. *Rubber Chem. Technol.* **1969**, *42*, 1058.
56. Gardiner, J. B. *Rubber Chem. Technol.* **1970**, *43*, 370.
57. Woods, M. E.; Mass, T. R. In "Copolymers, Polyblends and Composites"; ADVANCES IN CHEMISTRY Series No. *142*, Platzer, N. A. J., ed.; American Chemical Society: Washington, D.C., 1975, p. 386.
58. Horsley, R. A. *Appl. Polym. Symp.* **1971**, *17*, 117.
59. Wang, S. S.; Mandell, J. F.; McGarry, F. J. Research Report R76–1, "Fracture of Adhesive Joints", Department of Materials Science and Engineering, MIT.

RECEIVED for review January 20, 1983. ACCEPTED October 17, 1983.

13

Mechanical Behavior of Polyolefin Blends

JAISIMHA KESARI and RONALD SALOVEY

Departments of Chemical Engineering and Materials Science,
University of Southern California, Los Angeles, CA 90089-1211

Ternary blends of polyethylene (PE), polypropylene (PP), and ethylene-propylene-diene terpolymer (EPDM) showed variations in tensile and impact behavior after periods of compression molding. These variations resulted from changes in structure. The modulus of ternary blends decreased during 10-60 min of compression molding time and was thereafter constant and characteristic of composition. The energy to break determined from instrumented impact measurements on ternary blends depended linearly on modified fracture area in accordance with the behavior of brittle materials. The critical strain energy release rate derived from impact studies showed large variations after the early stages of compression molding. Changes in mechanical behavior are associated with rearrangements of phase morphology in the blends. Composite droplets of EPDM and PE, initially dispersed in a PP matrix, separate into PE droplets that are surrounded by EPDM and isolated from the PP matrix on subsequent compression molding.

TERNARY BLENDS of polypropylene (PP), ethylene-propylene-diene terpolymer (EPDM), and polyethylene (PE) are complex mixtures of immiscible phases, crystalline spherulites, and amorphous domains. The detailed morphology after processing results from kinetic factors, such as the rates of crystallization and vitrification, as well as thermodynamic considerations, such as the solubility and adhesion of phases. Accordingly, blend structure is sensitive to thermal and shear history.

Ho et al. (*1*) reported that the phase morphology of compression molded ternary polyolefin blends depends on molding time. A blend containing equal weights of EPDM and PE was mixed in a twin screw extruder, pelletized, dry blended, and injection molded with PP pellets. Injection molded samples were then compression molded for various periods of time at 213 °C. After short compression molding times, composite droplets containing EPDM and PE were dispersed in a PP matrix. After increasing the heating time under compression in the mold, EPDM separated from

0065-2393/84/0206-0211$06.00/0

the PE and concentrated at the edges of PE droplets. The PE was surrounded and isolated from the PP matrix by the EPDM. Rubbery polymers such as EPDM are frequently added to a glassy or semicrystalline polymer to enhance the impact resistance (*2*). Morphological changes that affect the size and composition of the dispersed phase should alter the mechanical properties. Here, we report on the variation with compression molding time of the tensile and instrumented impact behavior of compression molded ternary blends.

Experimental

Sample Preparation. MATERIALS. The polymers are commercial grades of isotactic PP (Profax 6523, Hercules), EPDM (Nordel 1070, Du Pont), and PE (Alathon 7030, Du Pont). An antioxidant [Ethyl 330, 1,3,5-trimethyl-2,4,6-tris-(3,5-di-*tert*-butyl-4-hydroxybenzyl) benzene, Ethyl Corp.] was added to retard oxidative degradation.

PROCESSING. A blend containing 50 wt% PE, 49.8% EPDM, and 0.2% of antioxidant Ethyl 330 was prepared in a 2.5-in. twin screw extruder (Werner and Pfliederer). This master batch was dry blended with PP pellets. Various compositions were prepared and injection molded (Arburg 200 U) at a melt temperature of 200 and mold temperature of 50 °C. Injection molded blends were compression molded at 213 °C and 5.5 MPa at various times from 10 min to 5 h.

Testing. TENSILE. The tensile behavior of ternary blends was examined with a floor model Instron Corporation Universal testing machine at room temperature and at a cross head speed of 0.2 in./min. Stress–strain curves were prepared, and the initial modulus was carefully determined by testing a minimum of five samples for each composition. Tensile bar geometry conformed to that specified in ASTM D-638 (*3*).

INSTRUMENTED IMPACT. The instrumented impact tester was developed (*4*), and measurements were performed at the Du Pont Experimental Station, Wilmington, Delaware. The polymer specimen was a Charpy impact bar, as specified in ASTM D-256 (*3*). The Charpy test uses a three-point loading of the specimen, which is notched to provide a stress concentrator. The notch angle and base radius were fixed at 40 ± 1° and 0.25 ± 0.05 mm, respectively. The sample was placed on a base beneath a hollow tube in which a striker (tup) was allowed to drop. The tup was attached to an accelerometer, a transducer that sensed the instantaneous acceleration as a function of time. The mass of the tup was controllably varied. Double integration yielded the instantaneous displacement, and force–displacement plots were automatically produced. From the force–displacement curve, we obtained the total energy to break (E_m), which is composed of initiation (E_i) and propagation (E_p) energies. The appearance of this curve depends on whether the failure is brittle or ductile. The E_m was determined for various notch depths and sample dimensions, and the data were treated as described below.

According to the method suggested in ASTM D-256 (*3*), the impact energy is reported as the E_m per unit length of notch. In the conventional Charpy test, a notch of 0.100 in. depth is made in the sample and the energy to break the sample is measured. This value of the energy defines a point on a plot of the E_m versus fracture area. This point is joined to the origin and the resulting line is extrapolated to a fracture area corresponding to 1-in. notch length. The corresponding energy defines a conventional impact number in terms of foot-pounds per inch or Joules per millimeter. There is no a priori reason for assuming that the energy/fracture area line passes through the origin. The conventional impact number is not a fundamental toughness parameter and depends on sample geometry. By using methods based on fracture mechanics (5–7), more sophisticated measures of impact strength were developed. The fracture toughness (K_c) can be determined by measuring the stress required to cause the fracture of specimens. Each specimen contains a central flaw of specified length (5). Another parameter indicating impact strength is G_c, the critical strain energy release rate needed to initiate a crack. This parameter is related to the fracture toughness K_c by

$$G_c = \frac{K_c^2}{E}$$

where E is Young's modulus.

For brittle materials, the energy at fracture is plotted against the modified area ($BD\phi$) for different crack geometries. This plot yields a straight line of slope G_c; B and D refer to the thickness and width of the Charpy bar, respectively, and ϕ may be determined experimentally by measuring the compliance of the sample for various notch depths or may be calculated theoretically (6).

Ductile materials undergo plastic yielding before fracture and a simple elastic analysis cannot be applied. Here we define a parameter J_c, which is equal to G_c, as the critical strain energy release rate for the purely elastic case (6). In general, J_c is applicable to all degrees of plasticity. For ductile materials, E_m is plotted directly against fracture area for various notch depths to give a straight line whose slope is $J_c/2$. Both G_c and J_c are critical strain energy release rates and are fundamental toughness parameters that are material properties, independent of specimen geometry.

Results

The tensile modulus of compression molded ternary blends is plotted as a function of compression molding time for various blend compositions in Figure 1. At each composition, the modulus of injection molded blends exceeded that of compression molded blends. The dependence of E_m, which we determined from instrumented impact tests, on the fracture area for a ductile material is illustrated in Figure 2. We derived G_c from instrumented impact measurements and plotted G_c versus compression molding time for three ternary blend compositions (Figure 3).

Discussion

The initial modulus of all blends decreases with an increase in molding time during the first hour of compression molding (Figure 1). After the first hour, the blends achieve a constant modulus independent of the molding time. Because morphological changes in the blends were observed during this interval (*1*), we infer that the origin of the decrease in modulus lies in the blend structure. Following injection molding and after the initial stages of compression molding, composite droplets of EPDM and PE are dispersed in a PP matrix. This distribution of phases produces the maximum tensile

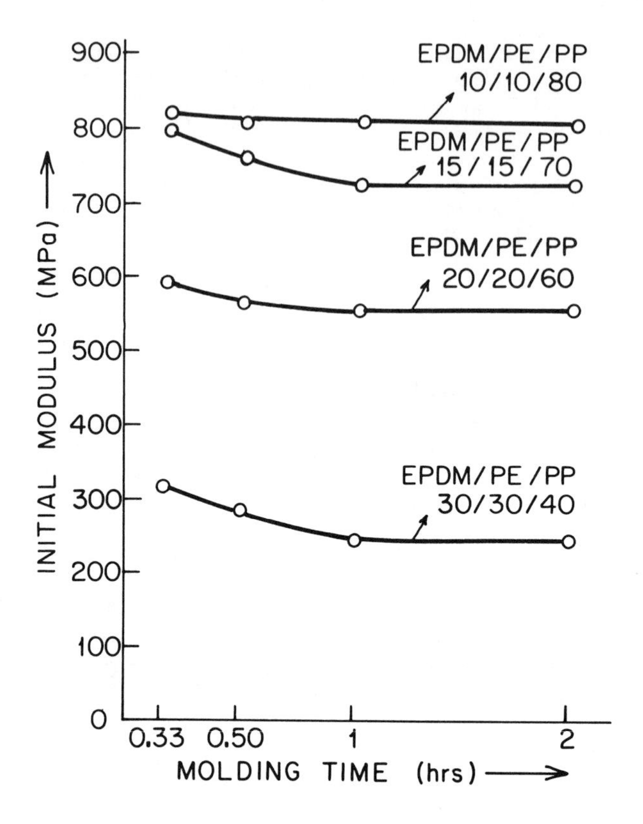

Figure 1. The variation of tensile modulus with compression molding time for ternary blends of specified composition.

modulus for ternary blends. When the compression molding time is increased, EPDM migrates out of the PE phase and occupies the interface between the PE droplet and PP matrix. Because the inclusion of low modulus EPDM reduces the modulus of the PP matrix, it is not surprising that the effect of the EPDM on PP is enhanced by phase separation from PE. Also, most likely the modulus decreases because of a relaxation of molecular and phase orientation produced by injection molding.

The appearance of force/displacement diagrams from instrumented impact tests differentiates between ductile and brittle materials. The total area under the force–displacement curve is E_m. Ductile materials yield before fracturing. The initially linear dependence of force on displacement deviates from linearity during the initiation stage and defines a yield point. Also, ductile materials have much higher propagation energies than brittle

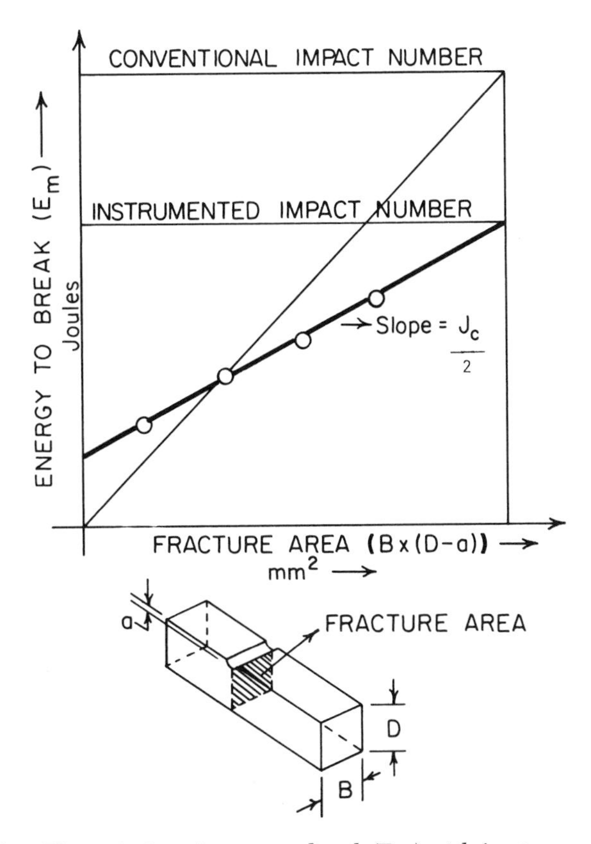

Figure 2. The variation of energy to break (E_m) *with fracture area as measured by instrumented impact testing for a ductile polymer.*

materials. Thus, pure PP evidences brittle behavior by not showing a yield point and by giving a relatively low propagation energy. On the other hand, PE shows a yield point and a large propagation energy. The force–displacement curves for ternary blends exhibit yielding and show some propagation energy. Accordingly, such blends were treated as ductile materials. To determine the impact behavior of the ternary blends, E_m was plotted against the fracture area for four notch depths. Unfortunately, poor straight lines and negative intercepts on the energy axis resulted for ternary blends, unlike the typically ductile behavior illustrated in Figure 2. However, plots of E_m versus $BD\phi$ (modified fracture area) yielded straight lines having excellent correlation coefficients and positive intercepts. This result suggests that the fracture of ternary blends is not really ductile, and that we are fracturing the brittle matrix of PP. Although the ternary blends do not break at very high rubber (EPDM) contents, ternary blends of EPDM/PE/PP are apparently "semibrittle" and can be treated by methods applied to brittle materials. From the slopes of the plots of E_m versus $BD\phi$,

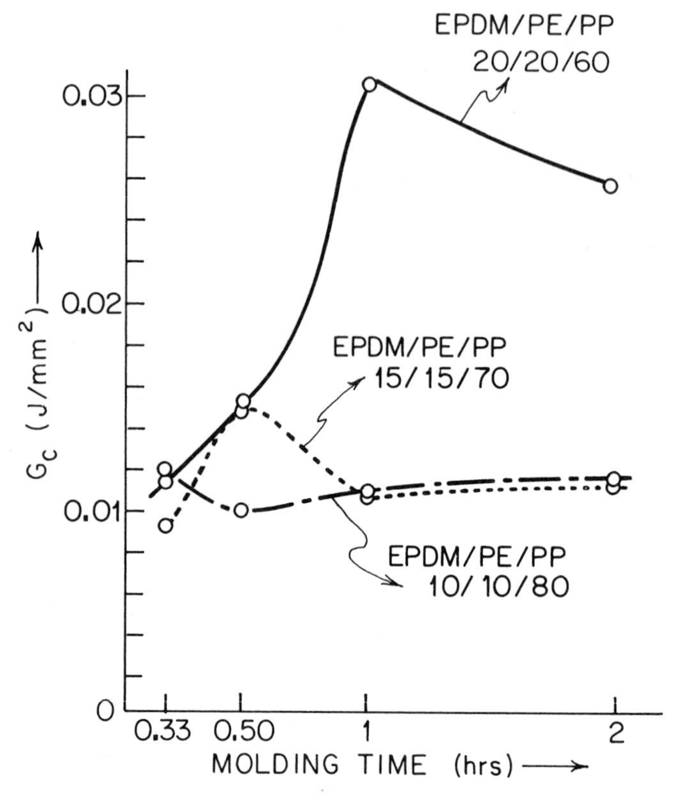

Figure 3. The critical strain energy release rate (G_c) *as a function of compression molding time for ternary blends of specified composition.*

values of G_c are derived for plane strain fracture. Figure 3 is a plot of G_c as a function of compression molding time for various compositions of ternary blends. Changes in G_c occur for short compression molding times. After 1 h of compression molding, G_c levels off.

Stehling et al. (*8*) suggested that a fracture surface might not be representative of the internal morphology of blends and that surfaces for scanning electron microscopy should be generated alternatively by microtoming at low temperatures. Because, in a binary blend of a rubbery ethylene–propylene copolymer in PP, the bulk concentration of the rubber exceeds its concentration on the fracture surface, they conclude that the fracture tends to propagate through the PP matrix rather than through the frozen rubber particles (*8*). Nonetheless, the internal morphology of ternary blends deduced from studies involving microtoming at low temperature is in basic agreement with the structure inferred from low temperature fracture. Stehling et al. suggested (*8*) that, in these ternary blends, PE and ethylene–propylene copolymer form a composite dispersed phase in a PP matrix in which, after heating to the molten state, PE droplets become surrounded by the rubbery copolymer. If the ethylene–propylene copolymer and PE are premixed before addition to PP, the composite droplets show an interpenetrating structure (*8*). However, the smaller dispersed particles tend to phase separate into PE droplets surrounded by the ethylene–propylene copolymer.

In agreement with our observations, D'Orazio et al. reported (*9*) that scanning electron micrographs of fracture surfaces from ternary blends of PP, PE, and ethylene–propylene copolymers show composite droplets of PE and ethylene–propylene copolymer in a PP matrix. Similarly, studies (*10, 11*) of the toughness of blends of PP and EPDM by instrumented impact and dynamic mechanical measurements support many of our observations. Load–displacement curves from instrumented impact are characteristic of brittle fracture. Notched specimens are elastic bodies and are amenable to the application of linear elastic fracture mechanics. A fracture energy or critical strain energy release rate, G_c, was calculated (*10, 11*). Also, the impact strength of the blends was shown to increase with dynamic mechanical loss (*10, 11*). The internal morphology of these binary blends of PP and EPDM was established by fracturing at − 40 °C, etching with heptane vapor, and examining by scanning electron microscopy (*12*). The average diameter of the EPDM particles was approximately 0.6 μm, as long as the melt viscosity of the EPDM did not exceed that of the PP by more than a factor of 2.7 (*12*).

Although the polyolefin phases in the ternary blends are immiscible, considerable interaction occurs between the phases. For example, the effect of amorphous ethylene–propylene copolymers on spherulitic crystallization in isotactic PP has been described (*13*), and considerable synergism in mechanical behavior has been reported (*14*) for PP–PE binary blends. Moreover, the mechanical properties of such binary blends are very sensitive to processing (*14*). Unlike binary blends of PE and PP, which do not evidence adhesion between dispersed PE droplets and PP matrix, ternary blends with a random ethylene–propylene copolymer indicate adhesion between the composite droplets and the matrix (*9*). The copolymer may act as an emulsifier by promoting adhesion between PE and PP (*1, 9*) and thereby facilitating multicraze formation and shear yielding in fracture (*2*). The rubber particle size and interfacial adhesion are important in determining the impact strength (*8*). Indeed, the mechanical properties of rubbery blends are particularly sensitive to interfacial adhesion (*15*). Evidence in support of our observation that ethylene–propylene copolymers prefer a PE environment over that of PP was also reported in binary blend studies of these polyolefins (*9*).

Conclusions

Ternary blends of PP, EPDM, and PE are composite droplets of EPDM and PE dispersed in a PP matrix. During compression molding at elevated temperature the dispersed phase separates so that EPDM concentrates at the edges of PE droplets. As a result of these structural changes in the phase morphology of ternary polyolefin blends during the initial hour of compression molding at constant temperature, the tensile modulus decreases. The fracture of these blends is amenable to treatment by linear elastic fracture mechanics if we assume brittle fracture. The critical strain energy release rate (G_c) varies for short compression times and levels off after 1 h of compression molding.

Nomenclature

B Specimen thickness
D Specimen width
E Young's modulus
E_m Total energy to break
E_i Initiation energy
E_p Propagation energy
G_c Critical strain energy release rate
J_c Plastic work parameter
K_c Fracture toughness
ϕ Calibration function (*6*)

Acknowledgments

The financial support and encouragement of the Hydril Company and of the Los Angeles Rubber Group (TLARGI) are acknowledged with gratitude. We sincerely appreciate the assistance and guidance of researchers from the Du Pont Company as well as the opportunity to use experimental facilities at the Du Pont Experimental Station, Wilmington, Delaware, during the summer of 1981.

Literature Cited

1. Ho, W.; Salovey, R. *Polymer Eng. Sci.* **1981**, *21*, 839.
2. Bucknall, C. B. "Toughened Plastics"; Applied Science Publishers Ltd.: London, 1977.
3. "ASTM Standards, Part 35."; American Society of Testing and Materials: Philadelphia, 1981.
4. Adams, G. C., presented at the 21st Meeting of the Rubber Division, ACS, Philadelphia, Penn., 1982.
5. Williams, J. G. *Polymer Eng. Sci.* **1977**, *17*, 144.
6. Plati, E.; Williams, J. G. *Polymer Eng. Sci.* **1975**, *15*, 470.
7. Williams, J. G.; Birch, M. W. *Fracture* **1977**, *1*, ICF 4, Waterloo.
8. Stehling, F. C.; Huff, T.; Speed, C. S.; Wissler, G. *J. Applied Polym. Sci.* **1981**, *26*, 2693.
9. D'Orazio, L.; Greco, R.; Mancarella, C.; Martuscelli, E.; Ragosta, G.; Silvestre, C. *Polymer Eng. Sci.* **1982**, *22*, 536, Figure 6C.

10. Karger-Kocsis, J.; Kiss, L.; Kuleznev, V. N. *Acta Polym.* **1982**, *33*, 14.
11. Karger-Kocsis, J.; Kuleznev, V. N. *Polymer* **1982**, *23*, 699.
12. Karger-Kocsis, J.; Kallo, D.; Kuleznev, V. N. *Acta Polym.* **1981**, *32*, 578.
13. Martuscelli, E.; Silvestre, C.; Abate, G. *Polymer* **1982**, *23*, 229.
14. Bartlett, D. W.; Barlow, J. W.; Paul, D. R. *J. Applied Polym. Sci.* **1982**, *27*, 2351.
15. Hamed, G. R. *Rubber Chem. Technol.* **1982**, *55*, 151.

RECEIVED January 20, 1983. ACCEPTED July 18, 1983.

14

Model Studies of Rubber Additives in High-Impact Plastics

MAURICE MORTON, M. CIZMECIOGLU,[1] and R. LHILA[2]
Institute of Polymer Science, The University of Akron, Akron, OH 44325

Polymer latex blends were used to prepare dispersions of rubber, ranging from 5% to 20% by weight, in rigid thermoplastics. The three thermoplastics used were polystyrene (PS), poly(vinyl chloride) (PVC), and styrene–acrylonitrile copolymer (SAN), and the three types of rubber used were polybutadiene (PBD), styrene–butadiene rubber (SBR), and polychloroprene (PCP). We studied the effect of particle size and rubber–plastic bonding on the impact resistance of the thermoplastic. With PS, which fractures by a crazing mechanism, no optimum was found in the rubber particle size below 1.2 μm, which represented the largest particle size possible by emulsion polymerization; the impact strength increased monotonically with particle size. With PVC, which fractures by a shear yielding mechanism, an optimum was found at a particle size of about 200 nm. For the SAN, a broad plateau maximum was found, ranging from 300 to 800 nm. This plateau was presumably a result of the presence of both types of fracture mechanisms. The presence of chemical bonds between the rubber and the plastic greatly enhanced the impact strength in all cases.

THE EFFECT OF FINELY DIVIDED RUBBER INCLUSIONS on the impact resistance of plastics has been known for some time and represents a well-developed technology (*1*). However, the morphology of these heterophase materials is usually quite complex, and has made unequivocal establishment of any relationship between the physical properties and the morphology difficult. Thus, for example, high-impact polystyrene (HIPS) is usually prepared by polymerization of a styrene monomer containing about 5–10% of dissolved rubber. As the polystyrene forms, phase separation of the rubber occurs as

[1] Current address: Jet Propulsion Laboratory, Pasadena, CA 91103
[2] Current address: Tuck Tapes, Inc., New Rochelle, NY 10801

0065-2393/84/0206-0221$06.00/0

a dispersion in the polystyrene solution. The dispersed rubber particles are swollen with styrene monomer, however, that continues to polymerize and forms a dispersion of polystyrene *within* the dispersed rubber particles. Hence this system ultimately has a three-phase morphology. As an added complication, during the polymerization of the styrene, considerable cross-linking of and grafting to the rubber occurs.

Some consensus has been reached that an optimum in the particle size of the rubber (*2–4*) exists and depends on the mechanism of fracture. Thus, where failure occurs through craze formation, this optimum in rubber particle size is higher than when a shear yielding mechanism is operative. Some studies have also been carried out on the influence of cross-linking of and grafting to the rubber. The conclusions concerning the effect of rubber cross-linking are rather ambiguous (*6–8*), but a small extent of such cross-linking should give optimum properties (*5–9*). Similarly the improved properties resulting from chemical bonding at the rubber–plastic interface have also been noted (*1, 4, 10–12*).

Some time ago we thought that it might be possible to throw more light on these systems by using as a model blends of polymer latex. In these blends the particle size of the rubber inclusions could be established a priori, and cross-linking of the rubber and/or rubber–plastic bonding could be deliberately introduced and measured. This model involved the preparation, by emulsion polymerization, of various synthetic rubber latices, the blending of these latices with the latex of the desired plastic, the coagulation of the blended latex (in isopropyl alcohol to remove emulsifier), and the molding of the dry polymer blend. Three different plastics were used: polystyrene (PS), styrene–acrylonitrile copolymer (SAN), and poly(vinyl chloride) (PVC). Three different rubbers were also used: polybutadiene (PBD), styrene–butadiene rubber (SBR), and polychloroprene (PCP). When cross-linking of the rubber was desired, the rubber latex was treated with a peroxide at elevated temperature, prior to blending. When bonding between the rubber and plastic was desired, a peroxide was incorporated in the latex blend, thus carrying out the grafting (and the unavoidable concomitant rubber cross-linking) during the molding step.

Polystyrene (PS)

The PS used in this work was Dow 788 latex. For this study a series of PBD latices was prepared, with particle sizes in the range of 50 to 400 nm, by using a "seeded" emulsion polymerization technique. Because this method is limited in attaining larger particles, a commercial PBD of particle size 700 nm was also obtained. The latices we prepared had relatively uniform particle sizes. The commercial latex, however, showed a rather wide distribution of sizes. Similarly, a series of SBR latices was prepared with particle sizes ranging from 80 to 450 nm, and a commercial sample with a particle size of 865 nm was also used. The fact that these rubber latex particles became incorporated into the PS without distortion or coalescence was clearly evidenced by transmission electron microscopy of a thin microtomed sample of the molded composite (*see* Figure 1).

The effect of these rubber inclusions on the Izod impact test for PS is shown in Figures 2 and 3, which show clearly that the larger particle sizes of the rubber have a greater effect on impact resistance. An optimum had been suggested in the rubber particle size of about 1–2 μm for high-impact PS, but the range of sizes shown in Figures 2 and 3 is not high enough to corroborate this suggestion. Some other work here with PCP latex has

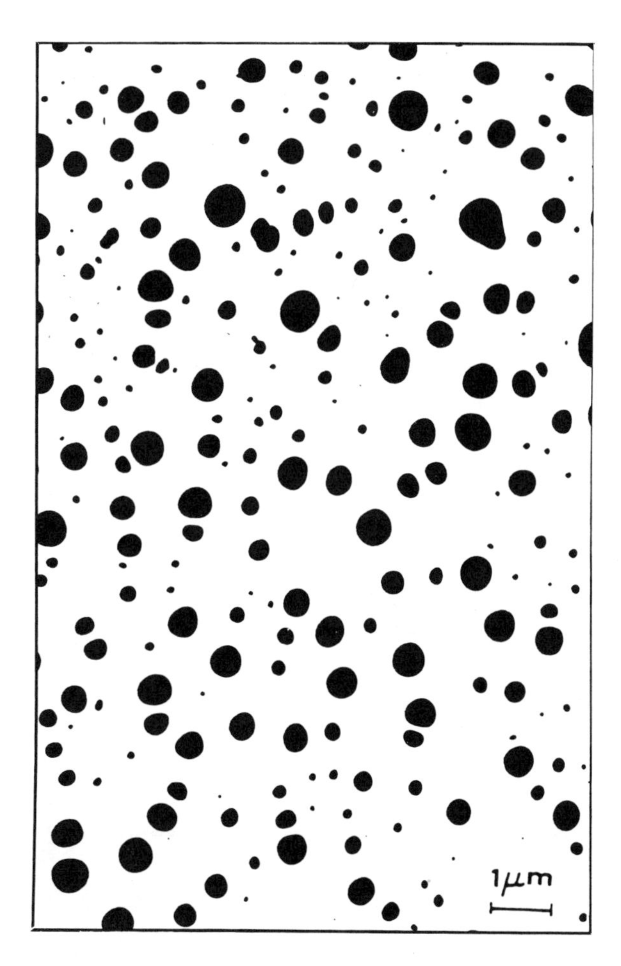

Figure 1. Transmission electron microphotograph of an ultra-thin film of PS (~50 nm), stained with O_2SO_4 and containing 10% SBR (865 nm).

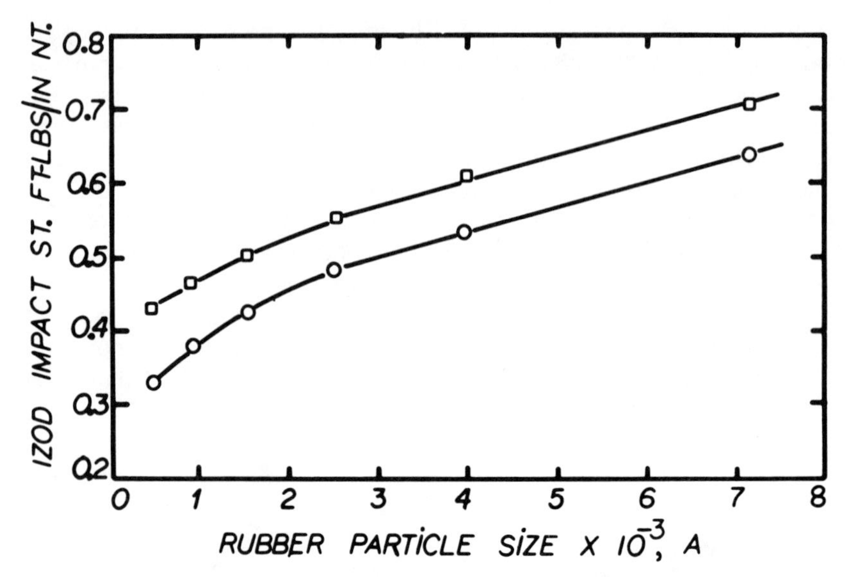

Figure 2. Effect of polybutadiene particle size on impact strength of PS (Impact strength of PS = 0.25 ± 0.01 ft-lbs/in notch). Key: □, PS/PBD 90/10; and ○, PS/PBD 95/5.

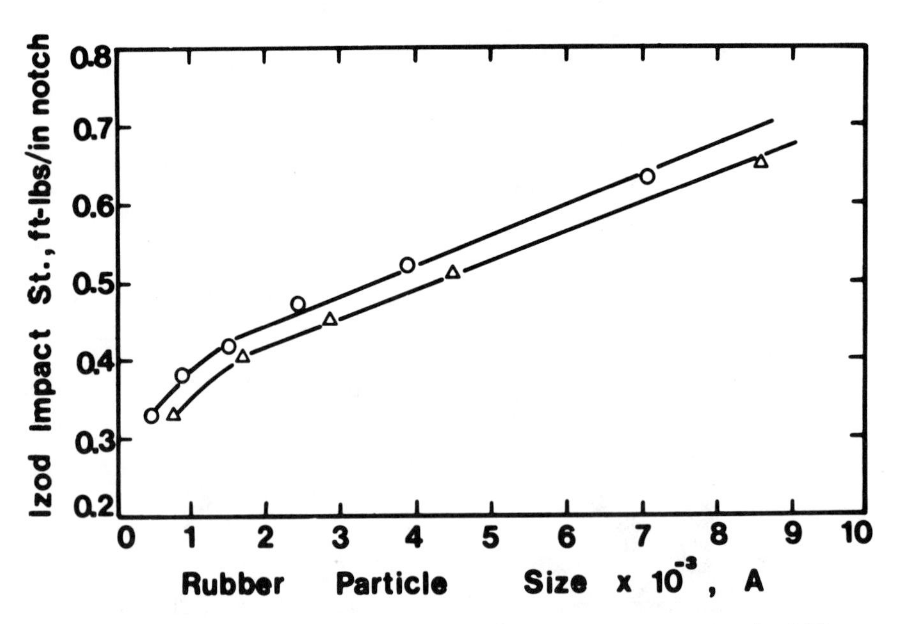

Figure 3. Effect of PBD and SBR particle size on impact strength of PS (Impact strength of PS = 0.25 ± 0.01 ft-lbs/in notch). Key: ○, PS/PBD 95/5; and △, PS/SBR 95/5.

shown that the impact strength of PS is still rising even when the rubber particle size reaches 1.2 μm. Figure 3 also suggests another conclusion: that a rubber having a lower glass transition temperature (T_g), such as PBD, has a slightly greater effect than the SBR.

Figure 4 shows that these rubber-toughened PS samples, prepared by latex blending, also exhibit the expected decrease in stiffness. The softer PBD apparently exerts a greater softening effect than the stiffer SBR.

Figure 5 shows the effect of both grafting and cross-linking of the rubber on the impact strength of PS. Chemical bonding of the rubber to the plastic apparently has a marked effect in improving the impact resistance; thus, at the highest grafting level shown (a "grafting index" of 22, i.e., 22% by weight of PS attached to the rubber), a 50% increase in impact strength is found. On the other hand, cross-linking of the rubber has only a minor effect in improving impact resistance, and excessive cross-linking actually decreases it drastically. Bucknall's conclusions (*5*, *9*) about the desirability of *light* cross-linking may be correct.

Poly(vinyl chloride) (PVC)

PS fractures by a crazing mechanism: formation of microvoids precedes the actual development of cracks (*2*, *14*). Hence the role of the rubber inclusions is both to initiate and to terminate such crazes; these effects are ac-

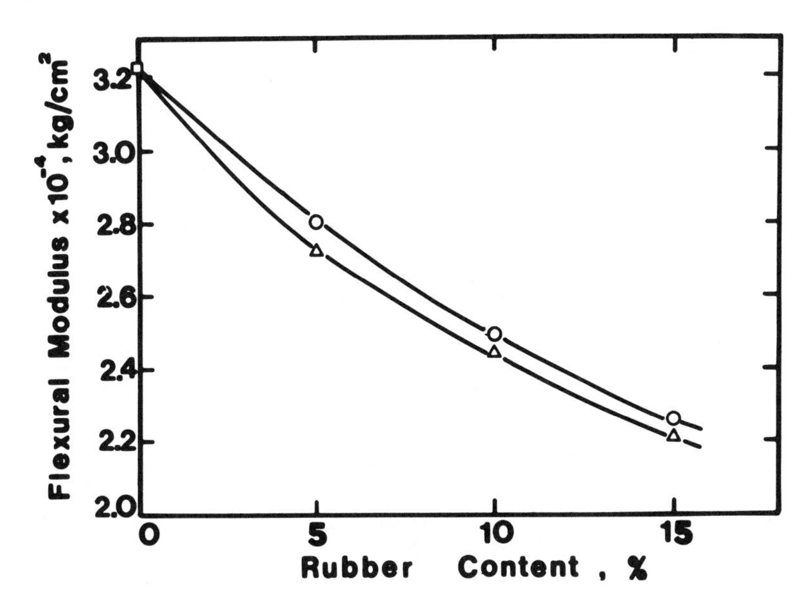

Figure 4. Effect of rubber content on flexural modulus of PS. Key: □, PS; ○, PS/SBR; and Δ, PS/PBD.

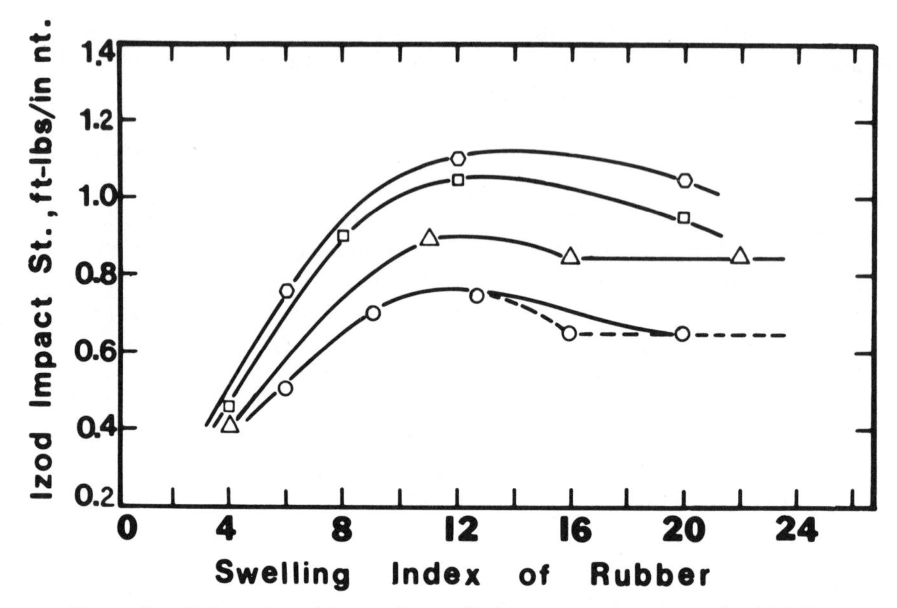

Figure 5. Effect of grafting and cross-linking on impact strength of PS (PS/SBR 90/10). Grafting index (G.I.) = (wt of grafted PS/wt of rubber) × 100. Key: ⬡, G.I. = 22; □, G.I. = 15.5; △, G.I. = 8; and ○, G.I. = 0.

complished most efficiently when the size of the rubber particles is in the range of 2 to 4 μm (*2–4, 11*).

A polymer like PVC, which is more ductile than PS, fails by an entirely different mechanism, known as shear yielding, (i.e., "cold drawing") as first proposed by Newman and Strella (*15*). The optimum particle size for impact resistance should *decrease* with increasing ductility of the plastic (*1, 16*) and this optimum for PVC should be in the range of 0.1 to 0.2 μm (*5–17*).

To investigate the factors affecting the rubber-toughening of PVC by using our model systems, a series of PBD latices was prepared with particle sizes in the range of 70 to 450 nm. In addition, a commercial SBR latex of particle size 940 nm was also used, because such a large particle size could not be produced in our laboratory. The PVC latex was Geon 150X20 made by the Goodrich Chemical Co. The effect of rubber particle size on the Izod impact test is shown in Figure 6. As expected, a maximum in impact resistance was obtained in the rubber particle size range of 150 to 200 nm. A hint of another possible maximum occurs at much higher particle size, possibly a result of the existence of a crazing mechanism in this system, but the data are too limited for any definite conclusion.

The effect of rubber–plastic bonding for this system is shown in Figure 7 by using 10% PBD in the optimum range of 250 nm. Again, a grafting index of 20 led to an increase of almost 50% in impact strength. The effect

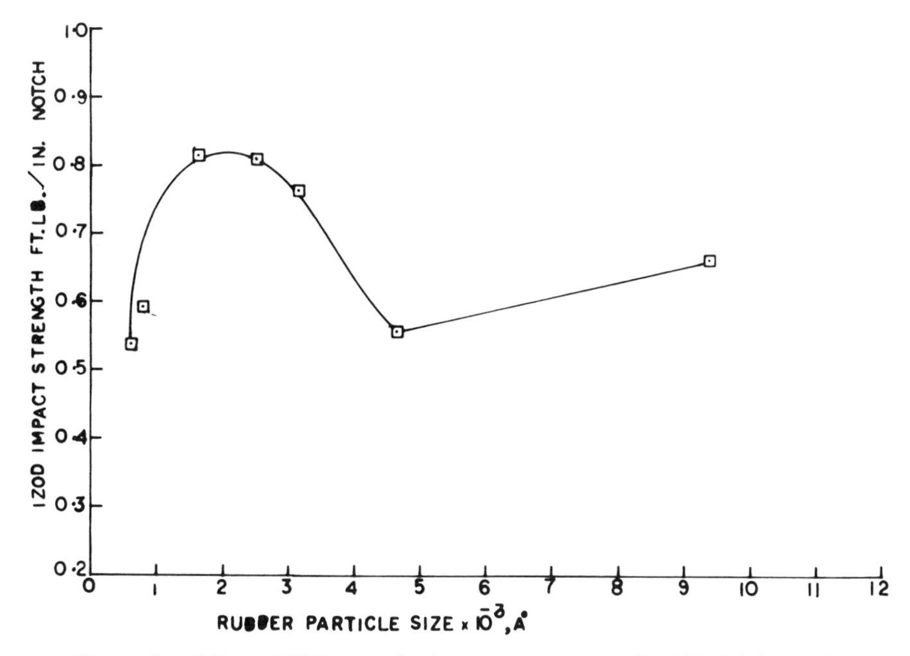

Figure 6. Effect of PBD particle size on impact strength of PVC (10 parts).

of bonding was also examined with a lower particle size with a PCP at 77 nm. In this case, the effect of grafting was only minor, presumably because these small rubber particles had only a limited effect in raising the impact strength from the value of 0.4 ft-lb/in. notch of the unmodified PVC.

Styrene–Acrylonitrile Copolymer (SAN)

The SAN latex used was an experimental lot of a 70/30 copolymer obtained from The Goodyear Tire & Rubber Co., Chemical Division. The unmodified polymer had an Izod impact strength of 0.28 ft-lb/in. notch. For this work a series of PCP latices was prepared, with particle sizes in the range of 77 to 1290 nm, and these latices were used as the rubber modifiers for the plastic matrix. The effect of rubber particle size on the impact strength of this system is shown in Figure 8. A very broad maximum of impact strength in the particle size range of 300 to 800 nm is obvious. These results agree with those of other investigators (*18*, *19*), who studied a very similar system, that is, acrylonitrile–butadiene–styrene graft copolymers (ABS). Apparently the SAN copolymers can fail by both mechanisms, crazing and shear yielding (*9*, *16*), so that both the smaller and larger rubber particles can be effective in preventing fracture.

The effect of rubber–plastic bonding on the impact properties of the SAN copolymer is shown in Figure 9, when 10% PCP with a particle size of

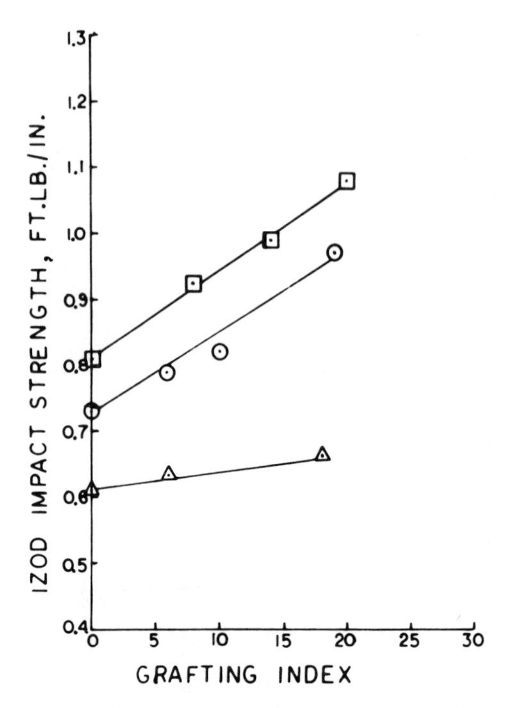

Figure 7. Effect of rubber–plastic bonding (grafting) on impact strength of PVC (10 parts PCP). Key: ⊡, PBD (249 nm); ⊙, PCP (152 nm); and △, PCP (77 nm).

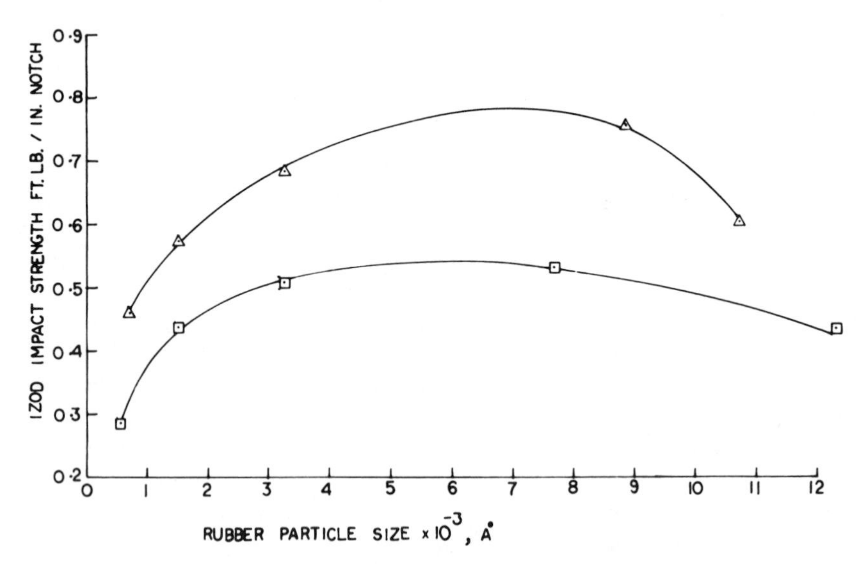

Figure 8. Effect of PCP particle size on impact strength of SAN. Key: △, 20 parts PCP; and ⊡, 10 parts PCP.

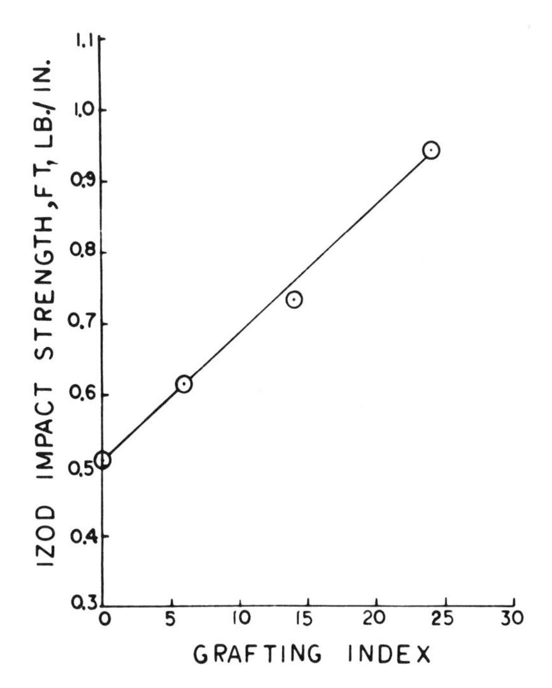

Figure 9. Effect of rubber–plastic bonding (grafting) on impact strength of SAN (10 parts PCP, 332 nm).

332 nm was used as the rubber dispersion. Here the effect is quite dramatic; a grafting index of 25 resulted in doubling the impact strength of the non-grafted material and more than tripling the impact resistance of the unmodified SAN.

Acknowledgment

This work was supported in part by a grant from The BFGoodrich Co.

Literature Cited

1. Bucknall, C. B. "Toughened Plastics"; Applied Science Publishers, Ltd.: London, 1977.
2. Merz, E. H.; Claver, G. C.; Baer, M. *J. Polym. Sci.* **1956**, *22*, 325.
3. Rose, S. L. *Polym. Eng. Sci.* **1967**, *7*, 115.
4. Newman, S. In "Polymer Blends"; Paul, D. R.; Newman, S., Eds.; Academic Press: New York, 1978; Vol. 2, Chap. 13.
5. Ref. 1, Chap. 7.
6. Wagner, E. R.; Robeson, L. M. *Rubber Chem. Technol.* **1970**, *43*, 1129.
7. Keskkula, H.; Turley, S. G. *Polymer* **1978**, *19*, 797.
8. Saam, J. C.; Mettler, C. M.; Falender, J. R.; Dill, T. J. *J. Appl. Polym. Sci.* **1979**, *24*, 187.
9. Bucknall, C. B. In "Polymer Blends"; Paul, D. R.; Newman, S., Eds.; Academic Press: New York, 1978; Chap. 14.

10. Keskkula, H. *Appl. Polym. Symp.* **1970,** *15*, 51.
11. Chang, E. P.; Takahashi, A. *Polym. Eng. Sci.* **1978,** *18*, 350.
12. Paul, D. R. "Polymer Blends"; Paul, D. R.; Newman, S., Eds.; Academic Press: New York, 1978; Chap. 12.
13. Boyer, R. F.; Keskkula, H. "Encycl. Polym. Sci. Technol"; John Wiley and Sons: New York, 1970; p. 392.
14. Bucknall, C. B.; Smith, R. B. *Polymer* **1965,** *6*, 437.
15. Newman, S.; Strella, S. *J. Appl. Polym. Sci.* **1965,** *9*, 2297.
16. Bucknall, C. B.; Street, D. G. *SCI Monog.* **1967,** *26*, 272.
17. Purcell, T. O. *Polymer Prepr. Am. Chem. Soc.* **1972,** *13(1)*, 699.
18. Parsons, C. F.; Suck, E. L. In "Multicomponent Polymer Systems"; Platzer, Norbert A. J., Ed.; ADVANCES IN CHEMISTRY No. 99; American Chemical Society: Washington, D.C., 1971; p. 340.
19. Schott, N. R.; Dhabalia, D. *Org. Coat. Plast. Chem.* **1974,** *34(2)*, 380.

RECEIVED for review January 20, 1983. ACCEPTED August 29, 1983.

POLYMER COMPOSITES

Reinforcement of Butadiene–Acrylonitrile Elastomer by Carbon Black

W. V. CHANG, B. WIJAYARATHNA, and RONALD SALOVEY

Department of Chemical Engineering, University of Southern California, Los Angeles, CA 90089-1211

Reinforcement and rupture mechanisms of filled nitrile elastomers were investigated. The initial modulus of filled elastomers could be adequately represented by the Frenkel and Acrivos equation for volume fractions above 0.2. The strain amplification concept and the structure concentration equivalence principle are applicable only at low strain. Above a critical strain, which depends on the concentration of filler, filler particles in the vulcanizate can not deform affinely. Stress softening in highly filled vulcanizates also was investigated. Existing mechanisms could not adequately represent all the facts associated with stress softening. On the basis of lubrication theory, a bonded rubber disk was used to simulate highly filled vulcanizates. Similar mechanical behavior was observed in both systems. Fractography, mechanical tests, stress analysis, and acoustic emission indicate that microcavitation is mainly responsible for stress softening in such systems.

THE REINFORCEMENT OF ELASTOMERS BY FILLERS is still poorly understood, in spite of its practical importance. To a great extent, this lack is due to the various morphological structures and chemical activities of different fillers, the distribution of fillers in a polymeric matrix, and the nonlinear viscoelastic behavior of the matrix itself. Rigorous and satisfactory theories only exist for small deformations and, even so, only to the regions of very low concentration (*1*) and extremely high concentration (*2*). Outside these regions the picture is much less clear, although many useful ideas have been proposed (*3*). In many cases our understanding is retarded by incomplete, inadequate, or even inaccurate data and analyses.

Our objective in this chapter is to assess the validity of some of the

0065-2393/84/0206-0233$07.75/0

proposed concepts in reinforcement and their applicability to a carbon black–nitrile elastomer system. In particular, initial modulus, finite and failure stress, and hysteresis will be studied as functions of the concentration and structure of the carbon black. We will examine the applicability of the structure–concentration equivalence principle (*4–6*) and the strain amplification concept (*7*) to our system at large deformations. We will critically evaluate the different mechanisms proposed for stress softening in elastomers (*8–10*). We will propose and demonstrate that, at least for highly filled systems, microcavitation in the matrix is an important source for stress softening. The toughest polymers are those that exhibit the highest hysteresis and the largest stress softening (*11*); new understanding of stress softening will provide us with insight into reinforcement.

Experimental

Mixing Experiments. Rubber and carbon black were mixed as shown in Table I in the Brabender Plasticorder. The optimum conditions for mixing to achieve an acceptable degree of dispersion were 60 rpm and 180–200 °F (*12*). Thus with the special mixing chamber developed in earlier work, we made all compounding work at these conditions. As the batch size was limited to 50 g in the Brabender, the two-roll mill was used for mixing larger batches. Ingredients were mixed into nitrile rubber as customarily done; however, 3–5 min of extra mixing were carried out to ensure good mixing.

In studies carried out with unfilled gum, the gum was also mixed in the two-roll mill with the following ingredients and compositions (parts by weight): Krynac-800, 100; sulfur, variable; zinc oxide, 3; stearic acid, 2; Thermoflex A, 2; and dop oil, 10.

Molding and Sample Preparation. The sheet made from the two-roll mill was cut into slabs approximately 6.5 × 6.5 in. that were compression molded in a Peco press in a tensile slab mold (6 × 6 in.) at 30,000 pounds ram pressure for 35 min at 350 °F. Specimens for tensile modulus and other experiments were obtained by punching out dumbbells with a 1-in. dumbbell cutter. The specimens were then bench marked at 1-in. intervals.

Table I. Composition and Mixing Procedure

Ingredient	*Composition*	*Loading Sequence Time (min)*
Krynac-800	0.500	0
Sulfur	variable	0
Zinc oxide	0.035	0
Agerite resin	0.035	0
HAF (Statex-R) or MT	variable	1.0
Stearic acid	0.0035	2.5
Dop oil	0.0565	3.0
ISAF-N19 or MT	0.115	4.5
Dump		6.5

Tensile Experiments. An Instron (Model TM) tensile testing machine was used for tensile tests. The thickness and width of the dumbbell specimen were measured with a dial gauge (accurate up to 0.001 in.) at several places along its length. The average measurement was used in estimating the cross-sectional area of the tensile piece. Dumbbells were then held between the jaws of pneumatic grips.

MODULUS EVALUATION. The initial secant modulus was evaluated by using the most sensitive setting of the load cell of the Instron machine, namely at a full scale of 500 psi. The cross-head speed used was the slowest available (0.2 in./min). A fast chart speed of 20 in./min was used to expand the infinitesimal deformation part of the stress–strain curve. These experiments were performed in a room where the air temperature was controlled and where neither wind nor heavy machines could cause vibrations. A tangent drawn to the stress–strain curve at zero strain yielded the initial modulus of the specimen.

FINITE AND FAILURE STRESS–STRAIN EXPERIMENTS. In these tests, we used the same type of dumbbells as in the modulus experiment. An extensometer was attached to the dumbbell to obtain the elongation of the specimen. Specimens were stretched at 5 in./min. The stress history during extension was recorded in the Instron strip chart recorder. Finite deformation and ultimate tensile properties were obtained from these curves. Usually five to six specimens were tested in each case and the average and standard deviation were calculated. The errors in most cases were less than 3–4%.

ENERGY AND HYSTERESIS TO BREAK. From the results of the tensile experiments, we have obtained the extension ratio to break and the engineering stress at break for each sample. Hysteresis and energy to break were determined as follows: A fresh sample (in each case) was stretched at a given speed (5 in./min) to a given extension and the cross head was returned at the same speed. The area under the stretching curve and the area between the stretching and retracting curve were evaluated at this given strain. This process was repeated many times and each time the degree of strain was increased; the maximal strain was a few percent points short of the breaking strain for the given specimen. After five to six such experiments, the ratio of hysteresis to energy stored (H/U) was plotted against the ratio of strain to strain at break (ϵ/ϵ_b). By extrapolating the curve to one, we obtained the ratio of hysteresis to energy at break $(H/U)_b$. By calculating the total area under the stress–strain curve up to break, we obtained U_b, and, by using the ratio $(H/U)_b$ and U_b, we obtained the hysteresis at break H_b.

Triaxial Experiments. The test piece consisted of vulcanized unfilled-rubber cylinders, 0.75 in. and 2 in. in diameter (D) and varying in thickness (L). These parameters lead to aspect ratios ($D/4L$) of 1/4 to 2.5. The rubber compound was compression molded into cylindrical metal pieces. The metal surface was sand-blasted and cleaned by a suitable solvent. A primer coat was then applied (Chemlock 205) and left overnight to dry at room temperature, 70 °F. The bonding agent (Chemlock 220) was then applied and similarly dried.

The inside of the mold was initially sprayed with a silicon mold-release to ease the removal of the bonded-rubber cylinder, called hereafter a poker-chip specimen. The specimen was very slowly pushed out with a mechanical ram.

The specimens so prepared were stretched in the Instron tensile testing machine at a cross-head speed of 0.2 in./min. The stress–strain curve was obtained from the Instron chart.

Morphology Studies. Rubber samples containing different volume-fraction loadings of carbon black and previously subjected to different stress conditions were examined under the scanning electron microscope (SEM).

The fracture surfaces of poker chips were examined visually and under lens power ($\times 5$) in an optical microscope to study the cavitation and crack growth.

Results and Discussion

Infinitesimal Deformation Region. The deformation behavior of the carbon black filled elastomers in this region is characterized by the initial slope of the stress–strain curve (E_0), which for a constant stretch-rate experiment can be represented mathematically as

$$\lim_{\epsilon \to 0} \frac{\sigma(t)}{\epsilon} = \lim_{\epsilon \to 0} E_s(t) = \lim_{t \to 0} \frac{\sigma(t)}{\dot{\epsilon} t} = E_0 \tag{1}$$

where $\sigma(t)$ is the time-dependent stress, $\dot{\epsilon}$ is the strain rate, ϵ is the strain, $E_s(t)$ is the secant modulus, and t is the time to achieve a given strain. The linear viscoelastic theory states that

$$E_s(t) = \frac{1}{t} \int_0^t E_R(t)\, dt \tag{2}$$

where $E_R(t)$ is the relaxation modulus. In our experiment t is approximately 0.1 min.

The relaxation modulus of our system can be represented by

$$E_R(t) \propto t^{-m} \tag{3}$$

Essentially, m is independent of strain (*13*) and the volume fraction of filler (*14*) over certain ranges of strain, time, and concentration. For our system m is approximately 0.09.

The modulus of a filled elastomer depends on the structure and morphology of the carbon black, the polymer, the degree of cross-linking, and on several significant interactions between filler and elastomer. The most important interactions are the degree of rubber–filler bonding, the rigid, immobilized adsorption layer on the particles, and the possible effects of the carbon black on the vulcanization reaction, leading to characteristic changes in the number and distribution of cross-links in the network. The first two effects become more significant with decrease in the size of the filler and increase in filler concentration. The third effect depends more on the chemical nature of the fillers and the curing system. According to Bueche (*15*), only strong, filler–elastomer bonds contribute to the reinforcement. Rivin et al. (*16*) calculated the number of chemisorptive attachments per gram of elastomer for a 100 phr HAF (80 m^2/g) as at least one order of magnitude less than the number of cross-links in a typical rubber network. Therefore, we ignore the first two effects.

The possible effects of fillers on the curing were studied by evaluating the initial modulus of an MT carbon black filled nitrile elastomer containing different quantities of sulfur as a function of filler concentration. A similar experiment was performed with nitrile rubber without any filler, but with varying degrees of sulfur. By dividing the modulus of the filled system by that of the unfilled system of corresponding sulfur content, the original individual curves of modulus as a function of volume concentration of fillers seem to condense into a single master curve (Figure 1). This result indicates that the effects of MT carbon black on the efficiency of curing is not important in our study.

The same conclusion can be drawn from the corresponding data of HAF carbon black filled nitrile rubber. This result is consistent with the finding of Cotten (*17*) who showed that the difference in reinforcement between a graphitized carbon black (zero acid concentration) and a regular carbon black filled SBR-1500/CBS system was typically less than 10%, and the difference decreases with an increase in carbon black structure.

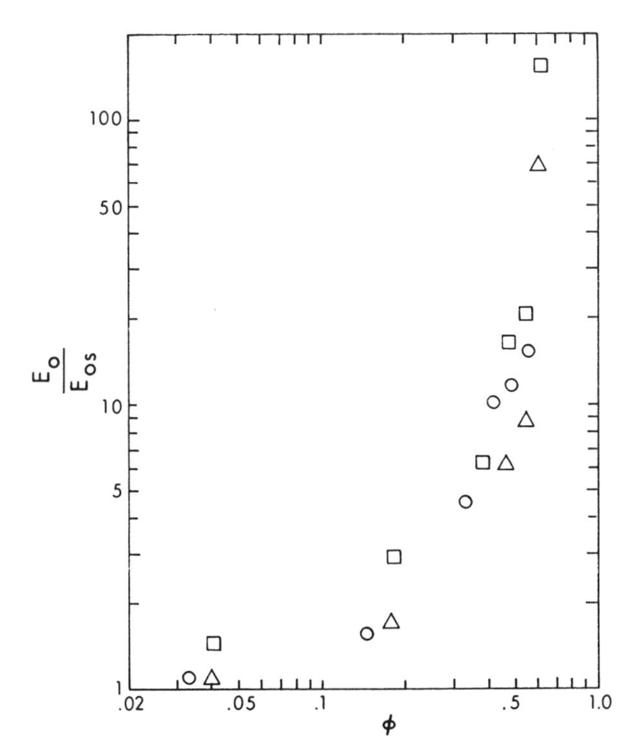

Figure 1. Relative initial modulus of the MT-filled nitrile elastomer as a function of the volume concentration of MT carbon black. Key (phr): ○, 1 part sulfur; □, 2 parts sulfur; and △, 6 parts sulfur.

Frankel and Acrivos (*2*) developed a relationship between the relative viscosity η^r and the volume fraction of filler based on lubrication theory,

$$\eta^r = \frac{9}{8} \frac{(\phi/\phi_m)^{1/3}}{1 - (\phi/\phi_m)^{1/3}} \tag{4}$$

Because the matrix is incompressible and dynamic effect is negligible, the equation can be used for the relative Young's modulus as well. If ϕ_m = 0.65, the packing density for random packing of equal spheres, then Equation 4 predicts E as 2.34, 3.8, 6.4, 12.3, 52.6, and 72 for ϕ of 0.2, 0.3, 0.4, 0.5, 0.61, and 0.62, respectively. The agreement with experimental data in Figure 1 is remarkably good; the theory is supposed to be valid only in the vicinity of ϕ_m. The Mooney equation (*18*) was also used but failed to represent our data.

Figure 2 shows a plot of the relative modulus E_r versus the volume fraction of filler for two types of carbon black, medium thermal (MT) and high structure furnace (HAF). Medium thermal (N990) is a more or less spherical carbon black and high structure furnace (N330) is a carbon black with structure. Carbon black N990 had a surface area [cetyltrimethylammonium bromide (CTAB)] of 8 m^2/g and an oil absorption (DBP 24M4) of 40 mL/100 g; carbon black N300 had a surface area of 79 m^2/g and an oil absorption of 89 mL/100 g (*4*). Both compounds contained the same amount of curing agent, 2 phr of sulfur.

As shown in Figure 2, the experimental data for the two compounds are not perfectly parallel and hence are not horizontally shiftable. This result is not completely unexpected because at very small deformations secondary aggregation and dispersion effects become dominant in governing the modulus; both effects depend on the volume concentration and type of fillers (*19*). However, with a little sacrifice in accuracy, two parallel curves can be drawn through the data points. A horizontal shift factor defined as

$$a_\phi = \left. \frac{\phi_{\mathrm{MT}}}{\phi_{\mathrm{HAF}}} \right|_{\text{const stress}} \tag{5}$$

is obtained from the curves. Its value is 1.785. The structure-dependent shift factor can be estimated by the dibutyl phthalate (DBP) absorption test. According to Kraus (*4*)

$$a_\phi = \frac{24 + A}{24 + A_0} \tag{6}$$

where A and A_0 are the DBP 24M4 values of the subject carbon black and the reference carbon black, respectively. In our case, A is 89, A_0 is 40, and therefore, a_ϕ is 1.766, which agrees well with the experimental value. Because $\phi_{\mathrm{MT.\,max}}$ is 0.65, $\phi_{\mathrm{HAF,\,max}}$ is 0.368.

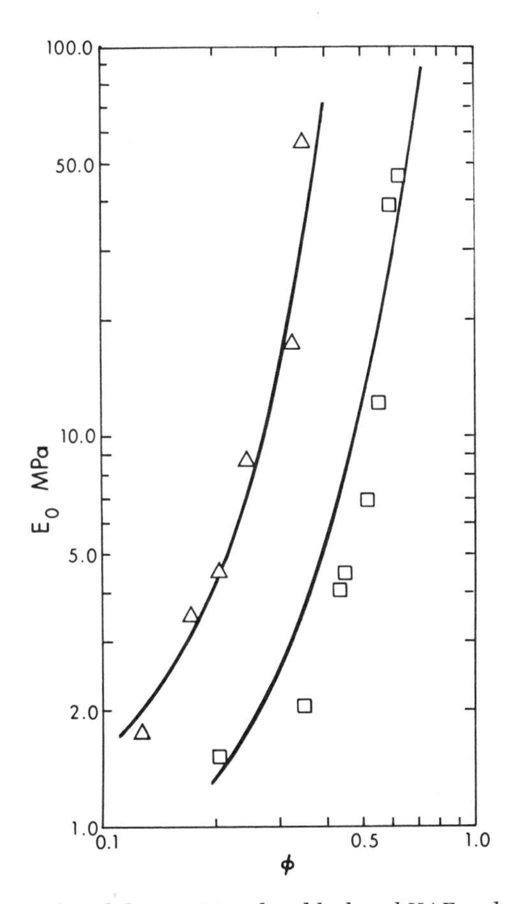

Figure 2. Initial modulus of MT carbon black and HAF carbon black filled nitrile elastomer as a function of the volume concentration of filler. Key: △, HAF-loaded BAN; and □, MT-loaded BAN.

Finite Deformation Region. Unlike the small deformation region where the stress–strain behavior can be assumed to be linear for all practical purposes, such analysis in the finite deformation region is complicated by the nonlinear deformation of filled rubber. In the absence of a rigorous theory we will be mainly evaluating the applicability of the concepts of strain amplification, of Smith's scheme of separating time and strain effects, and of the structure–concentration equivalence principle.

The strain-amplification concept recognizes that the microscope strain, ϵ_{mic}, existing in the elastomeric phase of a filled system is much greater than the applied macroscopic strain, ϵ_{mac}, because the filler particle are not deformable. Mathematically, a strain-amplification factor, $f(\phi)$, can be defined as

$$f(\phi) = \epsilon_{mic}/\epsilon_{mac} \tag{7}$$

and the macroscopic stress

$$\sigma = g(\epsilon_{mic}) \tag{8}$$

where g defines the constitutive equation of the rubber. The basic assumption of the strain-amplification concept is that $f(\phi)$ is independent of strain.

Mullins and Tobin (7) represented $f(\phi)$ by the Gath–Gold equation, and g followed the Mooney–Rivlin equation. This representation unnecessarily limited the applicability of the concept. In our discussion we will let both $f(\phi)$ and g be completely arbitrary. Equations 7 and 8 imply that in a log–log plot of the engineering stress as a function of strain for different MT carbon black concentrations, curves for different filler concentrations are horizontally superposable. We conclude (*21*) that the curves for unfilled elastomer and 10% MT were horizontally shiftable only if the uniaxial stretch ratio is less than 3.4. The range of applicability is even smaller for higher concentration systems, that is, for $\phi = 0.2$ it is applicable only if $\lambda < 2.1$.

A simple geometric consideration is helpful for understanding the dependence of the range of applicability on the volume concentration. In many so-called molecular theories of reinforcement the positions of the filler particles are assumed to undergo an affine deformation. This assumption is often justified because the particles are quite large in comparison to atomic dimensions. Hence, even at very large stresses, the unbalanced force on any given particle will be unable to move it far through the matrix. However, this argument only has a limited range of applicability, as shown in Figure 3. When a uniaxial stretch ratio, λ, is applied, the horizontal distance between the centers of two filler particles will be reduced by $\sqrt{\lambda}$, according to the incompressibility of the material and the affine assumption. However, as λ reaches a critical value, λ_c, the two neighboring horizontal particles will touch each other, which makes a further affine deformation of filler particles impossible. Any small unbalanced force on a particle such as cavitation in position *c* or detachment of rubber from one side of particle B will cause the particle to rotate. Because the stress fields before and after λ_c are totally different, the same strain amplification factor is an unrealistic assumption.

Therefore, how does λ_c depend on ϕ? The average original distance between the centers of neighboring filler particles is proportional to $\phi^{-1/3}$; the distance at λ_c is proportional to $\phi_m^{-1/3}$. Therefore,

$$\left(\frac{\phi_m}{\phi}\right)^{1/3} = \sqrt{\lambda_c} \tag{9}$$

We have already determined that $\phi_{max} = 0.65$ for MT. Hence, we obtain λ_c as 3.48, 2.19, 1.67, and 1.38 for ϕ equal to 0.1, 0.2, 0.3, and 0.4, respectively.

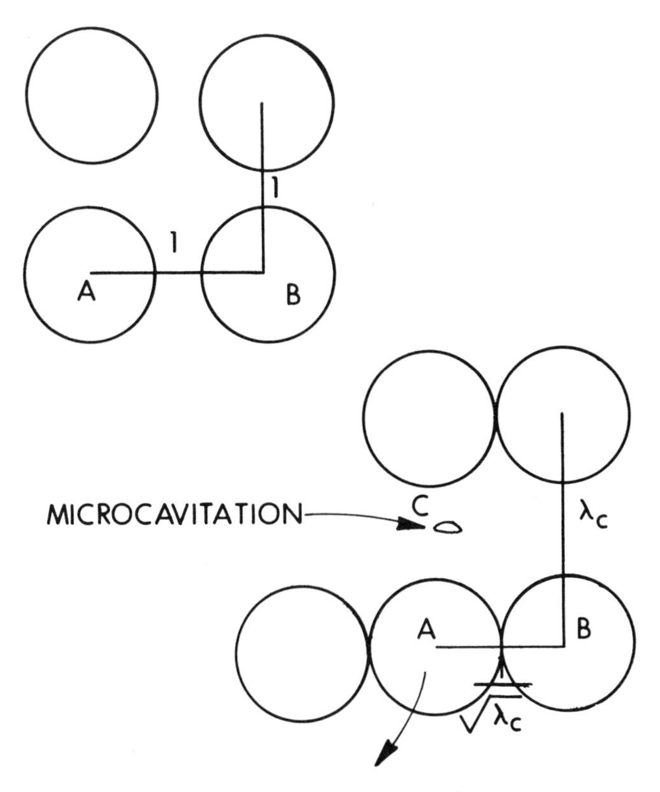

Figure 3. Schematic representation for microdeformation near critical strain.

An empirical equation was proposed by Smith (22) to represent the stress–strain data from constant strain-rate tests.

$$\sigma(t, \epsilon) = F(t) H(\phi) f(\epsilon) \tag{10}$$

where $F(t)$ is the constant strain-rate modulus. When the viscoelastic behavior is linear, $F(t)$ is the secant modulus. Because the strain rate is constant in our experiments, Equation 10 suggests that the stress–strain response plotted as $\log \sigma$ versus $\log \epsilon$ will yield a family of parallel curves that can be superposed by a vertical shift. The magnitude of vertical shift is equal to $\log H(\phi)$ or $\log E_r$. However, our data indicate that a vertical shift is only possible when λ is less than 3. We believe the argument given in our discussion of Figure 3 is also applicable here.

Next, can we predict the stress–strain behavior of a carbon black filled system from a known behavior of another carbon black system? With this in mind, we plotted stresses at 50, 100, 150, and 200% strains against the volume-fraction loading for both MT and HAF carbon black filled systems on the same graph paper. Figures 4 and 5 show such plots for strain at 50 and 200%, respectively. In Figure 4, these two curves are essentially parallel to each other and are horizontally shiftable. The superposability decreases as strain increases. The strain (%) and corresponding shift factor values (a_ϕ) are as follows: 0, 1.785; 50, 1.73; 100, 1.74; 150, 1.84; and 200, 1.94. Thus, within experimental error, the shift factor can be considered more or less constant and can be estimated from the structure of a carbon black.

At strains higher than 200%, the two curves are no longer shiftable. Next, we will discuss what occurs microscopically in a specimen deformed beyond 200% strain.

Failure Region. Figure 6 shows a plot of energy at break and hysteresis to break as functions of the volume fraction of the MT carbon black.

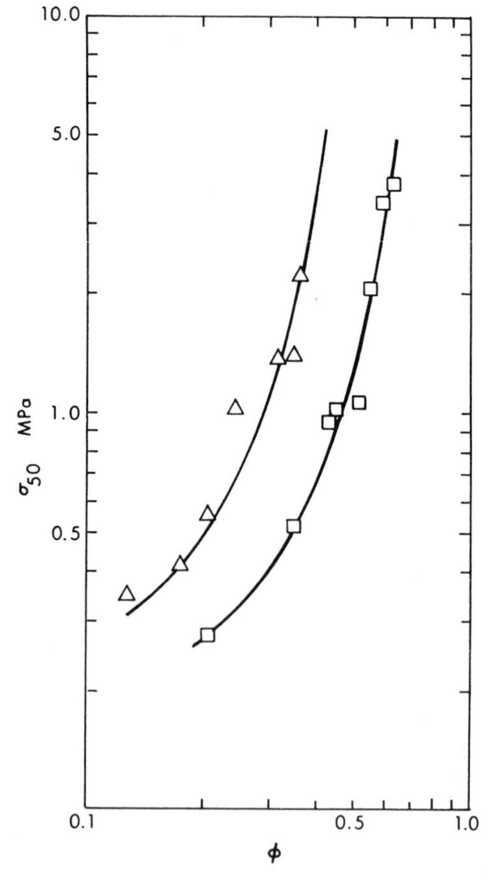

Figure 4. Stress at 50% deformation of MT and HAF carbon black filled nitrile elastomer as a function of the volume concentration of fillers. Key: see Figure 2.

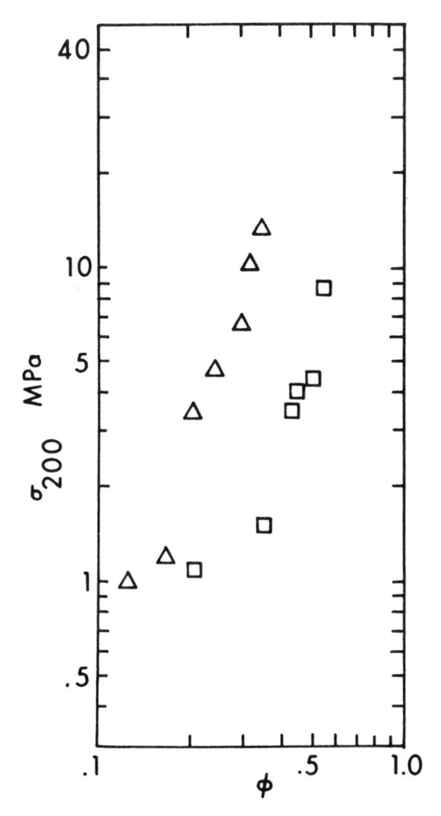

Figure 5. Stress at 200% deformation of MT and HAF carbon black filled nitrile elastomer as a function of the volume concentration of fillers. Key: see Figure 2.

Both U_b and H_b initially increase with the volume fraction of the filler and then decrease beyond a volume fraction loading of 20%. Can both U_b and H_b be accounted for by the viscoelastic properties of elastomer phase alone? Microscopic strain rate in the polymer matrix at a given ϕ can be estimated, for example, by Bueche's equations (8). The unfilled elastomer U_b was determined over a strain-rate range corresponding to the range of filler concentration and represented as a dot line in Figure 6. Thus, the addition of filler has increased the strength of the rubber above and beyond what could be offered by the viscoelastic properties alone. The presence of filler not only increases the ability of rubber to store energy but also increases the hysteresis. The ratio $(H/U)_b$ for a filled elastomer is interestingly almost double that for pure gum (Figure 7). The range of values for $(H/U)_b$ for a pure gum is 0.28–0.42 and for an MT-filled elastomer is 0.562–0.78, as determined from our work and the literature. The value for $(H/U)_b$ for pure gum was obtained from Harwood and Payne's data (23). The range of test data is not available; however, the minimum and maximum values of the energy density at break, 0.2 and 100 J/mL for pure vulcanized gum, indicate that various tensile tests have been performed. The value range for filled gum, 0.56–0.78, is for an MT black filled nitrile elastomer with volume fractions of filler that range from approximately 0.04 to 0.6. Thus, in a filled elastomer, hysteresis increased two to three times that of a gum vulcanizate per unit increase in the storage ability. What mechanisms are responsible for this increase in H_b? Is most of the energy dissipated in the elastomer phase or in the filler phase?

Perhaps the structure concentration equivalence principle will provide new insights. Figure 8 shows a plot of engineering stress at break versus the volume fraction of filler for MT and HAF carbon black. If this curve is replotted as $\sigma_b/(1 - \phi)$ versus ϕ/ϕ_m, we obtain the curve in Figure 9. Interestingly, the data are reduced to a single curve. The physical meaning of $\sigma_b/(1 - \phi)$ is not clear, but it suggests that at least most of the energy input and dissipation occur in the elastomer phase. But what mechanisms are responsible for this high dissipation in the elastomer phase?

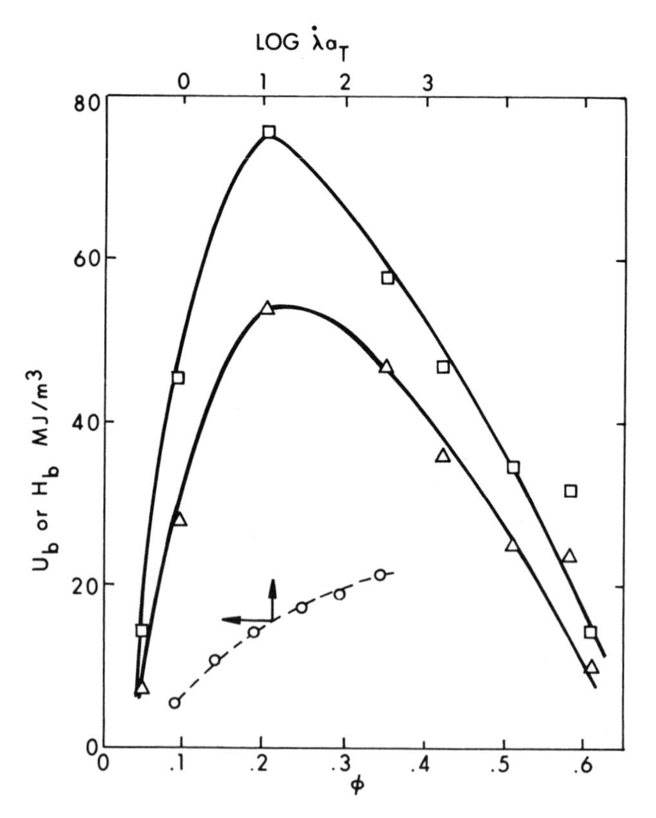

Figure 6. Energy density at break (□) and hysteresis at break (Δ) as a function of the volume fraction of MT carbon black.

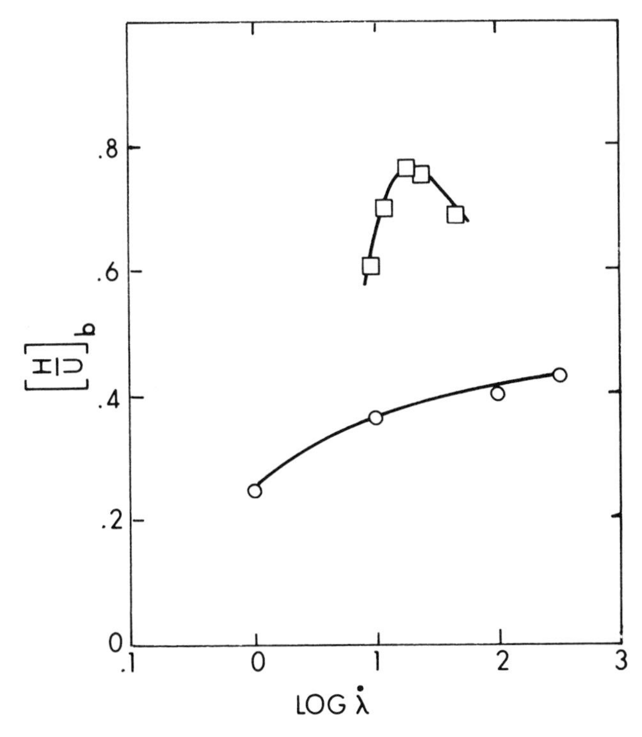

Figure 7. Ratio $(H/U)_b$ *of filled and unfilled nitrile elastomer as a function of deformation rate. The deformation rate of the elastomer phase in filled systems is adjusted according to the strain amplification factor. Key:* ○, *BAN; and* □, *BAN + MT.*

Energy dissipation in a filled system is often called stress softening. Several mechanisms have been proposed for stress softening:

1. Rearrangement of the rubber network associated with slippage of entanglements and nonaffine displacement of network junctions in the rubber matrix. This mechanism occurs in gum vulcanizates and in the gum phase of filled vulcanizates.
2. Structural changes of the carbon black aggregates. This mechanism is associated with the possible breakdown and reformation of filler domains, or breakage of the permanent structure of the carbon black particles. The latter mechanism was suggested by Kraus and others but no independent evidence has been presented. Furthermore, our data suggest that the contribution to H_b from this mechanism is not significant.
3. Slippage of rubber chains on the carbon black surface or the breakage of rubber-filled bonds. These are involved in the

Dannenberg–Boonstra and Bueche theories, respectively, but no independent evidence of these mechanisms exists.

4. Breakage of weak cross-links such as hydrogen bonds, ionic type links, and polysulfide links.

Among all the different mechanisms proposed for stress softening, Bueche's theory is most attractive because it has a clear molecular basis and an equation with parameters that can be experimentally determined and, to a certain extent, independently checked. Bueche attributed the softening primarily to the breaking of network chains extending between adjacent filler particles. In his model some strong bonds are assumed to be broken even at small elongations. Three parameters are defined in his equation: γ is a measure of the strength of the bonds that break and is given by $F_c a/kT$, where kT is the usual product of absolute temperature and Boltzmann constant; F_c is the tension in the chain when it breaks; and a is the length of a freely orienting segment in the chain. The quantities s and b characterize the filler and its dispersion. In particular, s is the average filler surface area per filler bond, and b is a measure of the separation of the filler particles.

Bueche's equation reproduces the softening curves for several systems (8, 9). Parameters have been used to evaluate different mixing practices and the efficiency of fillers. Poorly dispersed fillers show markedly increased

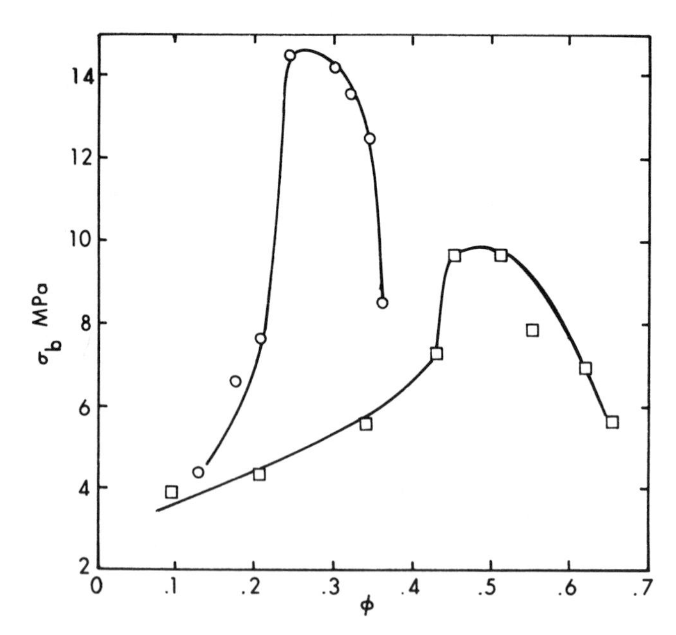

Figure 8. Engineering stress at break of MT and HAF filled nitrile elastomer as a function of the volume concentration of fillers. Key: ○, HAF-loaded BAN; and △, MT-loaded BAN.

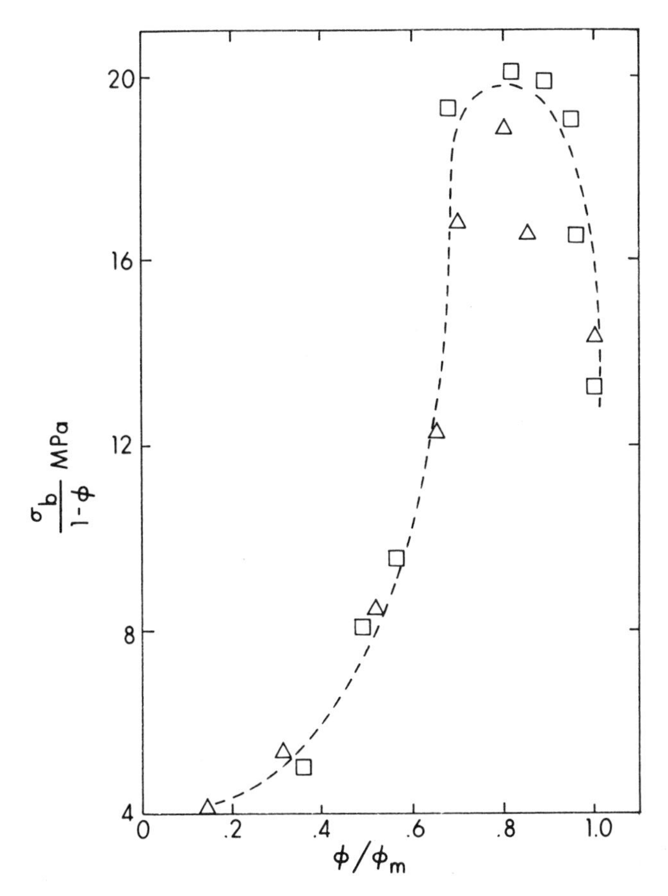

Figure 9. Same data as Figure 8 plotted against ϕ/ϕ_m after normalizing by the actual rubber volume. Key: □, HAF-loaded BAN; and △, MT-loaded BAN.

values of b. The poorly reinforcing fillers generally have large b and s values and small γ values. All of these findings seem physically reasonable.

Despite all these apparent successes we believe that Bueche's model suffers from some fundamental shortcomings. First, one of the most important assumptions in Bueche's theory is that filler particles deform affinely. However, beyond a critical strain this assumption can not be valid. And yet, the theory was often applied to the region well beyond that critical strain. Second, as pointed out by Bueche himself (8), the energy corresponding to γ is about 30 kcal less than the bond energy. This finding would imply that at $F = F_c$ the chain is far from being fully extended. Yet, in the original derivation a fully extended chain at $F = F_c$ was assumed. Third, the fact that poorly dispersed fillers show large values of b is not logical. A poorly dispersed system shows a higher initial modulus. Within the spirit of the original model, b, the average distance between particles, must be smaller.

In their study of the effect of milling on stress softening of an HAF–SBR vulcanizate, Wagner et al. (9) prepared the unmilled stock by dispersing the pigment and curing ingredients in a benzene solution of the rubber followed by freeze drying and curing without any intermediate milling. The milled stock was the same specimen passed through a two-roll mill (after drying and before curing) a sufficient number of times. Milling significantly reduced the total amount of softening. This fact was attributed to the pigment aggregation in the unmilled system. Similar experiments were also conducted on hydrated-silica filled styrene–butadiene rubber (SBR). Although the reduction of stress softening by milling was about the same for both cases, the changes in b values were significantly different. For HAF, b changes from 9.8 to 3.4 nm, but for hydrated silica, b reduced from 172 to 6.2 nm. No other independent measurement was provided to explain such a tremendous difference in dispersion between these two unmilled systems. Relating these numbers to the original physical meaning of b may not be meaningful.

We suspect that the high stress softening in the unmilled system may not be directly relevant to the dispersion state of fillers. Perhaps a freeze-drying process may cause some microphase separation that leaves minute defects such as voids in the matrix. The size and number of voids greatly contribute to stress softening of the elastomer.

Microcavitations. The stress–strain curve of highly filled elastomers shows a distinct yield point. The yield stress decreased initially with an increase of carbon black content. SEM pictures reveal that a large number of vacuoles are formed when the MT carbon black concentration is increased. This finding is shown in Figures 10 and 11 and by a similar work carried out by Takahashi on rubber containing different types and quantities of carbon black (*24*). Figure 12 shows SEMs of the fracture surface of a highly filled sample before and after yielding. The yield point may be a consequence of substantial microcavitation, and microcavitation happens at lower stress levels as the carbon black concentration is increased from ϕ = 0.25 to 0.60. Microcavitation in an elastomer reinforced by high structure-reinforcing filler has also been reported (*24*). Although less pronounced in such systems than in the corresponding MT-filled systems, we believe the mechanism of microcavitation is of similar nature. Therefore, we focused our study on the MT-filled systems.

Minute voids with a diameter of 1 μm or less always exist in rubber parts. Under negative hydrostatic pressure, vacuoles grow in size and would grow irreversibly (microcavitation) if the applied negative pressure field exceeded a critical value (*25–27*). This critical value depends on the size of the initial voids, Young's modulus, the tearing energy, the thermo-

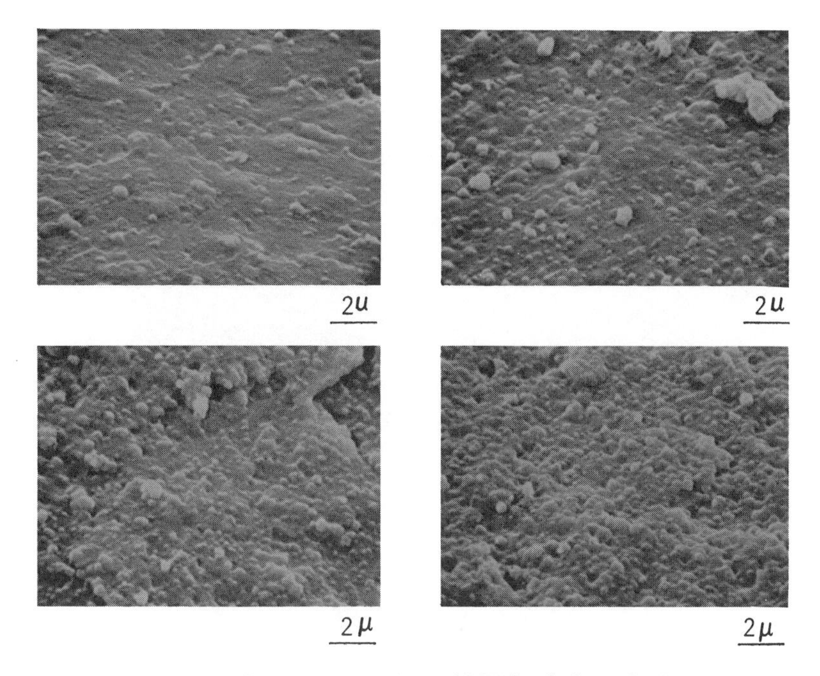

Figure 10. SEM of the fracture surface of MT-loaded nitrile elastomer with ϕ from 0.25 to 0.62, showing vacuole formation during deformation.

Figure 11. SEM of the fracture surface of MT-loaded nitrile elastomer with ϕ from 0.25 to 0.62, showing vacuole formation during deformation.

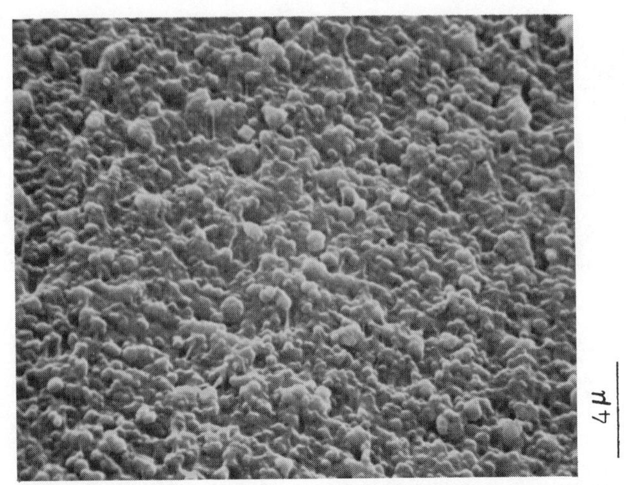

Figure 12a. SEM of fracture surface before yield (ϕ = 0.62).

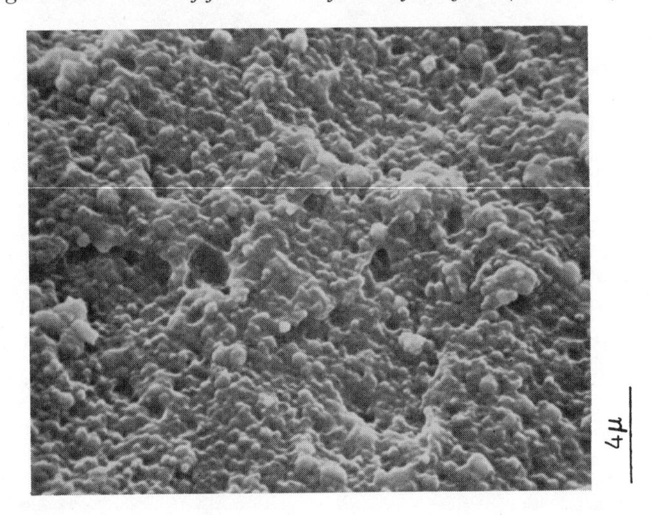

Figure 12b. SEM of fracture surface after yield (ϕ = 0.62).

dynamic surface energy, and the extensibility of the matrix (*27*). The critical negative pressure increases with a decrease in the initial size of the voids. A consequence of microcavitation is the softening of the material.

Perhaps the yield point can be related to the critical negative pressure generated in the rubber phase of a filled system. Thus, the negative hydrostatic pressure in the matrix increases with the concentration of the filler.

A linear relationship exists between the yield stress and the failure stress of filled vulcanized rubber in the region of concentrations under study (Table II).

Also, the yield stress at a given concentration of carbon black increases

Table II. Relationship Between Yield Stress and Failure Stress in Vulcanized Rubber

Volume Fraction of Carbon Black in Rubber[a]	*Yield Stress (psi)*	*Failure Stress (psi)*
0.47	1750	2165
0.45	1450	1968
0.51	1350	1804
0.55	1300	1758
0.59	1250	1748
0.62	800	1172

[a]Poly(butadiene-*co*-acrylonitrile) with 2% sulfur.

with the concentration of curing agent (sulfur). Because the curing agent results only in an increase in the modulus of the matrix, an increase in the matrix modulus retards the vacuole formation and thereby the yield process. With the increased matrix modulus, a linear relationship between the yield stress and failure stress is observed, and, interestingly enough, the two lines (for 2 and 6% sulfur cured) have the same slope. This result indicates that the yield process depends on both the modulus of the matrix and the magnitude of the negative hydrostatic pressure due to the type of stress field in the matrix.

In a highly filled system in which the particles are very close to each other, such that the gap between two particles is small compared to the radius of the spherical particle, we may consider the surfaces of those particles facing each other to be essentially flat. In such a situation, the lubrication approximation could be used to calculate the stress distribution in the matrix because the local displacement may be treated as a function of only the local gap distance, the local pressure gradient, and the local boundary conditions. The lubrication approximation represents our initial modulus data very successfully even when ϕ is as low as 0.2. We believe that the approximation will also be useful in understanding the fracture process.

In this sense, the poker-chip specimens discussed earlier can be used to simulate the stress field in the matrix between two spherical filler particles. Thus, we were inspired to investigate any possible similarities between the stress–strain behavior of poker chips under deformation with those of very highly filled rubber vulcanizates. In doing so, we molded nitrile elastomer (without filler) containing the same amount of sulfur as in the filled system, between two steel disks with properly prepared surfaces as described in the experimental section. We have already shown that the presence of filler does not affect the cross-link density of the rubber, and, therefore, the stiffness of the matrix between the carbon black particles is probably the same as that between the two poker chips in the present experiment.

The stress–strain curve of the poker-chip experiment was essentially similar in shape to that of a highly loaded nitrile elastomer sample. In addition, a definite dependence of yield point on the aspect ratio of the elastomer disk of the poker-chip specimen was seen. In Figure 13 the poker-chip specimen shows tremendous stress softening if the previous deformation is beyond the yield point. Chang (28) has shown that the stress-softening behavior can be accounted for, from the linear elastic analysis point of view, by the small increase in the void content of the matrix during deformation. This early stage microcavitation does not result in a significant increase in volume but surely affects the mechanical behavior of the specimen significantly. Conventional techniques for monitoring microcavitation such as dilatometry are not sensitive enough to detect the early stage of microcavitation. Therefore, acoustic emission has been used by Chang (29) to monitor the microcavitation. This work shows that microcavitation happens well before the yield point.

In our experiments, the poker chips separated with a pop sound similar to that reported by Gent and Lindley (25). They also reported that as the aspect ratio was increased the yield stress decreased and went through a minimum and increased again. Our results on filled vulcanizates are in agreement with theirs. Bubbles or vacuoles also formed at or near the longitudinal axis of the rubber cylinder closer to one of the bonded surfaces, as shown in Figure 14. As the aspect ratio (D/L) decreases (i.e., when the

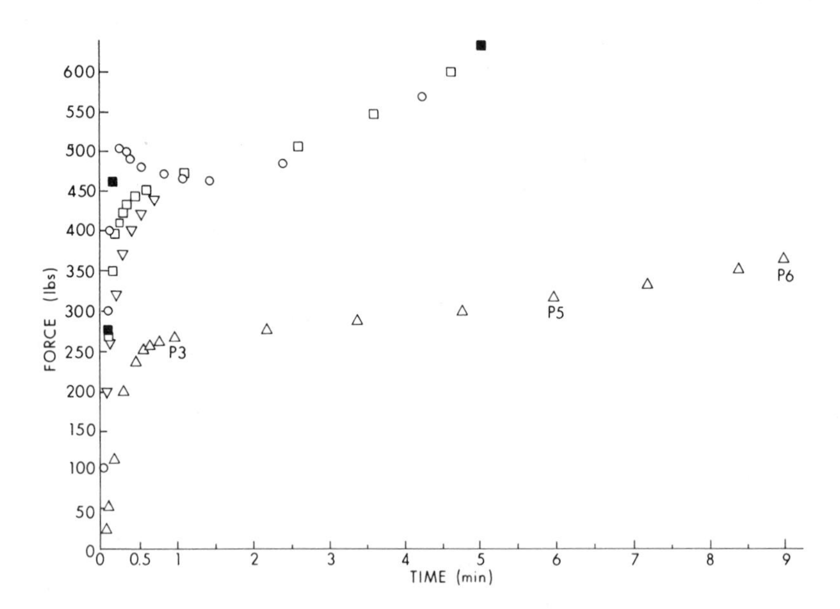

Figure 13. Load extension relations for poker-chip specimens. Pulling speed = 0.2 in./min. Specimen 5 and 7, D/L = 10. Specimen 2, D/L = 6.667. Key: ○, 5; ■, 7 (first extension); ▽, 7 (7th extension); □, 7 (after 9 days); and △, 2.

Figure 14. Fracture surface of poker-chip specimens.

surfaces were further apart), the size of the bubbles decreased and the area in which the bubbles formed spread radially outwards. This result shows that the severity of the negative hydrostatic pressure decreased as the distance between the surfaces increased and the stress distribution gets increasingly uniformed in the radial direction. This radial distribution has been mathematically accounted for by Chang (*28*). Gent and Lindley (*30*) showed that the yield point in the load-extension curve is related to the aspect ratio as

$$\sigma_y = E_m\left[1 + 8\left(\frac{L}{D}\right)^2\right] \qquad (11)$$

where E_m is the Young's modulus of the elastomer in the poker-chip specimen. Equation 11 represents our poker-chip data. The value of E_m calculated is in the range of 17–19 psi.

Furthermore, a linear correlation between the rupture stress and yield stress seems to exist (Figure 15). The slope of this curve appears to be similar to that of the highly filled nitrile rubber. Although the slopes are similar, the magnitude of the stresses in the filled rubber is about 10 times higher. This finding suggests that the failure in the two systems is governed by the same mechanism. The difference in magnitude may be a reflection of the viscoelastic nature of the rubber. The stress–strain relationship for a filled- or unfilled-rubber vulcanizate is dependent on the strain rate and temperature. For filled rubber, the actual deformation taking place in the binder, commonly known as the microscopic strain, is much higher than the applied macroscopic deformation. Thus, to compare the poker-chip experiment data with those of filled rubber, we have to consider test rate, sample dimension, and strain amplification on the basis of the concentration of the filler in the case of filled rubber, or geometry, in the case of poker chips.

Our filled-rubber experiments were carried out with tensile samples 1 in. long at a rate of 5 in./min. On the other hand, the poker-chip experiments were carried out at an extremely slow speed of 0.2 in./min, and the poker chip samples were 0.1 in. thick. To consider the rate effects, we use a correction for strain amplification. In the absence of an exact theory that would tell us the exact strain rate in the process, we propose to use the initial modulus to account for the increased microscopic strain. Thus we normalized the σ_y and σ_b values by the corresponding initial moduli. For poker chips, Gent derived mathematically a relationship for the apparent modulus of the poker chip on the basis of its geometry as

$$E_{\mathrm{app}} = E\left[1 + 2\left(\frac{D}{4L}\right)^2\right] \tag{12}$$

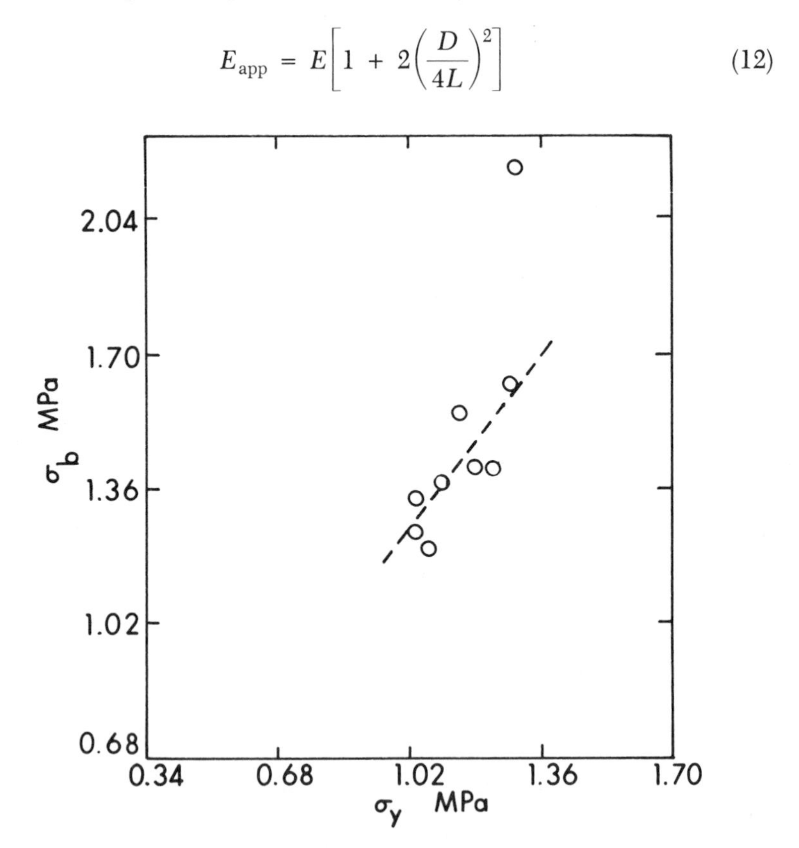

Figure 15. Stress at break as a function of the yield stress for poker-chip specimens.

On the basis of this method, we normalized the σ_y and σ_b values of poker-chip specimens by dividing with $E_{app} = 112$ psi. Now, if we plot σ_b/E_{app} (or σ_b/E_0 for filled specimens) versus σ_y/E_{app} (or σ_y/E_0 for filled specimens), we find that data of filled specimens of different filler concentrations and different amounts of sulfur, and data of poker-chip specimens can be condensed onto a narrow band (Figure 16). This finding provides strong evidence in support of our conclusions that the deformation and fracture processes in a highly filled elastomer can be simulated and represented adequately by poker-chip specimens, and that the microcavitation plays a vital role in the deformation and fracture processes of both systems.

If we plot σ_b/E_{app} for both filled-elastomer and poker-chip specimens against ϕ and D/L, respectively, we obtain Figures 17 and 18, respectively. They both show a linear relationship, indicating again that the mechanism of failure is similar. An increase in filler concentration, according to our argument, is represented by an increase in the aspect ratio, D/L, in the

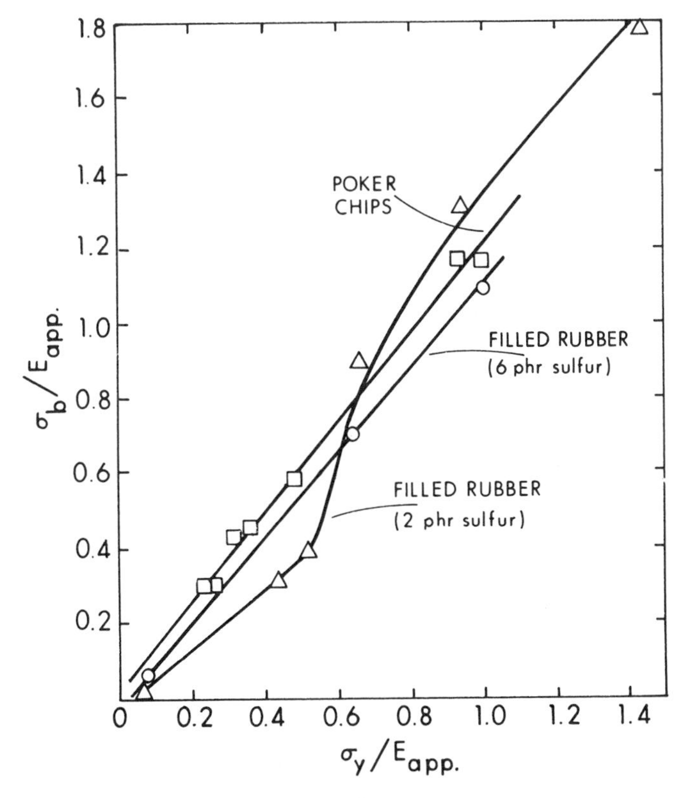

Figure 16. Stress at break normalized by the apparent modulus as a function of yield stress normalized by the apparent modulus for both carbon black filled nitrile elastomers and for poker-chip specimens. Key: △, filled rubber (2 phr sulfur); ○, filled rubber (6 phr sulfur) and □, poker chips.

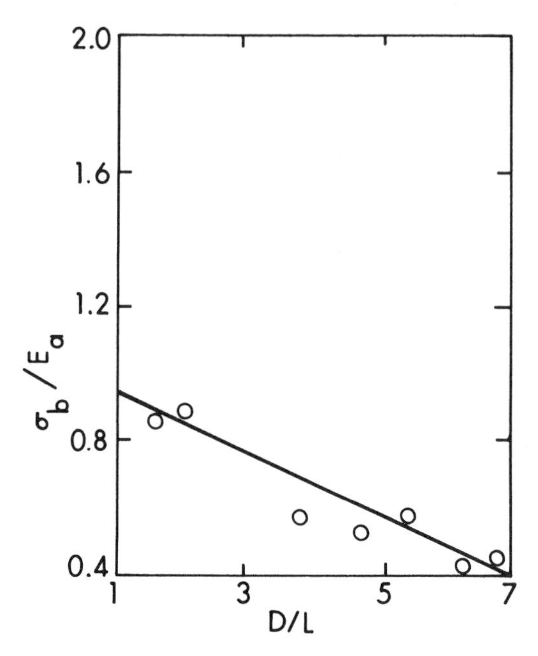

Figure 17. Stress at break of poker-chip experiments normalized by the apparent modulus as a function of the aspect ratio D/L.

poker-chip experiments. Although they behave similarly, a rigorous, theoretical relationship between the aspect ratio D/L and ϕ is not available.

Summary

The relative initial modulus of carbon black filled nitrile elastomers can be represented by equations derived from elastic or hydrodynamic theory. For MT-filled elastomers Frankel's equation based on lubrication approximation successfully represents data for $\phi > 0.2$. A stress–concentration equivalence principle is applicable to initial modulus and for strain up to an initial value, which depends on the concentration of the filler. The shift factor can be estimated from the structure of the filler as determined by the DBP 24M4 test. The strain amplification concept and Smith's scheme are workable if the strain is less than a critical value, which depends on the concentration of the vulcanizate. Above the critical strain, affine deformation of the filler particles is not valid. Energy dissipation in a filled elastomer is much larger than that of the unfilled gum. Almost all the dissipation occurred in the elastomer phase. Stress softening can not be adequately described by the existing mechanisms such as Bueche's theory. A model specimen based on the lubrication theory shows similar mechanical behavior to that of highly filled elastomers. On the basis of the results of mechanical

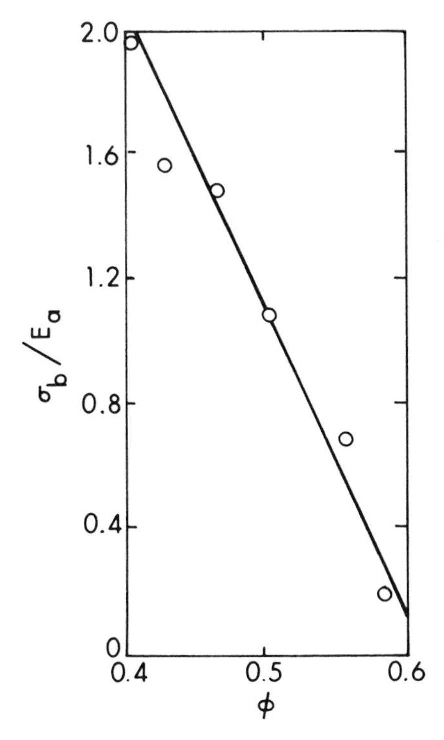

Figure 18. Stress at break normalized by the apparent modulus of highly filled vulcanizate as a function of the volume fraction of the filler.

tests, fractography, and stress analysis, microcavitation is concluded as the main cause of stress softening and as responsible for the fracture behavior of highly filled elastomers.

List of Symbols

a	Length of a freely orienting segment in the chain
a_ϕ	Structure concentration equivalence shift factor
b	Separation of neighboring filler particles
m	Slope of logarithmic relaxation modulus as a function of logarithmic time
k	Boltzmann's constant
s	Average filler surface area per elastomer-filler bond
t	Time
A_0	DBP 24M4 value of the subject black
A	DBP 24M4 value of the reference black
D	Diameter of a poker chip specimen
E_0	Initial (zero time) Young's modulus

E_r Relative Young's modulus, i.e., the ratio of the Young's modulus of a filled system to that of an unfilled system
$E_s(t)$ The secant Young's modulus
$E_R(t)$ The relaxation Young's modulus
E_m Young's modulus of elastomer in a poker chip specimen
E_{app} The apparent modulus (also referred as spring stiffness)
F_c Tension in a chain when it breaks
H_b Hysteresis at break
L Thickness of elastomer disk in a poker-chip specimen
U_b Total energy input at break
γ Strength of chemical bonds
ϵ Strain
ϵ_{mic} Microscopic strain, i.e., strain of the elastomer phase in a filled system
ϵ_{mac} Macroscopic strain
η^r Relative viscosity, i.e., the ratio of the viscosity of a filled system to that of the unfilled system
λ_c Critical stretch ratio beyond which affine deformation of filler particles cannot exist
λ Stretch rate or deformation rate
σ Stress
σ_b Stress at break
σ_y Yield stress
ϕ Filler volume concentration
ϕ_m Maximum filler volume concentration

Acknowledgments

The authors express their gratitude to the National Science Foundation (ENG 77–00458) and to the Hydril Company for their support of this project.

Literature Cited

1. Smallwood, W. M. *J. Appl. Physiol.* **1944**, *15*, 758.
2. Frankel, N. A.; Acrivos, A. *Chem. Eng. Sci.* **1967**, *22*, 847–853.
3. Eirich, F. R.; Smith, T. L. In Fracture: Liebowitz, H., Ed.; Academic Press: New York; 1972; Vol. 7, Ch. 7.
4. Kraus, G. *Rubber Chem. Technol.* **1971**, *44*, 199.
5. Medalia, A. I. *J. Colloid Interface Sci.* **1970**, *32*, 115.
6. Medalia, A. I. *Rubber Chem. Technol.* **1973**, *46*, 877.
7. Mullins, L.; Tobin, N. R. *J. Appl. Polym. Sci.* **1965**, *9*, 2993.
8. Bueche, F. *J. Appl. Polym. Sci.* **1960**, *4*, 107.
9. Wagner, M. P.; Wartmann, H. J.; Sellers, J. W. *Kautsch. & Gummi Kunstst.* **1967**, *20*, 407.
10. Harwood, J. A. C.; Mullins, L.; Payne, A. R. *Inst. of Rubber Ind. J.* **1967**, *1*, 17.
11. Harrow, J. A. C.; Payne, A. R. *J. Appl. Polym. Sci.* **1968**, *12*, 889.

12. Wijayathna, B.; Chang, W. V.; Salovey, R. *Rubber Chem. Technol.* **1978**, *51*, 1006.
13. Smith, T. L. *Trans. Soc. Rheol.* **1962**, *6*, 61.
14. Bagley, E. B.; Dixon, R. E. *Trans. Soc. Rheol.* **1974**, *18*, 371.
15. Bueche, F. In "Reinforcement of Elastomers"; Kraus, G. Ed.; Wiley: New York; **1965**, Ch. 1.
16. Rivin, D.; Avon, J.; Medalia, A. I. *Rubber Chem. Technol.* **1968**, *41*, 330.
17. Cotten, G. R. *Rubber Chem. Technol.* **1972**, *45*, 129.
18. Lewis, T. B.; Nielsen, L. E. *Trans. Soc. Rheol.* **1968**, *12*, 421–443.
19. Payne, A. R. *J. Polym. Sci.: Polym. Symposium 48* **1974**, 169.
20. Gruth, E. *J. Appl. Phys.* **1945**, *16*, 20.
21. Wijayathra, B. Ph.D. Thesis, Department of Chemical Engineering, University of Southern California, CA, **1979**.
22. Smith, T. L. Technical Documentary Report No. ASD-TDR-63-430; Wright Patterson Air Force Base, OH, May, 1963.
23. Harrow, J. A. C.; Payne, A. R. *J. Appl. Polym. Sci.* **1968**, *12*, 889.
24. Shinomura, T.; Takahashi, M. *Rubber Chem. Technol.* **1970**, *43*, 1015.
25. Gent, A. N.; Lindley, P. B. *Proc. Roy. Soc.* A249 **1958**, 195.
26. Williams, M. L.; Schapery, R. A. *Int. J. Fract. Mech.* **1965**, *1*. 64.
27. Chang, W. V. "Cavitation in Elastomers", submitted to *Inter. J. Fracture Mech.*
28. Chang, W. V. "Detection of Cavitation of Acoustic Emission in Bonded Rubber", submitted to *Polymer Science and Engineering.*
29. Chang, W. V. "Internal Rupture of Bonded Rubber Disks in Tension", submitted to *Polymer Science and Engineering.*
30. Gent, A. N.; Lindley, P. B. *Proc. Inst. Mech. Eng.* **1959**, *173*, 111–122.

RECEIVED for review January 20, 1983. ACCEPTED November 14, 1983.

16

Fatigue Crack Propagation in Short-Fiber Reinforced Plastics

R. W. LANG, J. A. MANSON, and R. W. HERTZBERG
Materials Research Center, Lehigh University, Bethlehem, PA 18015

Some of the problems and limitations associated with the application of linear elastic fracture mechanics (LEFM) techniques, as well as the usefulness of this approach to study fatigue crack propagation (FCP) behavior in short-fiber composites, are discussed. Recent work on the influence of several material parameters on the FCP resistance of these materials is reviewed, and various aspects related to the micromechanisms of failure are addressed. On the basis of existing data and considerations of the various energy-absorbing mechanisms involved in the micro-modes of crack advance, a qualitative model is presented to describe the effects of such material variables as fiber orientation, fiber content and aspect ratio, and matrix and interfacial properties.

SHORT-FIBER REINFORCED (SFR) PLASTICS are of increasing technological interest because of improvements in properties relative to those of the unreinforced materials. In addition, these materials can be processed into finished parts of complex shapes by injection molding, compression molding, or extrusion, techniques generally similar to those used for unreinforced polymers. Because many of these materials are being used for structural components that are subjected to fatigue loading, the behavior of these materials under this condition is of prime interest. The purposes of this chapter are to discuss some limitations in the linear elastic fracture mechanics (LEFM) approach to fatigue crack propagation (FCP) in sfr plastics, and to analyze the effects of several material variables on FCP resistance to obtain a better understanding of the failure process.

General Considerations

In spite of the need to understand fatigue in short-fiber (sf) composites, the state of the art is not nearly as advanced as it is for unreinforced polymers

0065-2393/84/0206-0261$06.00/0

(*1*) or metals (*2*). Although the S–N response (stress amplitude versus number of cycles to failure) has received some attention (*1, 3–6*), studies of this type generally do not distinguish between the initiation of cracks and their propagation to catastrophic failure. And yet, as a result of the stress concentrations introduced by the discontinuities associated with the presence of short fibers, failures may easily initiate (*6*). Therefore, the total fatigue life of these materials may well be dominated by the kinetics of crack and damage propagation in many applications.

Because of the limitations of the S–N approach, many efforts have concentrated on describing the kinetics of FCP in metals and polymers by using the concepts of fracture mechanics (*1, 2*). Crack growth rates per cycle, da/dN, are most commonly related to the prevailing stress intensity factor range, ΔK, according to a relationship proposed by Paris (*7*)

$$\frac{da}{dN} = A \Delta K_I^m \tag{1}$$

where A and m depend on the material and test conditions. The applied ΔK_I-range (subscript I stands for opening mode conditions) is given by the cyclic load range, ΔP, the crack length, a, the width, W, and the thickness, B, of the specimen, and a geometrical factor, Y, as

$$\Delta K_I = \frac{\Delta P}{B\sqrt{W}} Y\left(\frac{a}{W}\right) \tag{2}$$

Solutions for $Y(a/W)$ are available from the literature for many specimen configurations (*8*).

Materials and Fatigue Testing

To obtain a more generalized understanding of FCP in sf composites, previous results will be cited throughout this chapter. The thermoplastic matrix materials in the composites investigated so far include various types of nylon (*9–14*), poly(ethylene terephthalate) (PET) (*15–17*), polysulfone (PSF) (*18*), and polystyrene (PS) (*19*)[1]; short fibers include glass and carbon fibers and Kevlar. In addition, discontinuous fibers of glass and aluminum have been used in FCP studies of reinforced polyester (*20*) and epoxy resins (*21*), respectively.

All materials included in this study for illustrative purposes were obtained in the form of injection-molded plaques. A list of these materials

[1]In Reference 19, the polymer material was mislabeled as a styrene–acrylonitrile copolymer (SAN) instead of polystyrene (PS). However, this error does not alter any of the conclusions in the publication.

along with their designations and sources is given in Table I. Compact type specimens were machined to be loaded either longitudinally (L) or transversely (T) to the main processing direction. Although the material characteristics and testing variables pertinent to the discussion will be mentioned, more detailed information is available from the references listed in Table I.

The LEFM Approach to FCP in SF Composites

Whether or not FCP in sf composites can be characterized and accurately modeled by using LEFM techniques developed for isotropic materials will depend on several factors. One problem noted previously (*1*, *20*, *22*) is that in general with a fiber composite the "crack" is not a simple entity that can be readily characterized in terms of length. Even in cases where a crack is intentionally grown from a notch, one cannot always unambiguously define the position of the crack tip in an analogous simple way as in homogeneous materials. Rather, fatigue failure in notched sf composites occurs by propagation of a main crack surrounded by innumerable side cracks diverting from the plane of the main crack. Instead of a clear-cut crack tip, a diffuse damage zone consisting of many microcracks exists. Thus, the accumulation of damage in notched sf composites occurs by a damage propagation process rather than by a well-defined crack propagation mechanism. Of course this situation also implies that the term "crack growth rate" should be replaced by the physically more meaningful term "damage growth rate." Although both expressions are commonly used, this distinction should be kept in mind.

Whereas the crack-tip damage zone may be conceived as being much like a plastic zone in an elastic–plastic material (*23*), local areas may occur along the main crack where single fibers or fiber bundles bridge the crack

Table I. Materials and Specimen Designation

Material Designation[a]	*Commercial Designation*	*Matrix Type*	*Glass Fiber*[b] *Content, Vol (wt) %*	*Ref.*
A-N66	Zytel 101	N66	—	*13*
A-N66/18G	Zytel 70G33	N66	18(33)	*13*
B-N66	R-1000	N66	—	*13,14*
B-N66/16G	RF-1006	N66	16(30)	*13,14*
B-N66/31G	RF-10010	N66	31(50)	*13,14*
B-PS	C-1000	PS	—	*19*
B-PS/18G[c]	CF-1007	PS	18(35)	*19*

[a]Series A (Du Pont) and B (LNP-Corporation) were supplied as end-gated and side-gated injection molded plaques, respectively; symbols L and T added to the material designation refer to the direction of applied load relative to the major flow direction.
[b]E-glass.
[c]Supplied with fibers 3.2, 6.4, and 12.7 mm long before processing.

surfaces. In this manner stresses will be transferred across the cracked surfaces by either frictional forces or, as long as it is not already destroyed, by the interfacial bond between fiber and matrix itself. This condition certainly violates the conventional assumption in LEFM of the integrity of a crack.

Other complicating factors encountered in some sf composites are related to the anisotropy resulting from fiber orientation, which develops under some processing conditions. Although solutions for stress intensity factors exist for some special cases (*24*, *25*), fiber orientation and, therefore the degree of anisotropy, may be difficult to predict or to determine and may well vary with position within a component or even within a test specimen. Such factors make an accurate determination of K difficult, if not impossible.

Another problem experienced in some cases is the departure of the propagating crack from the desired horizontal fracture plane because of fiber orientation. Figure 1 shows characteristic examples of failed specimens of short glass-fiber reinforced (sgfr) Nylon 66 (N66) and of sgfr PS. Although the crack in sgfr N66 essentially grew in the desired plane perpendicular to the loading direction, excessive departure from this plane is evident in the PS composite. Crack growth in the latter case occurs under combined Mode I/Mode II (opening/in-plane shear) conditions and Equation 1 is no longer applicable.

This discussion implies that an interpretation of K for sf composites (based on calculations that assume a uniform, isotropic, elastic material) as a parameter that provides an accurate description of the stress field in the vicinity of the crack tip is by no means self-evident. Hence the results of an FCP test may not always provide a quantitative and specimen-independent description of the material behavior that ultimately could be used for component design. In such cases the significance of ΔK is restricted to that of a

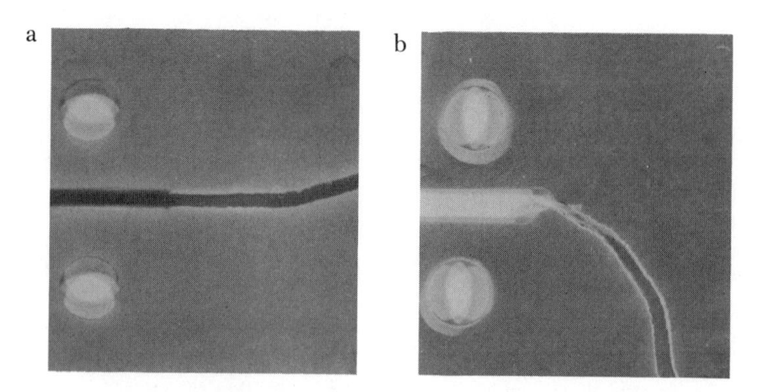

Figure 1. Typical examples of failed FCP specimens: (a) sgfr N66 (B-N66/16G-T), (b) sgfr PS (B-PS/18G-L).

convenient parameter to describe the combined conditions of loading and specimen geometry. Nevertheless, FCP experiments are still extremely useful for comparative purposes (if specimen geometry and loading conditions are kept the same) as an aid for the material selection for both notched and unnotched components. Valuable information on the micromechanisms of fatigue failure can be obtained; this information may ultimately lead to the development of materials with improved FCP resistance through a better understanding of the various energy-absorbing mechanisms associated with the fracture process.

Effects of Material Variables

The material parameters can be related to the three main elements present in an sf composite, namely the fibers, the matrix, and the fiber–matrix interface. Variables associated with the reinforcing fibers include the fiber content, fiber orientation, fiber aspect ratio (length-to-diameter ratio, l/d), and the kind of fiber used.

Effect of Fiber Content. Analogous to the substantial improvements in other mechanical properties (*6*, *26*, *27*), FCP resistance generally increases with fiber content at least over the range investigated (*9*, *12–14*, *16*, *19*). For example, a significant increase in FCP resistance resulting from the addition of glass fibers is clearly visible in the comparison of A-N66/18G with pure N66 in Figure 2. The value of ΔK necessary to drive a crack at a given growth rate is about twice as high for the composite than for the unreinforced matrix. Similar behavior is shown in Figure 3 for Nylon 66 of Series B (*see* Table I) in an alternative representation of FCP data. Crack growth rates at constant values of ΔK decrease continuously with an increase in fiber volume fraction, v_f, at least up to $v_f = 0.31$.

The improvement in behavior can be interpreted in terms of (1) the stress transfer from the matrix to the much stronger fibers along with the overall increase in specimen stiffness; (2) the additional energy dissipation mechanisms associated with debonding, fiber breakage and pullout, and local plastic deformation in the matrix around fibers; and (3) crack blunting due to the complex damage zone that acts to reduce the stress intensity level (i.e., lower effective ΔK) and hence to decrease the severity of the crack tip stress field.

Effects of Fiber Orientation. The fiber orientation in any sf composite will depend primarily on the melt flow characteristics as affected by the processing technique and the processing conditions (*27*). For example, parts with a high degree of orientation in the processing direction may be obtained by extrusion, whereas more or less random fiber orientation could be achieved by compression molding. In the case of injection molding, fi-

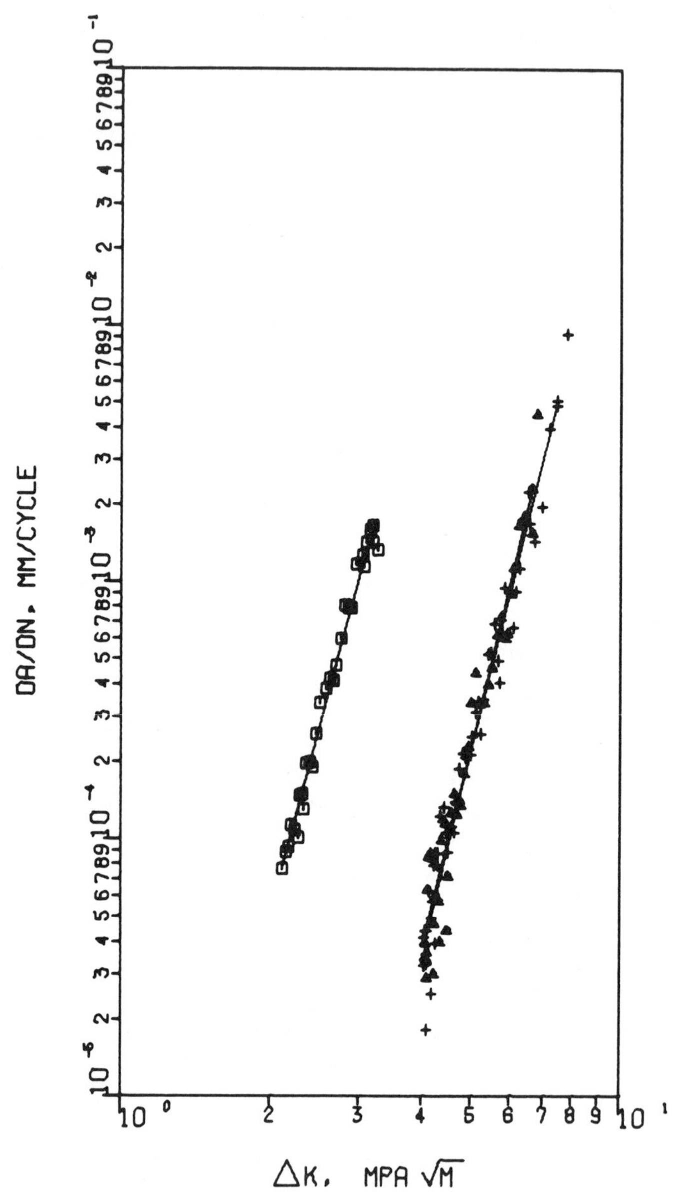

Figure 2. Fatigue crack propagation behavior of Nylon 66 (A-N66) and sgfr Nylon 66 (A-N66/18G) at 10 Hz and a minimum/maximum load ratio of 0.1. Key: ◫, A-N66; ▲, A-N66/18G-T; and +, A-N66/18G-L.

ber orientation will depend on injection speed, mold geometry, and geometry and position of the gate. In addition, the nature of the matrix, fiber content and aspect ratio, and fiber–matrix bond strength may be important. Indeed, several different observations have been made for injection molded materials.

Although no effects or only minor effects of specimen orientation have been reported in some cases (*12, 13*), pronounced effects have been found in others (*15–17*). For example, macroscopic growth rates in Figure 2 are essentially the same for specimens tested longitudinally and transversely to the mold fill direction. On the other hand, Figure 3 reveals that the influence of orientation may well vary with fiber content at least to some degree. Although no effect of orientation was found for B-N66 and B-N66/16G, some dependence on orientation was observed for the composite with the highest fiber content. An even more significant effect of orientation has been reported by Friedrich (*15–17*) for sgfr PET.

In all these instances, the influence of specimen orientation on FCP behavior has been related directly to the fiber orientation distribution

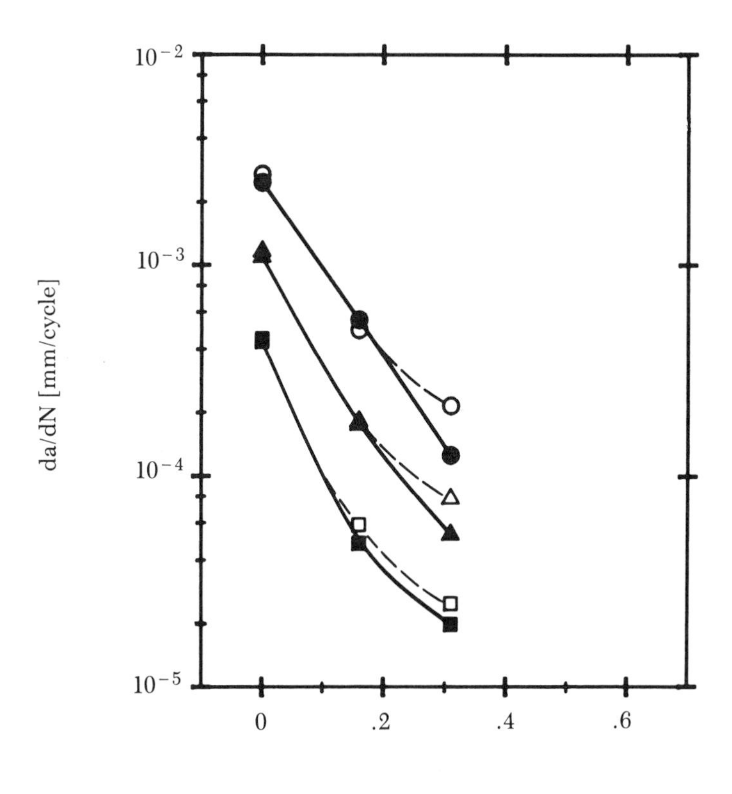

Figure 3. Effect of fiber volume fraction, v_f, on fatigue crack growth rates in sgfr Nylon 66 (B-N66) at various levels of ΔK for two specimen orientations (frequency = 10 Hz, minimum/maximum load ratio = 0.1). Key (ΔK = 4.5 MPa $\sqrt{m}$): ●, L direction; and ○, T direction; (ΔK = 4.0 MP$\sqrt{m}$) ▲, L direction; and △, T direction; (ΔK = 3.5 MPa$\sqrt{m}$) ■, L direction; and □, T direction.

within the specimens. In this context (*27, 28*) the fiber orientation in end-gated injection molded plaques varies over the plaque thickness, as shown schematically in Figure 4. Close to the mold surfaces where the shear gradient is high while filling the mold, skin layers will develop with fibers predominantly oriented in flow direction. In sharp contrast, fibers will be aligned transversely to the major flow axis in the core region because of divergent flow, which occurs when the melt stream passes from the narrow gate into the much wider mold.

This model might be somewhat oversimplified (*28*), however, it is sufficient for the purpose of our discussion and consistent with observations of fracture surfaces (*12, 13*). An example of the appearance of fracture surfaces is shown in Figure 5 for A-N66/18G-L. Indeed, the fiber orientation in skin and core layer corresponds roughly to what has been shown in Figure 4. That is, fibers for this specimen orientation are oriented mainly perpendicular to the fracture surface in the skin layer (Figure 5a) and are aligned predominantly in or at small angles to the fracture surface plane in the core layer (Figure 5b). Because the two skin layers together appeared to be approximately equal in size to the core layer, the orientation independent behavior of A-N66/18G in Figure 2 can be explained in terms of a quasi-isotropic fiber array, even though the specimens can be considered as laminates of plies having distinctly different fiber orientations.

As expected, microscopy (*16, 29*) and dynamic mechanical spectroscopy (*19*) have shown that a higher FCP resistance in a given direction is a

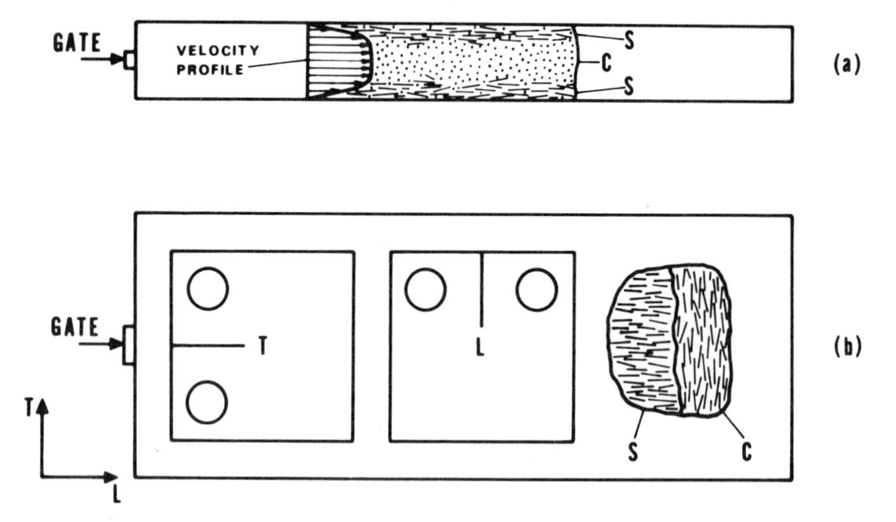

Figure 4. Schematic representation of an end-gated injection molded plaque showing (a) the "layer" structure in fiber orientation across the plaque thickness, and (b) the orientation of FCP specimens. Key: S, skin layer; C, core layer; and L, T, direction of applied load with respect to the mold fill direction.

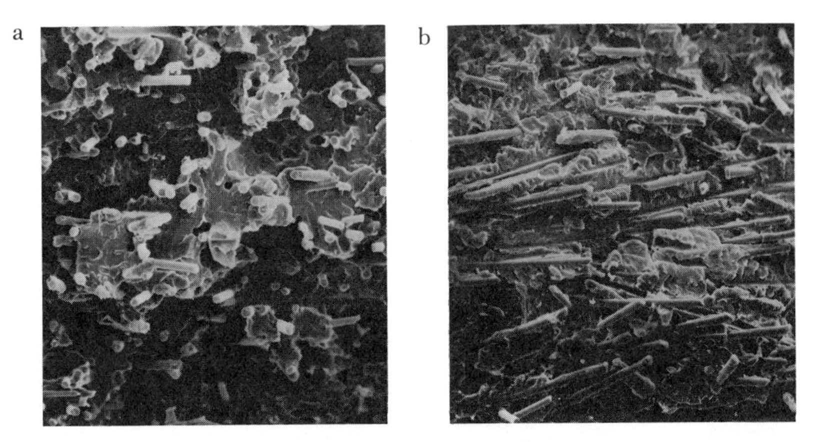

Figure 5. Scanning electron micrographs from the fast fracture region of an L specimen of sgfr Nylon 66 (A-N66/18G-L): (a) skin layer with fiber orientation at high angles to fracture surface and (b) core layer with main fiber orientation at small angles to fracture surface (fiber diameter ≈ 10 μm).

result of a smaller average angle between fibers and the loading axis. Again, this smaller angle increases the load bearing capability of the fibers and results in a more energy-consuming path of the crack.

One more aspect of the combined effects of fiber orientation and specimen thickness in injection molded plaques such as shown in Figure 4 deserves further comment. From simple rheological considerations one would expect that overall fiber orientation should be dominated by the effects of shear flow for small plaque thicknesses, whereas the relative contribution of orientation development associated with divergent flow should be prevalent at higher thicknesses. Accordingly one might speculate that L specimens should be more FCP resistant at low thicknesses and T specimens should be more FCP resistant at high thicknesses. Such conditions have indeed been verified experimentally by Friedrich.[2] Moreover, the differences in the results on the orientation effects in sgfr N66 (Series A) and sgfr PET (*15–17*) discussed are in agreement with studies on fiber orientation by Menges and Geisbüsch (*28*). From these considerations it follows that the FCP resistance in the L and T directions should decrease and increase, respectively, as the plaque thickness increases. However, this thickness dependence is solely a result of changes in the overall fiber orientation distribution and therefore must not be mistaken for a transition in stress state (from plane stress to plane strain) as has been observed in unreinforced polymers (*30*).

Other Effects Related to the Fibers. In sharp contrast to other mechanical properties (*6, 26, 27*) and to the fatigue resistance in unnotched

[2]Friedrich, K., personal communication, 1982.

specimens of boron-fiber reinforced epoxies (*31*), essentially no effect of fiber aspect ratio (before processing) on FCP resistance was found for sgfr PS. A modified testing procedure had to be used (*19*) because of excessive crack curvature (Figure 1). In a manner analogous to that of an S–N test for unnotched specimens, the initial stress intensity factor range, ΔK_o (given by the applied load range for specimens of constant geometry and initial crack length), is plotted in Figure 6 versus the number of cycles to failure, N_f, for compositions containing fibers of various length before processing. The lack of any noticeable effect of the fourfold change in fiber length (and hence aspect ratio) is somewhat surprising, but certainly related, at least in part, to the more severe reduction in fiber length by fiber breakage during processing in the composites prepared from the longer fibers. However, more work remains to be done to clearly establish the effect of fiber length and distribution. On the other hand, consistent with the results in Figure 3 for sgfr N66 (same plaque and specimen geometry) and with the findings from dynamic mechanical tests (*19*), the longer fatigue lives of the L speci-

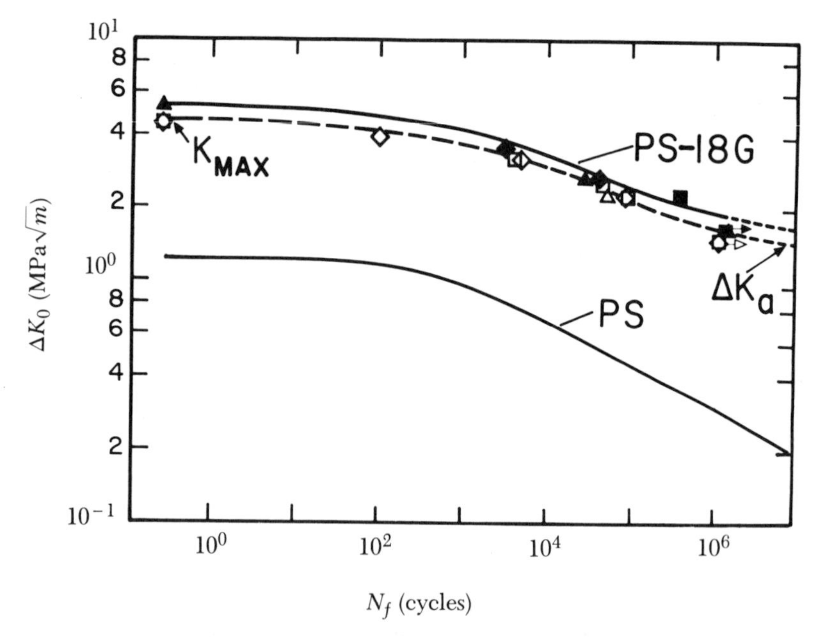

Figure 6. Plot of initial stress intensity factor range, ΔK_o, versus the number of cycles to failure, N_f, for PS (B-PS) and sgfr PS (B-PS/18G with different fiber length before processing). The limiting values K_{max} and ΔK_a correspond to a measure of quasi-static fracture toughness, and the value of ΔK_o below which the damage propagation rate became vanishingly small. Key (3.2-mm fibers): ■, L direction; and □, T direction; (6.4-mm fibers) ◆, L direction; and ◇, T direction; (12.7-mm fibers) ▲, L direction; and △, T direction.

mens are again believed to be related to the smaller average angle of fiber orientation with respect to the loading axis.

With regard to the kind of fiber used, Mandell et al. (*18*) have shown for PSF that the incorporation of carbon fibers yields a more FCP-resistant material than is obtained by using glass fibers. This comparison was based on composites with equal fiber weight content (40%); however, because of the lower density of carbon fibers, the difference in behavior should decrease when the physically more meaningful fiber volume content is used as the comparative parameter.

Another study using a total fiber content of 20 vol% of carbon fibers, polyaramid fibers (Kevlar), and a hybrid system of these two in a Nylon 66 matrix was performed by DiBenedetto and Salee (*11*). Whereas the hybrid and the polyaramid composite were approximately equal in FCP resistance, both were superior to the carbon reinforced material. Thus the improvement in overall properties of sf composites by using hybrid systems (*32*) may apply to FCP as well and offers many opportunities in the development of materials for specific needs.

Effects of Matrix and Interfacial Properties. Although hardly any information is available as to the influence of variations in matrix and interfacial properties, an increase in matrix ductility and interfacial bond strength leads to a decrease in FCP rates, at least for sgf-PET (*16*). Consistent with this observation is the higher FCP resistance of sgfr composites of the more ductile N66 in comparison to the rather brittle PS revealed in the course of this investigation for specimens cut from plaques of identical geometry. With regard to interfacial characteristics, a comparison between the sgfr versions of A-N66 and B-N66 suggested that the superiority in FCP behavior of the composite with the less fatigue resistant neat matrix (i.e., A-N66) was a result of a higher degree of fiber–matrix adhesion (*13*). Finally, for the ductile materials neat and short carbon-fiber reinforced poly-(ethylene) (PE), crack growth rates may in fact increase when fibers are added; this finding is in agreement with studies on the impact resistance of sgfr PE (*34*). All these observations clearly indicate the importance of matrix and interfacial properties on the FCP process.

Generalized Model

Clearly, all of these material parameters can be important in determining the overall FCP response in an sf composite. Although a rigorous and quantitative treatment of each variable is not yet possible, several conclusions may be drawn and predictions may be made that should aid research efforts in the future. The basis of this analysis is the combination of existing data with considerations on the various energy-absorbing mechanisms that may ultimately determine the fracture resistance in these materials.

Similarly to static or impact loading conditions (*6, 26, 27*), the crack propagation process may involve many energy-absorbing mechanisms such as fiber breakage, fiber debonding and pullout, cavitation, matrix fracture, microcrack development and coalescence, and crack branching. In this context, fracture of brittle fibers such as glass and carbon will contribute little to the overall fracture energy, whereas fiber debonding and pullout will have a high energy-absorbing tendency. In addition, the energy dissipated in matrix fracture may be quite significant whenever considerable plastic deformation (i.e., matrix drawing, craze and void formation along with void coalescence) is involved. Finally a high density of microcracks or secondary cracks around the main crack (including crack branching) will also act as an energy sink and blunt the crack by relaxation of the stresses at the crack tip.

On the basis of this information, the following qualitative model may provide some valuable insight as to the potential contributions of several material parameters to the FCP resistance of sf composites. By using the relative growth rate at a given value of ΔK

$$\left(\frac{da}{dN}\right)_{rel} = \left[\left(\frac{da}{dN}\right)_c \Big/ \left(\frac{da}{dN}\right)_m\right]_{\Delta K = \text{const}} \quad (3)$$

(subscripts c and m stand for composite and neat matrix, respectively) as the relevant normalization parameter, the upper and lower bounds in behavior may be defined as a function of the fiber volume fraction, v_f, as shown in Figure 7. The letters along with the arrows in this schematic graph indicate the expected shifts in behavior for an increase in matrix ductility, D, for improved fiber–matrix adhesion, A, and for an increase in fiber aspect ratio, F.

Lower Bound Behavior. A lower bound for the relative growth rate will exist for the case of good adhesion with fibers mainly oriented in load direction, that is perpendicular to the crack plane. In this case a maximum of the applied load will be transferred to the fibers and $(da/dN)_{rel}$ will initially decrease with an increase in fiber content for the reasons we have presented. A minimum in the relative growth rate may be expected at a fiber content of about 50% because of fabrication difficulties in obtaining a good quality composite at fiber concentrations above this value (*6, 16*).

Possibly a slight but random misorientation of the fibers from load direction might be beneficial. According to the maximum stress criterion proposed by Stowell and Liu (*35*), the dominant failure mode will change under these conditions from fiber failure to shear failure parallel to the fibers. Failure may then occur either in the matrix or along the fiber–matrix interface depending on the strength of the matrix relative to the interfacial bond. Although the loss in overall stiffness and strength of the composite by

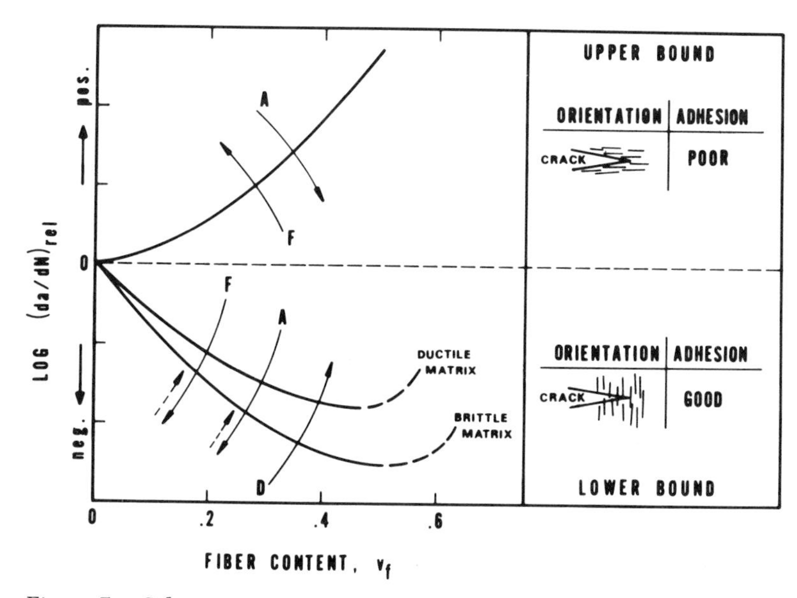

Figure 7. Schematic representation of the effect of fiber volume fraction, v_f*, on the relative crack-growth rate under upper bound and lower bound conditions. Key:* D*, matrix ductility;* A*, fiber–matrix adhesion; and* F*, fiber aspect ratio.*

slight fiber misalignment might still be rather insignificant (*36,37*), one might expect a larger damage zone at the crack tip because of an enhanced tendency for shear failure by debonding, fiber pullout, plastic deformation of the matrix, microcracking and crack branching, all mechanisms that increase energy absorption and crack-tip blunting.

An increase in matrix ductility should shift the lower bound behavior in Figure 7 toward higher *relative* growth rates. Thus, although an sf composite with a more ductile matrix may still be more FCP-resistant than one with a brittle matrix in an *absolute* sense, the *relative* improvement due to the incorporation of fibers will decrease as the matrix ductility increases. This decrease is consistent with the experimental observations on PS (brittle; strong improvement) (*19*), N66 (more ductile; less relative improvement) (*13*), and carbon-fiber reinforced PE (*33*) (very ductile; $(da/dN)_{rel} > 1$) discussed and also with results from fracture toughness (*16, 38*) and impact tests (*6*). The increase in the relative growth rate with matrix ductility may be explained as a result of (1) the generally higher FCP resistance of ductile versus brittle polymers in the unreinforced condition and (2) a relatively larger effect on toughness reduction in ductile matrices because of the triaxial constraint caused by the presence of the rigid fibers.

As mentioned, comparing the effects of two different fiber treatments showed, at least for sgfr PET, that improvement in the fiber–matrix bond strength was beneficial from the standpoint of fatigue crack propagation

(*16*). However, it is not clear yet whether the strongest possible interface will also yield a material with the highest FCP resistance. Excellent adhesion will certainly maximize the load transfer from the matrix to the fibers, but the presence of a somewhat weaker interface will tend to favor some of the important energy-dissipating mechanisms (i.e., debonding, fiber pullout, enhanced matrix deformability) and increase the degree of crack blunting by formation of shear cracks parallel to the fibers. Thus some intermediate value in adhesion may well lead to the highest FCP resistance, especially for brittle matrices in which the latter mechanisms are of utmost importance in improving toughness (*6*).

Fiber aspect ratio (before processing) did not seem to have any significant effect over the range investigated in PS composites (Figure 6). However, because of the detrimental effects on FCP reported for particulate fillers and glass spheres (corresponding to $l/d \approx 1$) (*13, 39*), we may expect that an increase in fiber aspect ratio will certainly decrease $(da/dN)_{rel}$ under lower bound conditions at least as long as the l/d ratio remains below some critical value. Thus, whereas stiffness and strength in sf composites continuously increase with an increase in fiber aspect ratio and approach an asymptotic value, energy considerations predict a maximum for fracture energy when the fiber length is equal to the minimum fiber length (also referred to as critical fiber length, l_c) in which the maximum allowable fiber stress can be achieved (*6, 27*). These contradictory trends indicate again, of course, that some intermediate value in l/d (or better, a certain l/d distribution) might exist for optimum performance.

Upper Bound Behavior. In sharp contrast to the lower bound conditions, an upper bound in behavior will exist for the case of poor (or no) adhesion with fibers predominantly oriented perpendicular to the load direction (Figure 7). Under these conditions the fibers cannot carry any of the load, and the relative growth rate can be expected to increase progressively with fiber content. This increase is a result of the additional stress concentrations introduced by the presence of the fibers and the increased contribution of weak interfaces over which crack growth can occur. An increase in fiber matrix adhesion will decrease relative growth rates by creating a more fatigue resistant path over the stronger interface or through the matrix itself. However, in some cases (*40*) the detrimental effect of a weak interface on the FCP resistance even under upper bound conditions may again be offset by the tendency to form a larger crack-tip damage zone, which enhances energy dissipation and acts to blunt the crack.

On the contrary, the relative effects of matrix ductility and fiber aspect ratio are much less obvious under these conditions. Although an sf composite with a ductile matrix may again be expected to be more FCP resistant than one with a brittle matrix in an absolute sense, a prediction of the relative growth rates is difficult and may well depend on the specific systems used. Similarly, whereas the influence of fiber length is somewhat uncertain, we might anticipate that $(da/dN)_{rel}$ will increase with an increase in aspect ratio, at least to some extent, because of the decrease in the average number of fiber ends at a given fiber volume content. In contrast to high-aspect ratio fibers where crack advance is facilitated and less disturbed over longer distances along the weak interface, the increased number of fiber ends associated with a lower l/d ratio will force the crack more frequently to take a more fracture- and fatigue-resistant path through the matrix. This expectation is, of course, consistent with the general concept of toughness enhancement by the presence of short fibers in continuous fiber composites (*27,41*). In any case, the contradictory effects of fiber aspect ratio on the improvement of upper and lower bound behavior are important considerations. Although an increase in l/d ratio may be beneficial under lower bound conditions, it will be detrimental under upper bound conditions.

Of course, the perfectly aligned model composite depicted in Figure 7 will be inappropriate for practical applications where multiaxial stress fields act on a component. In addition, the idealized, uniform fiber orientation assumed for the limiting conditions will be difficult if not impossible to achieve, depending on the processing conditions. Thus, extrusion techniques may yield components with fiber orientation approaching these idealized cases in and transverse to the extrusion direction. On the other hand, a "layer" structure is more characteristic for injection molded parts, and generally, isotropic composites with randomly oriented short fibers may be produced by compression molding. Hence, components produced by the latter two techniques should more or less approach the lower bound behavior in Figure 7 depending on the actual fiber orientation distribution. However, the fact that in many cases component failure may initiate where the local orientation is unfavorable with respect to the applied loads should emphasize the necessity to also study upper bound properties because they may well become the service-life determining factors.

Summary

In view of our discussion, it seems difficult to predict service performance and life times of sfr plastic components under fatigue loading conditions from simple molded test specimens by using conventional fracture mechanics techniques. Nevertheless FCP studies are extremely valuable to increase our knowledge of the mechanisms involved in the fatigue failure process. Although the problem of fatigue crack or damage propagation in sf composites is evidently rather complex, the large number of possible variations in material parameters offers great potential in the tailoring of materials with improved fatigue resistance. Our purpose was to outline some of the important variables involved in the crack and damage propagation process and to stimulate further efforts in the development of a better understanding of the principles governing fatigue failure in these materials.

Acknowledgments

The authors acknowledge partial support from the Office of Naval Research. We are also grateful to B. Epstein, E. I. du Pont de Nemours & Co., and J. Theberge, LNP Corporation, for supplying the material.

Literature Cited

1. Hertzberg, R. W.; Manson, J. A. "Fatigue of Engineering Plastics"; Academic Press: New York, 1980.
2. Hertzberg, R. W. "Deformation and Fracture Mechanics of Engineering Materials"; Wiley: New York, 1976.
3. Theberge, J.; Arkles, B.; Robinson, R. *Ind. Eng. Chem., Prod. Res. Div.* **1976**, *15*, 100.
4. Mandell, J. F.; Huang, D. D.; McGarry, F. J. *Polym. Compos.* **1981**, *2*, 137.
5. Dally, J. W., Carrillo, D. H. *Polym. Eng. Sci.* **1969**, *9(6)*, 434.
6. Agarwal, B. D.; Broutman; L. J. "Analysis and Performance of Fiber Composites"; Wiley-Interscience: New York, 1980.
7. Paris, P. C., Proc. 10th Sagamore Army Mater. Res. Conference 1964, Syracuse, NY; p. 107.
8. Sih, G. C. M. "Handbook of Stress Intensity Factors"; Lehigh University, 1973.
9. Suzuki, M.; Shimizu, M.; Jinen, E.; Maeda, M.; Nakamura, M. *Proc. Int. Conf. Mech. Behav. Mater, 1st.* **1971**, 5, 279, Kyoto.
10. DiBenedetto, A. T.; Salee, G. *Polym. Eng. Sci.* **1979**, *19(7)*, 512.
11. DiBenedetto, A. T.; Salee, G. *Tech. Rep. AMMRC CTR 77-12* **1977**.
12. Lang, R. W.; Manson, J. A.; Hertzberg, R. W. *Org. Coat. Plast. Chem.* **1981**, *45*, 778.
13. Lang, R. W.; Manson, J. A.; Hertzberg, R. W. *Polym. Eng. Sci.* **1982**, *22(15)*, 982.
14. Lang, R. W.; Manson, J. A.; Hertzberg, R. W. In "The Role of the Polymeric Matrix in the Processing and Structural Properties of Composite Materials," Nicholais, L. Eds.; Seferis, J. C.; Plenum Publishing Corporation: New York, 1983; pp. 377–396.
15. Friedrich, K. *Colloid Polym. Sci.* **1981**, *259*, 808.
16. Friedrich, K. *Kunstst.* **1982**, 72, 5, 290.
17. Friedrich, K. In "Deformation, Yield and Fracture of Polymers"; Plastics and Rubber Institute: London, 1982; p. 26.1.
18. Mandell, J. F.,; McGarry, F. J.; Huang, D. D.; Li, C. G. Res. Rep. R81-6, Dept. of Mat. Sci. Eng., MIT, Cambridge, MA 1981.
19. Lang, R. W.; Manson, J. A.; Hertzberg, R. W.; *Org. Coat. Plast. Chem.* **1983**, *48*, 816.
20. Owen, M. J.; Bishop, P. T. *J. Phys. D.* **1974**, *7*, 1214.
21. Thornton, P. A. *J. Compos. Mater.* **1972**, *6*, 147.
22. Reifsnider, K. *Int. J. Fract.* **1980**, *16(6)*, 563.
23. Gaggar, S. K.; Broutman, L. J. *Int. J. Fract.* **1974**, *10*, 606.
24. Delale, F.; Erdogan, F. *J. Appl. Mech., Trans. ASME* **1977**, *44*, 237.
25. Kaya, A. C.; Erdogan, F. *Int. J. Fract.* **1980**, 16, 171.
26. Manson, J. A.; Sperling, L. H. "Polymer Blends and Composites"; Plenum: New York, 1976.
27. Folkes, M. J. "Short Fibre Reinforced Thermoplastics"; Research Studies Press: Chichester, 1982.

28. Menges, G.; Geisbüsch, P. *Colloid Polym. Sci.* **1982**, *260*, 73.
29. Friedrich, K. *Praktische Metallographie* **1981**, *18(11)*, 513.
30. Pitman, G.; Ward, I. M. *J. Mater. Sci.* **1981**, *15*, 635.
31. Lavengood, R. E.; Gulbransen, L. E. *Polym. Eng. Sci.* **1969**, *9*, 365.
32. Richter, H. *Kunstst.* **1977**, *67*, 739.
33. Connelly, G., unpublished research, Lehigh University 1982.
34. Bernardo, A. C. *SPE J.* **1970**, *26(10)*.
35. Stowell, E. Z.; Liu, T. S. *J. Mech. Phys. Solids* **1961**, 9, 242.
36. Lees, J. K., *Polym. Eng. Sci.* **1968**, *8*, 186.
37. Lees, J. K. *Polym. Eng. Sci.* **1968**, *8*, 195.
38. Mandell, J. F.; Darwish, A. Y.; McGarry, F. J. In "Test Methods and Design Allowables for Fibrous Composites"; ASTM STP 734, C. C. Chamis, Ed.; 1981, pp. 73–90.
39. Friedrich, K.; Karsch, U. A. *Polym. Compos.* **1982**, *3(2)*, 65.
40. Attalla, G., unpublished research, Lehigh University 1983.
41. Short, D.; Summerscales, J. *Composites* **1980**, *33*.

RECEIVED for review April 11, 1983. ACCEPTED September 29, 1983.

17

Prediction and Control of Fiber Orientation in Molded Parts

WILLIAM C. JACKSON, FRANCISCO FOLGAR, and CHARLES L. TUCKER III

Department of Mechanical and Industrial Engineering, University of Illinois at Urbana-Champaign, Urbana, IL 61801

The problem of predicting the fiber orientation pattern in a part molded from a short-fiber reinforced polymer is formulated. The solution requires that a governing equation for the fiber orientation distribution function be integrated along the paths of fluid particles during mold filling. A set of simplifying assumptions is presented for thin compression-molded parts with fibers much longer than the part thickness. Example calculations agree with familiar qualitative rules for fiber orientation and with quantitative experiments on sheet molding compounds.

POLYMERS WITH DISCONTINUOUS FIBER REINFORCEMENT are attractive materials because they combine stiffness, strength, and light weight. In addition, they can be manufactured by low labor content processes such as injection molding, compression molding, and extrusion. Virtually all processing techniques for these materials can orient the fibers preferentially, a feature that controls many physical properties of the resulting composite. Stiffness, strength, thermal expansion, and thermal conductivity are among the many properties of a composite that change with fiber orientation. In many cases the changes are major. For example, the elastic modulus of a typical glass fiber–polyester matrix composite more than doubles in one direction and drops by a factor of four in the perpendicular direction as the fiber orientation changes from planar random to fully aligned (*1*).

Mechanical property variations of this order have significant impact on the performance of the materials, but this fact has largely been ignored by designers and processors. In many instances, fiber orientation has been treated as an undesirable source of mechanical property variation, and processors have sought to minimize any preferential orientation. A much more fruitful approach would be to control the fiber orientation to improve material performance. When this approach can be taken by judicious de-

0065-2393/84/0206-0279$06.25/0

sign of the part and careful choice of processing conditions, these improvements can be had with no increased costs. Even the ability to predict how fibers will be oriented in a part would be useful, because, otherwise, one must make a mold, manufacture prototype parts, and test them to be sure that they will work.

In fact, we currently have the ability to predict mechanical properties of short-fiber composites once the fiber orientation distribution is known. Elastic constants (*1–5*), thermal expansion coefficients (*1–3*), and tensile strength (*5, 6*) can be accurately predicted. Even the nonlinear stress–strain curve of a brittle matrix composite can be predicted (*7*). Other properties such as notch sensitivity and viscoelastic response have been shown experimentally to depend on fiber orientation (*1, 8*). Even a simple quantitative approach to account for the orientation distribution would improve design.

In this chapter, we consider the problem of predicting fiber orientation distributions as a function of processing variables for molded parts. We first present a general approach to the problem and then discuss its application to compression molding. The goal of the calculation is to predict the entire fiber orientation distribution at every point in the finished part. The suggestion has been made that one should try to predict or correlate two particular weighted integrals of the orientation distribution as a function of processing conditions, because, under certain symmetry conditions, one can predict stiffness constants and thermal expansion coefficients by knowing only the values of the two integrals (*1*). However, not all properties can be predicted from this information. We have chosen the more general approach of calculating the entire orientation distribution function.

Background

The motion of a single rigid ellipsoidal particle in a deforming viscous fluid was originally derived by Jeffrey (*9*). His equations have been shown (*10*) to apply to rigid particles with other shapes as well (e.g., cylinders). Although the particle motion is affected somewhat by viscoelasticity in the fluid (*11*), these equations can be applied with some confidence to polymer processing problems where few fibers are present. Certainly the high viscosity of polymer melts and resins and the small diameter of reinforcing fibers validate the assumption of negligible inertia.

Consider the simplified case of a rigid cylindrical particle whose axis lies in the x–y plane and a flow field for which the unperturbed flow is planar in the x–y plane. The fiber orientation is described by the single angle ϕ, as shown in Figure 1. Jeffrey's equations give the motion of this particle as

$$\dot{\phi} = [r_e^2/(r_e^2 + 1)]\{-\sin\phi\cos\phi\,(\partial v_x/\partial x) - \sin^2\phi\,(\partial v_x/\partial y) + \cos^2\phi\,(\partial v_y/\partial x) + \sin\phi\cos\phi\,(\partial v_y/\partial y)\} - [1/(r_e^2 + 1)]\{-\sin\phi\cos\phi(\partial v_x/\partial x) + \cos^2\phi\,(\partial v_x/\partial y) - \sin^2\phi(\partial v_y/\partial x) + \sin\phi\cos\phi\,(\partial v_y/\partial y)\} \quad (1)$$

where $\dot{\phi}$ is the angular velocity of the particle, and v_x and v_y are the fluid velocity components. The quantity r_e is the equivalent ellipsoidal aspect ratio (*10*) and is on the order of the length-to-diameter ratio of the cylinder.

This equation shows that fiber motions are controlled by the kinematics of the flow. In simple and planar extensional flows, the fiber rotates toward the direction of stretching; when the fiber is aligned in that direction it is in stable equilibrium. In simple shear flow, the fiber rotates with a periodic motion, but spends much more time oriented in the direction of the flow than in other directions. Hence, we have the familiar rules of thumb: (1) shear flows orient fibers in the direction of flow, and (2) extensional flows orient fibers in the direction of extension.

These equations of fiber motion have been used by Harris and Pittman (*12*) to analyze the fiber orientation of a dilute suspension of fibers in a converging nozzle. They solved for the velocity distribution in the nozzle, then used the deformation field from the solution to calculate fiber orientation changes. Their calculations were in excellent agreement with experiments.

Some calculations of fiber orientation in processing were also per-

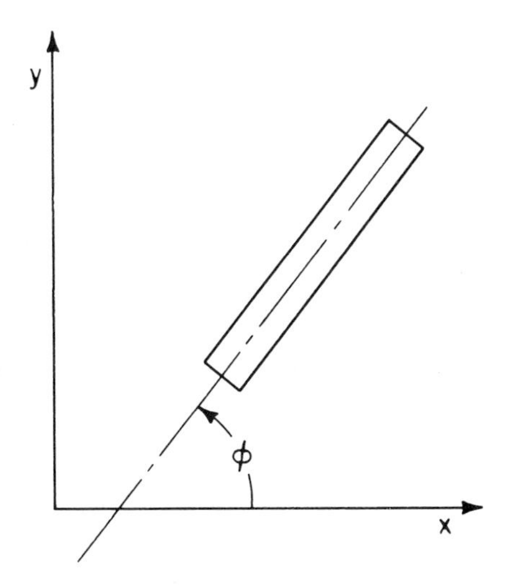

Figure 1. Definition of the angle ϕ that describes the orientation of a fiber lying in the x–y *plane.*

formed by Modlen (*13*), for fibers in a plastically deforming matrix, and by Nicolais et al. (*14*), for postextrusion drawing of a glass fiber/polystyrene composite. Both works involved primarily extensional strains of dilute suspensions, and both used a simple theory of fiber motion that assumes a fiber rotates like a fluid line. Their simple theory is in fact identical to Equation 1 as $r_e \rightarrow \infty$. This theory does not predict the observed periodic motion of fibers in simple shear flow, but is very close to the real motion of fibers in extensional flows. Hence, the agreement between theory and experiment in these two works is not surprising.

These works provide the pattern for predicting fiber orientation during processing: first find the deformation kinematics, then apply the laws of fiber motion. There are two problems with applying this pattern to molding of composite materials. One is that the well-known equations for the motion of a single fiber cease to apply once the concentration of fibers exceeds a very low level. At concentrations on the order of $(1/r_f)^2$ and above, r_f being the ratio of fiber length to diameter, fibers interact with one another, and each interaction causes a rapid change in fiber orientation (*15*). Although some information about interactions is available (*16*), no theory of the mechanics of interactions capable of solving fiber orientation problems is yet available.

The second problem is solving for the deformations accumulated by molding a part into a complex shape. In the most general case, the solution of the mold filling pattern is coupled with the fiber orientation problem, because the rheological properties of a fiber suspension are a function of fiber orientation. For example, sheet molding compound, which has nearly a planar random orientation distribution, has anisotropic viscosity. The extensional viscosity in the plane of the sheet can be 100 times greater than the shear viscosity across the thickness of the sheet (*17*). A detailed rheological model describing this phenomenon has been formulated by Dinh and Armstrong (*18*). Their model provides insight into the rheology of concentrated fiber suspensions. It does use a fiber orientation law that is in conflict with our recent experimental data. The solution of fluid mechanics problems using this type of model is very difficult.

Various experimental observations of fiber orientation in processing situations have been made (*19–28*). Notably, no conflicts occur between experimentally observed fiber orientations and qualitative predictions based on the ideas discussed so far. Where the overall flow is converging, such as at the entrance to an extrusion die or an injection mold gate, fibers align in the direction of flow in response to the stretching motion of the flow. Where the flow is diverging, such as where the flow enters an injection or transfer mold, fibers align transverse to the flow, in the direction of stretching. For simple shearing flows, such as flow in capillary tubes and in parts of mold cavities, fibers align in the direction of flow in response to the shearing deformation.

Fiber orientation was reported to have some dependence on the flow rate (and hence the strain rate). For instance, Crowson et al. (*29*) report that the degree of fiber orientation in a capillary tube increases as the flow rate increases. Changes in fiber orientation with filling rate in thermoplastic injection molding have also been reported by Bright et al. (*25*). At first this evidence might seem in conflict with the previous observation that changes in fiber orientation should depend on the strain accumulated by the material but not on the strain rate at which it occurs. However, these workers only observe a rate dependence of fiber orientation under conditions where changes in the flow rate also change the kinematics of the flow and cause different parts of the material to undergo different strains. For instance, in their capillary die equipment, Crowson et al. (*29*) report that the fiber orientation in the capillary changes as the flow moves from the Newtonian region to the power-law region and changes the velocity profile and the distribution of strain rate across the tube. The extensional flow at the capillary entrance produces more highly aligned fibers than a steady state shearing flow. The shear flow in the capillary tube actually disorients the fibers somewhat, and the flatter velocity profile at high flow rate causes less shear strain and has better orientation. For injection moldings, the velocity profile changes because of rheological properties that change with strain rate and temperature. As the flow rate is decreased, heat transfer becomes proportionally more important, and more orientation is "frozen in" near the mold walls. The differences in velocity profiles serve to explain the observed differences in fiber orientation. Thus, no conflict exists between this type of experimental evidence and the idea of a strain-dependent orientation pattern.

Although some rules for predicting orientation patterns based on the preceding ideas do exist (*26*), we need the ability to make quantitative predictions of fiber orientation; to predict the entire orientation distribution, not just the principal direction; and to do so on a firm fundamental basis. We present a method for these predictions, beginning with an appropriate model for fiber orientation behavior in concentrated suspensions. The specific application of this model to the compression molding of thin parts is then shown as an example of how this type of calculation can be done.

Theory

Basic Fiber Orientation Law. Experiments reveal that, in a concentrated suspension of fibers, the motion of an individual fiber is not determined, as in the dilute case (Equation 1), simply by its angle and by the flow field. Presumably the surrounding fibers also play a role through some type of interaction. In the absence of a detailed theory of interactions, we have chosen a probabilistic model that gives not the individual fiber orientations but the probability density function of fiber orientation. The suspension is viewed as a continuum, and the probability density function is taken to be a continuous function of space and time. For the case of a planar fiber orientation distribution, one can define the density function $\psi(\phi)$ such that $\psi(\phi)\,d\phi$ is the probability that a fiber will be oriented between the angles ϕ and $\phi + d\phi$.

This definition is consistent with the usual definition of a probability density function; the probability of a fiber lying between angles ϕ_1 and ϕ_2 (or the expected fraction of fibers oriented between ϕ_1 and ϕ_2) is

$$P(\phi_1 < \phi < \phi_2) = \int_{\phi_1}^{\phi_2} \psi(\phi)\, d\phi \qquad (2)$$

We will restrict this discussion to the case of planar fiber orientations for the sake of simplicity. A more complete discussion of this model and development of the three-dimensional equations has been given elsewhere (*30*).

The function $\psi(\phi)$ must meet several conditions. As a probability density function it must be non-negative. Because a fiber at angle ϕ is indistinguishable from one at angle $(\phi + \pi)$ then ψ must be periodic with period π:

$$\psi(\phi) = \psi(\phi + \pi) \qquad (3)$$

Also, each fiber must have an orientation in the range $(0, \pi)$, so

$$\int_0^{\pi} \psi(\phi)\, d\phi = 1 \qquad (4)$$

A suspension with uniform orientation has $\psi(\phi) = \text{constant} = 1/\pi$. This constant provides a convenient reference value when interpreting orientation functions.

When fibers are rotating, ψ is a function of time as well as ϕ. However, Equation 4 must always hold, so ψ is also subject to a conservation or continuity condition, (i.e., the area under the curve is constant). The average angular velocity of fibers is denoted by $\dot{\phi}$, which may be a function of ϕ and time, and the continuity equation is derived by considering a small angle increment, $d\phi$, and balancing the rate of change of the fraction of fibers in that increment against the motion of fibers into and out of that angle. The result,

$$\frac{\partial \psi}{\partial t} = \frac{-\partial}{\partial \phi}(\psi \dot{\phi}) \qquad (5)$$

is a form familiar from heat and mass transfer balance equations. Equation 3 provides the boundary conditions in ϕ for Equation 5.

Equations 3–5 are obvious consequences of the physical situation, but one still needs a particular form of ϕ to solve for ψ. We have used the phenomenological theory suggested by Folgar (*30, 31*), which was developed for rigid fibers of uniform length and diameter. The fiber motion, $\dot{\phi}$ is assumed to be the sum of two terms. The first term is identical to the motion of a fiber in a dilute suspension and is controlled by the bulk deformation of the suspension. The second term models the effect of interactions between fibers and is stochastic in origin. An interaction is presumed to occur when the centers of two fibers pass within one fiber length of each other. Each interaction causes a random reorientation of the interacting fibers and this reorientation causes a net motion of fibers from angles of high orientation toward angles of low orientation. That is, interactions tend to randomize the fiber orientation by eroding orientation peaks. The rate of interactions is controlled by the strain rate, so fiber orientation still depends on the strain experienced by the suspension just as in the dilute case. No detailed model of interactions is available, so individual fiber reorientations resulting from interactions are assumed to be independent, identically distributed random variables with zero mean. The net motion of fibers resulting from interactions $\dot{\phi}_i$, can then be derived (*30*) by using the same type of arguments that relate diffusion to Brownian motion (*32, 33*). The motion resulting from interactions is

$$\dot{\phi}_i = \frac{-C_I \dot{\gamma}}{\psi} \frac{\partial \psi}{\partial \phi} \tag{6}$$

Here, $\dot{\gamma}$ is the magnitude of the strain rate tensor, and C_I is a coefficient related to the frequency and strength (i.e., mean angle change) of interactions. This interaction coefficient C_I is hypothesized to be an intrinsic property of any given suspension, and to depend on quantities such as fiber volume fraction and fiber aspect ratio. It must be measured experimentally, because no theory for its prediction yet exists.

The total angular velocity for fibers in a concentrated suspension may then be written as the sum of the contribution from interactions (Equation 6) and the orienting effect of flow (Equation 1). Normally one is interested in fibers with large aspect ratios ($r_f > 10$). For these fibers, $r_e \gg 1$, and the contribution from Equation 1 is well approximated by taking the limiting case as r_e approaches infinity. The total angular velocity of the fibers then becomes

$$\dot{\phi} = \frac{-C_I \dot{\gamma}}{\psi} \frac{\partial \psi}{\partial \phi} - \sin \phi \cos \phi \frac{\partial v_x}{\partial x} - \sin^2 \phi \frac{\partial v_x}{\partial y} + \cos^2 \phi \frac{\partial v_y}{\partial x} + \sin \phi \cos \phi \frac{\partial v_y}{\partial y} \tag{7}$$

With the addition of an equation for $\dot{\phi}$, one can now write a complete equation for the changes in the distribution function ψ as a function of deformation by substituting Equation 7 into Equation 5. Individual fibers are convected with the fluid, so ψ is also a convected quantity. Thus, for systems in which ψ is also a function of x and y, we replace the left side of Equation 5 by the material derivative. For the case of two-dimensional fiber orientation, the governing equation becomes

$$\frac{D\psi}{Dt} = C_I \dot{\gamma} \frac{\partial^2 \psi}{\partial \phi^2} + \frac{\partial}{\partial \phi}\left[\psi\left(\sin\phi\cos\phi\frac{\partial v_x}{\partial x} + \sin^2\phi\frac{\partial v_x}{\partial y} - \cos^2\phi\frac{\partial v_y}{\partial x} - \sin\phi\cos\phi\frac{\partial v_y}{\partial y}\right)\right] \quad (8)$$

All fibers are assumed to lie in the x–y plane, and ϕ is the angle between the fiber axis and the x axis. In the strict sense, this equation applies only to flows that are two-dimensional, in the sense that there are no variations of velocity in the z-direction.

One interesting consequence of this theory of fiber motion is that there is a limit to the degree of orientation that can be achieved by deforming a suspension. For any steady homogeneous flow (simple shear, extensional, etc.) a corresponding steady state distribution function can be found by solving Equation 8 with $D\psi/Dt = 0$. This steady state is dynamic because individual fibers continue to rotate as long as the deformation continues, though $\psi(\phi)$ remains constant. The steady state distribution is dependent on the value of C_I because it represents a balance between the orienting effects of flow and the randomizing effects of interactions. Figure 2 shows the steady state distributions predicted for simple shear flow for several values of C_I. As C_I increases, the steady state condition becomes less oriented. The same is true of extensional flows, though the orientation distributions are slightly different.

In practice, C_I can be measured experimentally by subjecting a suspension to a large deformation so as to achieve the steady state distribution, then measuring the distribution function and finding the value of C_I that best fits the measured function. Figure 3 shows an example of an experimental distribution function fit this way. In this experiment, orientation angles of 1800 tracer fibers are measured to create a histogram of the orientation distribution. When normalized according to Equation 4, the histogram is an experimental measure of $\psi(\phi)$. Note how well the single choice of C_I fits both the height of the orientation peak and its location. Experimentally, C_I has been found to increase with both the fiber volume fraction and the fiber aspect ratio. As the suspension becomes more crowded, less perfect orientation results. This kind of evidence gives strong credence to

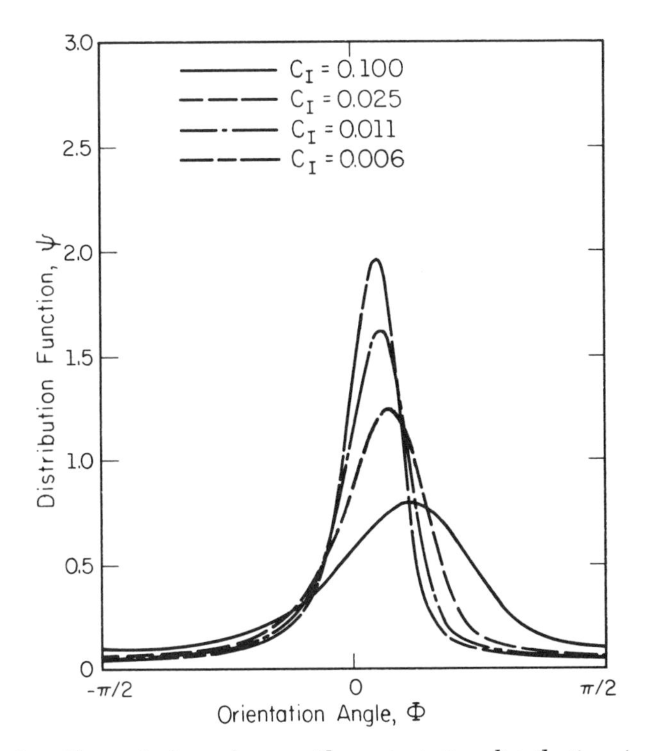

Figure 2. Theoretical steady state fiber orientation distributions in simple shear flow as a function of the interaction coefficient, C_I. *The velocity field is* $v_x = \gamma y$.

the theory. More experiments and discussion may be found in Reference *30.*

The existence of steady state orientation distribution functions has several interesting implications in processing. First, the steady state orientation for extensional flow is the state of maximal orientation for any given material; if the theory is correct, then greater orientation with more polarization of mechanical properties is not possible in a flow molding process. Any increases in orientation beyond this level can only be achieved by orienting the fibers before combining them with the resin and not disturbing that orientation subsequently. (There are processes for manufacturing short-fiber composites that do prepare aligned fiber mats and add the matrix resin in a subsequent step, e.g., Ref *34.*) Second, analysis of orientation problems may be simplified when the material reaches steady state at some point during the process. For example, in many injection molding situations the flow into the gate might have enough extensional deformation to reach the corresponding steady state orientation. If so, one could begin the orientation analysis from that point and with that initial condition, and ignore the previous history of the material in the screw and runner system.

Application to Molding Processes. The general procedure for calculating the fiber orientation distribution for a specific process is to start with the known (statistical) orientation distribution function of the molding compound, $\psi(t = 0)$, integrate it forward in time with Equation 8 and boundary conditions derived from Equation 2, and arrive at the orientation distribution. Obviously, the entire velocity history in the material during processing must be known to do this procedure. As mentioned, if the rheology of the molding compound depends on fiber orientation, the two problems may be coupled. Even without this complexity, the general case leaves one with a complicated differential equation for the variable ψ in four space variables, x, y, z, and ϕ (plus an additional orientation variable for nonplanar distributions) and in time. The solution of such an equation is clearly a difficult task. Any simplification that can be gained when analyzing a specific problem would be welcome.

As an example of the method we show calculations for compression molded parts that are thin and sheet-like, similar to the part shown in Figure 4, and that have fibers much longer than the part thickness. These conditions are typical of parts made from sheet molding compound (SMC).

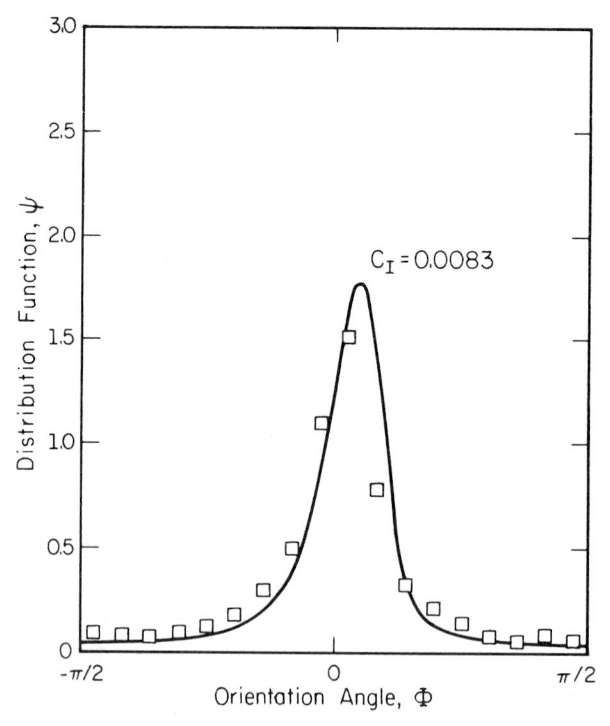

Figure 3. Comparison between experimental (□) and theoretical (—) steady state fiber orientation distributions in simple shear flow. The experiments use 8% by volume of nylon monofilaments with an aspect ratio of 16 in 12,500-cs silicone oil.

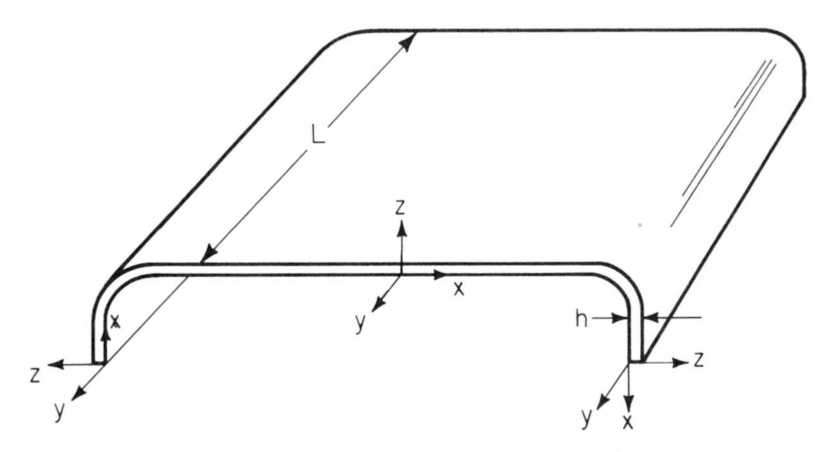

Figure 4. Coordinate system used for mold filling and fiber orientation analysis of thin compression molded parts. The true shape is "laid flat" in the x–y *plane.*

Mold Filling Model. To solve the fiber orientation problem, one must know the velocity distribution during mold filling. For compression molding of thin parts, the Generalized Hele–Shaw flow model has proven appropriate. Details of the theoretical development and a finite element scheme for solving compression molding problems have been given elsewhere (*35*, *36*). Some important features of the model are given here for clarity.

The part is modeled (Figure 4) by "unfolding" or "laying flat" in the x–y plane, so that z is everywhere the thickness direction. The Generalized Hele–Shaw model is appropriate for parts with $h \ll L$, where h is the characteristic thickness, and L is the characteristic lateral dimension.

During compression molding, a charge of molding compound somewhat smaller than the part in its x and y dimensions and larger than the part in its z dimension is placed in the mold. The two mold halves are then brought together and squeeze the charge so that it flows laterally to fill the mold. In thin parts, the vertical component of velocity v_z is much smaller than the horizontal components v_x and v_y and can be neglected. In many cases, the stresses that arise from in-plane stretching, τ_{xx}, τ_{xy}, and τ_{yy}, are small compared to the cross-thickness shearing stresses τ_{xz} and τ_{yz}, so the former can be neglected in the equations of motion. When this case is true, the Generalized Hele–Shaw flow model is valid.

One important feature of Hele–Shaw flows is that, at any location in the x–y plane, the direction of the velocity vector is not a function of z. The magnitude of velocity does change; it is zero at the surfaces of the upper and lower mold and has a single maximum in between. For non-Newtonian fluids and for the nonisothermal conditions typical of SMC molding (cold molding compound in a hot mold), the velocity profiles tend to be very flat in the z direction with thin regions of high velocity gradient near the upper and lower mold surfaces.

An interesting consequence of the Generalized Hele–Shaw model is that the rheological response of the molding compound to stretching deformations in the x–y plane has no effect on the filling pattern, which is instead dominated by shearing stresses across the thickness direction. SMC has anisotropic rheological properties with much greater resistance to stretching than to shearing (*17*). These properties are a consequence of the planar orientation of the fibers and would be expected to change as the fibers became preferentially oriented during flow. To account for this expectation in the mold filling model would greatly increase the difficulty of the problem. For our case, however, this effect should be negligible because fiber orientation primarily affects the in-plane stretching response and has little affect on the dominant cross-plane shearing properties.

Fiber Orientation Calculations. To calculate the fiber orientation in compression molded thin SMC parts like those shown in Figure 4, we make the following assumptions:

1. The fiber orientation distribution is planar in the x–y plane.
2. The mold filling pattern is independent of the fiber orientation distribution and is governed by the Generalized Hele–Shaw flow model.
3. The x and y velocity distributions across the z direction of the mold are nearly flat, showing plug flow.
4. The fiber orientation distribution does not vary in the z direction, and its rate of change is determined by the velocities in the x–y plane.

The first assumption is excellent because in SMC the fibers are typically 1 in. long and the parts are seldom more than 1/8 in. thick. This condition constrains all the fibers to lie very nearly in the plane of the part.

The second and third assumptions have been discussed. The fourth assumption is reasonable if the fiber orientation in the initial charge is independent of z and if the first and third assumptions are satisfied. Significant variations in orientation across the thickness of the part exist in injection molded thermoplastics (*22*, *23*, *27*), and are thought to arise from the fountain effect at the flow front and become frozen in as the mold fills. This variation does not seem to occur in SMC because the fibers are too long to rotate out of the plane of the part and cannot respond to the fountain effect at the flow front.

The effects of these assumptions are to decouple the mold filling problem from the fiber orientation problem and to reduce the number of space dimensions in the problem.

The finite element mold filling solution starts with a definition of the shape of the charge that is placed in the mold and the shape of the mold. Subsequent charge shapes are stepped out in time by imposing a finite element mesh on the current charge, solving for the current pressure and velocity distributions subject to the current boundary conditions, and using the velocities to predict the charge shape a short time later. This process is repeated until the mold has been filled. Because our assumptions decouple the filling and fiber orientation problems, we run the entire mold filling simulation first, and write the appropriate results to an output file that is subsequently used for fiber orientation calculations.

Rather than solve for ψ as a function of x and y everywhere within the charge at every time step, we solve for it only at specific points. This step is done by integrating Equation 8 along the paths of fluid particles, so that the convective terms $v_x\ (\partial\psi/\partial x)$ and $v_y\ (\partial\psi/dy)$ on the left side of Equation 8 do not come into play. One is left with a partial differential equation of ψ in ϕ and t. This equation is solved numerically with a commercially available routine that uses Galerkin's method with B-splines in space and a variable order, variable time-step extrapolation procedure in time (*37*, *38*). Stable solutions to Equation 8 were obtained by using $k = 4$, where k is the order of the B-spline used to fit $\psi(\phi)$.

In practice, the solution begins by specifying an initial condition for $\psi(\phi, t = 0)$ in the charge. A random orientation would be a reasonable first guess for SMC, though the SMC machine can introduce some fiber orientation, usually in the machine direction (*28*). One then chooses a point of interest in the original charge. The solution proceeds by evaluating the velocities and velocity gradients at that point from the finite element results, solving Equation 8 over the interval of the first time step, and computing the position of the point at the end of the time step. From that position, the velocities and velocity gradients at the next time step are computed, and the process is repeated. The cycle continues until the end of the final time step. The result is the complete fiber orientation distribution at a given point in the final part. The entire process can be repeated as many times as desired with different starting points.

This approach is chosen to reduce computational effort, because one would seldom require the orientation distribution as a complete function of space. For example, when doing a finite element structural analysis the part would be divided into many elements, each of which could have different mechanical properties. These could be calculated from the fiber orientation distribution at a single representative point in each element. One could easily calculate the initial positions of these points in the charge from the results of the mold filling simulation, and then solve for the orientation distributions.

One inherent deficiency in this model lies in the way it handles the edges of the part. The Generalized Hele–Shaw flow model cannot account for the no-slip boundary conditions on the vertical edge boundaries of the mold. This deficiency does not usually disturb the mold filling calculations, because this boundary condition only affects the flow over a region about one part thickness wide in the x–y plane. However, the strong x–y velocity gradients in this region can and do affect fiber orientation, and, because the velocity gradients change over a distance less than the fiber length, the effect is complex. Accordingly, one would not expect this model to provide accurate results within one to two fiber lengths of the edge of the part. This condition is not particularly restrictive for SMC parts, which are often several feet in width.

Example Calculations

Some compression molding experiments that provide a convenient test for this model have been performed by Chen and Tucker (7). They molded a special sheet molding compound with 65% by weight of 1-in. long glass fibers and no filler into plaques 18 in. × 12 in. × 1/8 in. Different charges, occupying 100, 67, 50, and 33% of the mold, were used in the configuration shown in Figure 5. The 100% charge plaques have almost no deformation, but the other charges experience an extensional strain in the direction of flow that increases as the charge size decreases. About 1.5% of fibers made from a high lead content glass were included in the special SMC so that the fiber orientation distribution could be observed by radiography. Portions of two of the radiographs are shown in Figure 6. Fiber orientation in the region within about one fiber length of the edge of the part is very

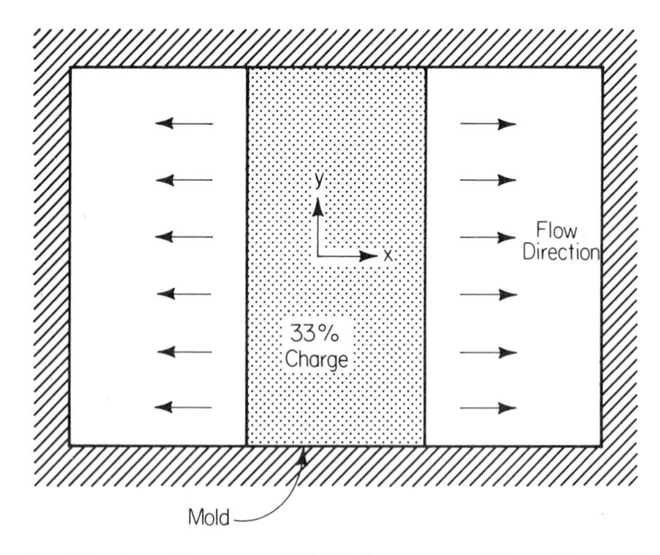

Figure 5. Shape and location of SMC charge in mold used to produce the data shown in Figures 6–10.

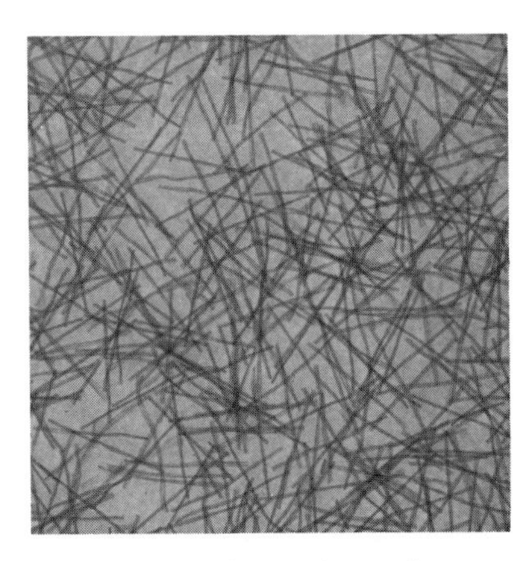
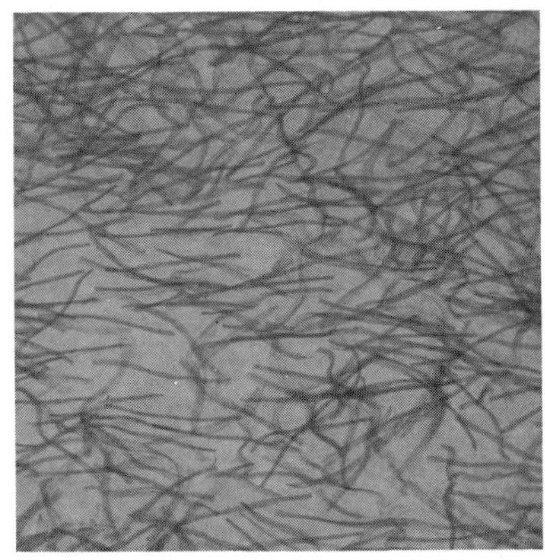

Figure 6. Radiographs of fiber orientation patterns in compression molded specimens. Left: 100% charge, nearly random orientation. Right: 33% charge, highly aligned fibers.

different from the rest, because of the edge effects just mentioned. However, the rest of the piece appears to have a uniform orientation field (ψ not dependent on x or y) as is predicted by our model. Hence, the fiber orientation distributions were compiled from samples taken over the entire central region of the plaques. Totals of 500 fibers were sampled for each distribution.

Two other unknowns exist in this experiment: the initial orientation distribution and the interaction coefficient C_I. Measurements of the 100% charge samples, which underwent no deformation during molding, provided the initial distribution function. The data and a smooth curve used to fit it are shown in Figure 7. The slight preferential orientation is presumably a result of the SMC machine, so all charges were cut with the same orientation relative to the SMC machine direction. (The machine direction corresponds to $\phi = 0$ in Figure 7.) With the 100% charge data as an initial condition, trial computations were made by varying the value of C_I until good agreement with the 33% charge data was obtained. This value and the initial conditions were then used to predict the orientation distributions for the 50% and 67% charges to test the model. The predictions are compared with the experimental results in Figures 8–10, where it is seen that the predictions are quite accurate in spite of the simplifying assumptions.

In the preceding example the mold filling flow is very simple, because the flow fronts simply move outward as straight lines and every point in the charge experiences the same strain. Figure 11 shows a slightly more complicated geometry. Although comparisons with experiments are not yet available, this geometry was selected to demonstrate the effects of converging and diverging flow. Fiber orientation distributions were calculated for the three points indicated in Figure 11; the results are shown in Figure 12. This charge undergoes a large stretching deformation along the right-hand diag-

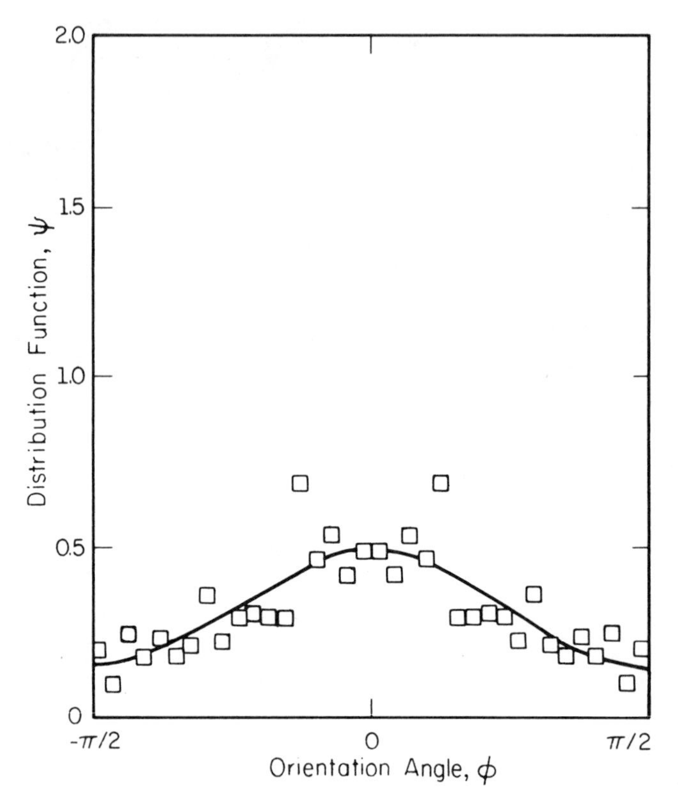

Figure 7. Measured orientation distribution for 100% charge and fitted curve used as initial condition in subsequent calculations.

onal, similar to the previous example, so that primary orientation is along $\phi = \pi/4$ everywhere. However, other features of the flow affect the orientation as well. The point labeled 1 in Figure 11 experiences a converging flow in the x–y plane that accentuates the stretching and produces a highly aligned state. Point 2, on the other hand, experiences a diverging flow in the x–y plane that reduces the degree of alignment, though it is not strong enough to alter the direction of principal orientation. The tendency for flow fronts to become round in compression molding causes the principal orientation direction at Point 3 to be deflected somewhat counterclockwise.

These results are in accord with the empirical rule of thumb about the effects of converging and diverging flow. However, it is not possible to test the rule of thumb about shear flow under the assumptions we have made. A Hele–Shaw flow in a gap of uniform thickness is irrotational. That is, $\partial v_x/\partial y$ equals $\partial v_y/\partial x$ everywhere, so there can be no location where the flow in the x–y plane is simple shear flow. The shearing across the thickness (z)

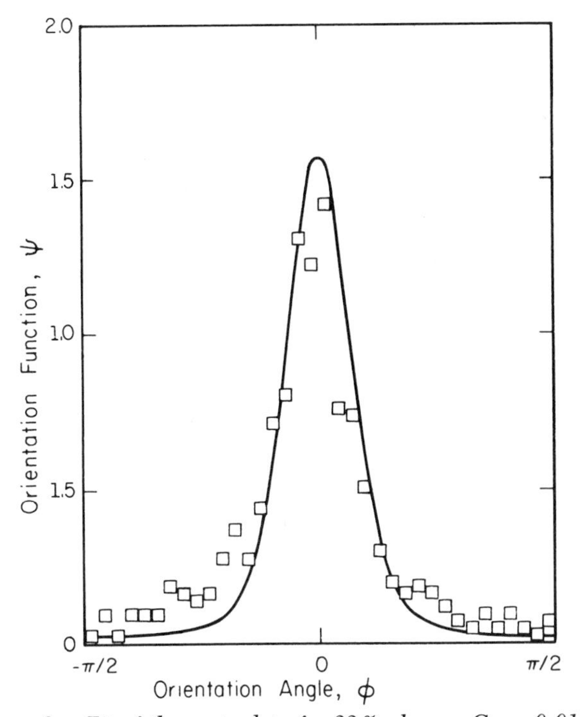

Figure 8. Fit of theory to data for 33% charge, $C_I = 0.0185$.

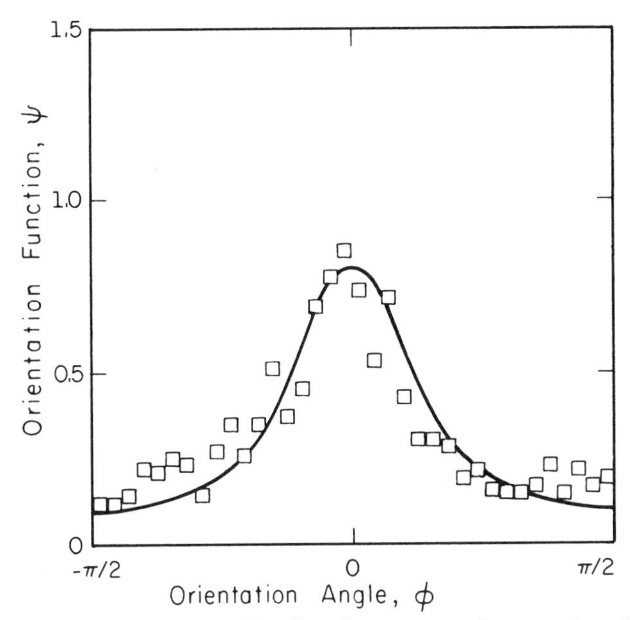

Figure 9. Comparison of predicted and experimental orientations for 67% initial mold coverage, $C_I = 0.0185$.

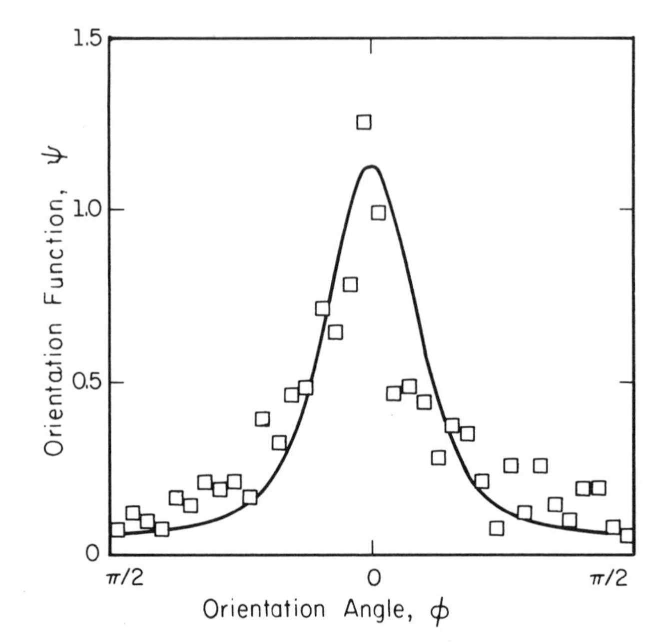

Figure 10. Comparison of predicted and experimental orientations for 50% initial mold coverage, C_I = *0.0185.*

direction that does occur in Hele–Shaw flow has been assumed not to affect orientation because the fiber length far exceeds the gap width.

Discussion

We have showed that for a given set of molding conditions and a given initial condition, the final fiber orientation pattern can be predicted. But how may the orientation pattern be *controlled*? According to the model, fiber orientation depends only on the strain experienced by the molding compound. Thus, factors such as mold temperature and fill rate do not affect orientation directly. Instead, they affect fiber orientation only in that they can affect the kinematics of the mold filling flow. The primary variables controlling mold filling kinematics include cavity shape, gate location and shape (for injection molding), and charge location and shape (for compression molding). Not surprisingly, these factors have long been identified as the primary variables affecting fiber orientation in molded parts. No new techniques for controlling orientation have been revealed by our theoretical development. However, by knowing how to deal with fiber orientation in a quantitative and predictive fashion one can make much better use of the available techniques.

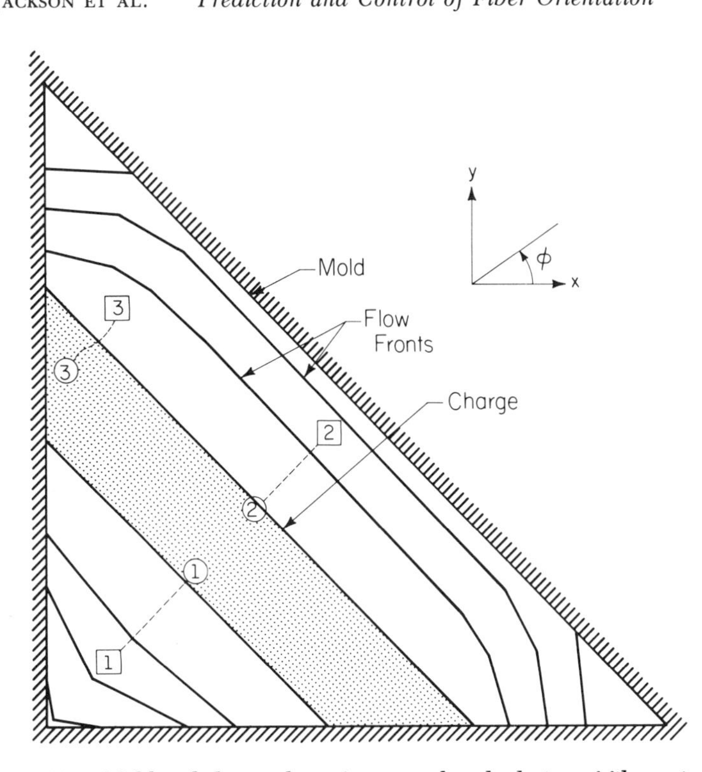

Figure 11. Mold and charge shape for example calculation of fiber orientation. The solid lines represent locations of the charge front during filling. Initial locations (○) path lines and final locations (□) of three points at which the fiber orientation distribution was calculated are shown.

Summary

The problem of predicting the fiber orientation pattern in a part molded from a short fiber reinforced polymer has been formulated. The solution requires a governing equation for the fiber orientation distribution function to be integrated along the paths of fluid particles during mold filling. Hence, a solution to the mold filling flow is also needed. A set of simplifying assumptions for the case of thin compression molded parts with fibers much longer than the part thickness has been presented and shown to provide useful results. Example predictions are in accord with well-known qualitative rules for fiber orientation. The ability to make quantitative predictions of fiber orientation can enhance the reliability and usefulness of short-fiber reinforced composite parts.

Nomenclature

C_I Interaction coefficient describing a fiber suspension
t Time

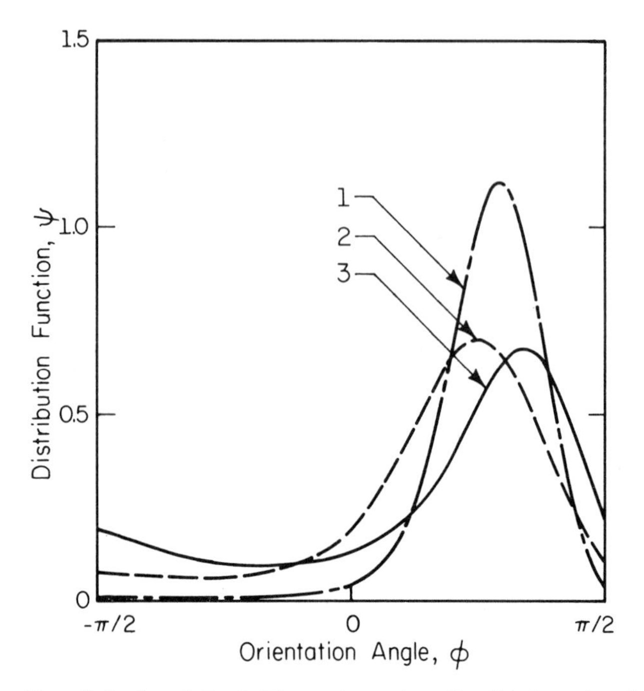

Figure 12. Calculated final fiber orientation distributions for the part shown in Figure 11. Numbers refer to the points indicated in Figure 11.

v_x, v_y, v_z	Bulk velocity components
x, y, z	Cartesian coordinates
$\dot{\gamma}$	Bulk strain rate
ϕ	Angular coordinate for fiber in the x–y plane
$\dot{\phi}$	Average angular velocity of fibers
$\dot{\phi}_i$	Angular velocity component due to interactions between fibers
ψ	Fiber orientation distribution function

Acknowledgments

This work was sponsored by the Owens-Corning Fiberglas Corporation and by the National Science Foundation, Grant No. MEA 81–12037. The authors thank Che-Yang Chen of the University of Illinois and Don Sage and Doug Denton of Owens-Corning Fiberglas for their help with the molding experiments cited. Advice and encouragement from the staff of OCF and the Industrial Advisory Board of the NSF project is gratefully acknowledged.

Literature Cited

1. Pipes, R. B.; McCullough, R. L., Taggart, D. G. *Polym. Compos.* **1982**, *3*, 34.
2. Halpin, J. C. *J. Compos. Mater.* **1969**, *3*, 732.

3. Halpin, J. C.; Pagano, N. J. *J. Compos. Mater.* **1969**, *3*, 720.
4. Halpin, J. C.; Jerina, K.; Whitney, J. M. *J. Compos. Mater.* **1971**, *5*, 36.
5. Loughlin, P. T.; Chen, C.-Y.; Tucker, C. L. *SPE Tech. Pap.* **1981**, *27*, 61.
6. Halpin, J. C.; Kardos, J. L. *Polym. Eng. Sci.* **1978**, *18*, 496.
7. Chen. C.-Y.; Tucker, C. L. *SPE Tech. Pap.* **1983**, *29*, 337.
8. Darlington, M. W.; Gladwell, B. K.; Smith, G. R. *Polymer* **1977**, *18*, 1269.
9. Jeffrey, G. B. *Proc. R. Soc.* **1923**, *A102*, 161.
10. Mason, S. G. "The Microrheology of Dispersions," in "Rheology: Theory and Applications," Eirich, F. R., Ed.; Academic Press: New York, 1967; Vol. 4.
11. Gauthier, F.; Goldsmith, H. L.; Mason, S. G. *Rheol. Acta* **1971**, *10*, 344.
12. Harris, J. B.; Pittman, J. F. T. *Trans. Inst. Chem. Eng.* **1976**, *54*, 73.
13. Modlen, G. F. *J. Mater. Sci.* **1969**, *4*, 283.
14. Nicloais, L.; Nicodemo, L.; Masi, P.; DiBenedetto, A. T. *Polym. Eng. Sci.* **1979**, *19*, 1046.
15. Arp, P. A.; Mason, S. G. *J. Colloid Interface Sci.* **1977**, *59*, 378.
16. Okagawa, A.; Cox, R. G.; Mason, S. G. *J. Colloid Interface Sci.* **1973**, *45*, 303.
17. Lee, L. J.; Marker, L. F.; Griffith, R. M. *Polym. Compos.* **1981**, *2*, 209.
18. Dinh, S. M. Sc.D Thesis, Dept. of Chem. Engr., MIT, Cambridge, MA, 1981.
19. Bell, J. P. *J. Compos. Mater.* **1969**, *3*, 244.
20. Goettler, L. A. presented at the 25th SPI RP/C Conf. 1970, Sect. 14-A.
21. Goettler, L. A. *Mod. Plast.* **1970**, *47*, No. 4, 140.
22. Darlington, M. W.; McGinley, P. L., *J. Mater. Sci.* **1975**, *10*, 906.
23. Darlington, M. W.; McGinley, P. L., Smith, G. R. *J. Mater. Sci.* **1976**, *11*, 877.
24. Lee, W.-K; George, H. H. *Polym. Eng. Sci.* **1978**, *18*, 146.
25. Bright, P. F.; Crowson, R. J.; Folkes, M. J. *J. Mater. Sci.* **1978**, *13*, 2497.
26. Owen, M. J.; Thomas, D. H.; Found, M. S. presented at the 33rd SPI RP/C Conf., 1978, Sect. 20-B.
27. Taggart, D. G.; Pipes, R. B. Report No. CCM 79–12, Center for Composite Materials, University of Delaware, 1979.
28. Denton, D. L. presented at the 36th SPI RP/C Conf. 1981, Sect. 16-A.
29. Crowson, R. J.; Folkes, M. J.; Bright, P. F. *Polym. Eng. Sci.* **1980**, *20*, 925.
30. Folgar, F. Ph.D. Thesis, Dept. of Mechanical and Industrial Engineering, University of Illinois at Urbana-Champaign, 1982.
31. Folgar, F.; Tucker, C. L. "Orientation Behavior of Fibers in Concentrated Suspensions," Proc. of Symp. on Transport Phenomena in Materials Processing, ASME Winter Annual Meeting, Boston, November 1983.
32. Reif, F. In "Fundamentals of Statistical and Thermal Physics"; McGraw-Hill: New York, 1965; Chapt. 15.
33. Brenner, H. *Int. J. Multiphase Flow* **1974**, *1*, No. 2, 195.
34. Kacir, L.; Narkis, M.; Ishai, O. *Polym. Eng. Sci.* **1975**, *15*, 525.
35. Lee, C.-C.; Folgar, F.; Tucker, C. L. "Simulation of Compression Molding for Fiber Reinforced Thermosetting Polymers" in "Polymer Processing: Analysis and Innovation"; Suh, N. P.; Tucker, C. L., Eds.; ASME: New York, 1982, PED- Vol. 5.
36. Tucker, C. L.; Folgar, F. *Polym. Eng. Sci.* **1983**, *23*, 69.
37. Schryer, N. L. Computing Sci. Tech. Report No. 52; Bell Laboratories: Murray Hill, N.J., 1976.
38. Schryer, N. L. Computing Sci. Tech. Report No. 53; Bell Laboratories: Murray Hill, N.J., 1977.

RECEIVED for review January 20, 1983. ACCEPTED August 11, 1983.

18

Pulling Force and Its Variation in Composite Materials Pultrusion

HOWARD L. PRICE
Bendix Advanced Technology Center, Columbia, MD 21045

STEPHEN G. CUPSCHALK
Old Dominion University, Norfolk, VA 23508

The pulling force in a prototype pultrusion operation was affected by the volume of resin and fiber compacted into and pulled through the die, by the die temperature both below and in the resin reaction range, and by the pulling speed. For a constant temperature and pulling speed, the force increased exponentially with volume of material. For a given material volume and die temperature in the reaction range, the best quality composite material was obtained at the pulling speed that provided the highest pulling force. These findings were obtained by using epoxy resin and carbon fiber prepreg tape and a laboratory-scale pultruder constructed for this investigation.

FIBROUS COMPOSITE MATERIALS for mechanical and structural applications often are expensive because of high labor costs. One economical way of making composites, a way less labor intensive than press molding or autoclave technology, is pultrusion (*1*). Pultrusion is a manufacturing process in which polymeric resin-impregnated fibers are pulled at a constant speed through a heated die (Figure 1). The die shapes and compacts the resin-fiber mass into a predetermined constant cross section and cures the resin. Pultruded stock has been used in structural, electrical, and chemical processing applications.

Most of the published work on pultrusion (*2–59*) has been of an empirical nature, used full-scale (production) pultruders, and required large amounts of material (typically polyester resins and glass fibers). In the few cases in which process parameters have been investigated, the emphasis has been on the control variables of die temperature and pulling speed. The force required to maintain a constant pulling speed has been either ignored or treated as some upper bound that must not be exceeded. Although such

0065-2393/84/0206-0301$06.50/0

an approach may be useful for pultruder design, it does little to elucidate the pultrusion process itself. A more fundamental understanding of the process is needed if the inherent economies of pultrusion are to be realized.

Therefore, we investigated the pulling force and its variation in a prototype pultrusion operation. The objectives were to determine the origin and magnitude of the pulling force and to use the changes in that force to obtain a better understanding of the pultrusion process. A laboratory-scale pultruder was constructed to make thin, rectangular cross-section pultruded stock representative of the plates, angles, channels, and beams often made by pultrusion. The thin section minimized heat transfer problems so a short, economical die could be used to study the process. Epoxy resin-impregnated carbon fibers (prepreg tape) were pulled through the pultruder at constant speed while the die temperature and pulling force were measured. The volume of tape compacted into and pulled through the die, the die temperature both below and in the epoxy reaction range, and the pulling speed affected the pulling force.

Pultrusion Process Analysis

Pultrusion Technology. The process now called pultrusion can be traced to a way of making better fishing rods (*2*). The argument in favor of the process was that bamboo fly rods had variable flexibility and that the process of tying together bamboo splits to obtain the right flexibility was costly. Such an argument reflects a valuable composite material concept (tailor properties as needed) and a pultrusion process advantage (reduced manufacturing costs). Another advantage of pultrusion, which was recognized early (*4*), is that of having the fibers under tension as the resin cures. Then, when the composite material is put under a mechanical load, the load is taken immediately by the high strength fibers, not by the low strength matrix. This concept is similar to that used in prestressed concrete in which the steel reinforcing bars are loaded in tension while the concrete member is still in the mold.

The pultrusion process is shown in schematic form in Figure 1. In the simplest terms, fibers pass through a bath of (typically) thermosetting resin, some of which adheres to the fibers. The resin-impregnated fibers continue through a heated die that shapes and compacts the resin–fiber mass into a predetermined cross section and cures the resin. The pultruded stock that emerges from the die is a composite material with a polymeric matrix and fibrous reinforcement. A traction mechanism applies a pulling force (which may vary) to the pultruded stock to keep the resin–fiber mass moving at a constant speed through the die. Generally, the highest speeds, consistent with the cure requirements of the resin, are sought to maximize output. The length of the stock is limited only by material supply, handling, and transportation considerations.

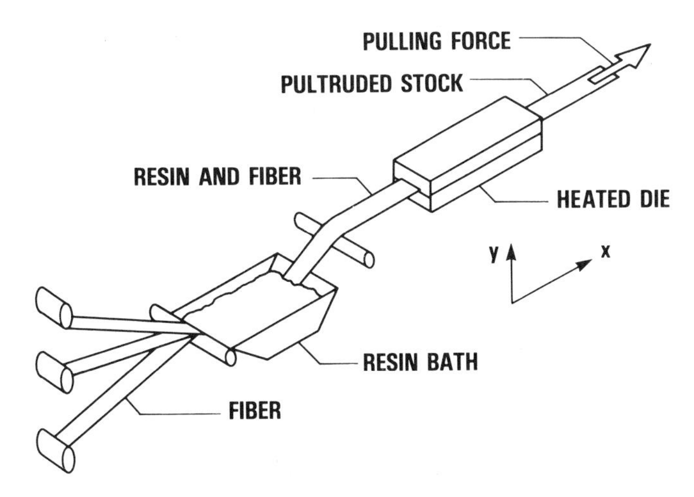

Figure 1. Schematic of pultrusion process showing reference axes for the heated die. (Reproduced with permission from Ref. 57. Copyright 1980, 1st Intl. Conf. on Reactive Processing of Polym.)

Pulling Force. A review of the published work on pultrusion (*1–59*) yielded little specific or unambiguous information on the origin and magnitude of the pulling force. In most investigations, the pulling force was either ignored or else thought of in terms of an upper bound: the force must not exceed the capacity of the pultruder, the tensile strength of the pultruded stock, or the shearing strength of the resin–fiber mass within the die (*11*, *30*, *34*). By contrast, the pultruder capacity may be so large [in the meganewton range (*9*)], or the required force so small [stock pulled through the die manually (*6*)], that the pultrusion force is of little concern. Even if the pulling *force* is small, high hydrostatic *pressures* generated by excess resin at the die entrance may cause fiber displacement (*14*), a condition that can compromise the composite mechanical properties (*60*). However, Garnett et al. (*12*) pointed out that such die entrance resistance can be useful in maintaining fiber tension and reported a pressure there of 140–310 kPa for a 0.5-fiber fraction. This work, done on a laboratory-scale pultruder, was carried further and included the construction of a pilot-scale pultruder described in Reference 23. This later pultruder had electrically heated aluminum dies 150 mm long. The die mount had a strain gauge so the total friction between the die and the resin–fiber mass could be measured. Actual force values were not reported, but several observations were made. Shorter dies developed less friction, but, for a given die length, the friction was higher if the resin cured early. Friction was higher at the higher carbon fiber loadings, and a fiber fraction above 0.6 was not practical. By contrast, glass fiber fractions may approach 0.75 (*33*).

In another pultrusion investigation (*22*) a full-scale (production) pultruder was fitted with strain gauges on the die and central mandrel mount. A 25-mm diameter tube with a 1.6-mm wall was pultruded at pulling speeds up to 30 mm/s. The total force (on die and mandrel) increased with pulling speed up to about 25 mm/s (1 in./s) and reached a value of about 1 kN. By using the dimensions given in Reference 22, the estimated *average* shear stress (between resin–fiber mass and the die and mandrel) was found to be a low 26 kPa. A special apparatus was constructed (*22*) to measure the shear stress between a curing resin–fiber mass and a steel die. The stress generally was low for most of the curing time and seldom exceeded 10 kPa. However, the stress showed two peaks during curing and reached values as high as 56 kPa for a few seconds. The investigation showed that, although the average pulling force on a full-scale pultruder might be quite low, the force could undergo some very large changes.

Resisting Forces. The review of pultrusion publications was useful in suggesting how and under what circumstances the resisting forces arise. The resisting forces are those that the pulling force must exceed to move the resin–fiber mass through the heated die at a constant speed. On the basis of that review, these resisting forces are identified as the collimation force, the bulk compaction force, and the temperature-induced force.

The collimation force results from the gathering and alignment of the fibers before they enter the die and from the viscous drag as the fibers pass through the resin bath. Such forces are strongly dependent on the physical arrangement of the pultruder, so few generalizations can be made about them. In an attempt to obtain a constant collimation force and reduce it as much as possible, we used prepreg tapes (discussed later).

The other two components of the pulling force, the bulk compaction and temperature-induced forces, are associated with the pultruder die. The bulk compaction force arises from the compaction of the resin–fiber mass from an unrestrained condition (high volume, low density) outside the die, to a highly restrained condition (lower volume, higher density) inside the die. Such forces are strongly dependent on the relative volumes of the die cavity and on the resin–fiber mass that is compacted into and passes through the die. These forces result from constraint normal to the fiber axis but are manifested parallel to that axis. The bulk compaction force is essentially a mechanically induced force, and although it is temperature sensitive, it is not primarily temperature induced.

Like the bulk compaction force, the temperature-induced force is associated with the heated die. This force arises from the heating and resultant chemical reaction of the resin during its passage through the die (*54*). Part of this force is a result of the increasing resin viscosity (a shear force), and part is a result of the attempted resin thermal expansion as the resin is initially heated. Some of the expansion is relieved by backflow toward the die entrance. Chemical expansion and contraction takes place in association with the resin-curing reaction. These actions generate normal forces on the die and thus influence the force required to move the resin–fiber mass at a constant speed. Shear forces arising from resin viscosity changes first decrease as resin temperature rises, then increase as the resin curing reaction takes place.

The three forces (collimation, bulk compaction, and temperature induced) act as resisting forces. Thus they are, in some respects, the origins of the pulling force in a pultruder. As conditions in the pultruder change, the pulling force would be expected to change also. Such changes might provide an indication of the characteristics of the process as the resin–fiber mass moves through the heated die.

Experimental

Materials. The materials used were commercially available epoxy resins and carbon fibers. The epoxy resins, which were tacky semisolids at room temperature, were based on diglycidyl ether of bisphenol A. The carbon fibers, made from polyacrylonitrile precursor, were 6000-filament tows. To minimize collimation forces and avoid problems associated with the resin bath (Figure 1), the tows were melt impregnated with the resin. These preimpregnated collimated fibers, or prepreg tapes, were prepared by two different manufacturers. Some properties of the prepreg tapes, designated A and B, are given in Table I.

Tests. A laboratory-scale pultruder, shown in Figure 2, was constructed. The two-piece, aluminum die was mounted in the movable crosshead of a mechanical testing machine that could measure loads to 900 N. Die temperatures to 550 K were provided by two thermocouple-controlled, quartz tube, radiant heaters. The die had five thermocouple wells so thermocouple beads could be located along the centerline within 0.5 mm of the die cavity wall. The cavity was 20 mm wide and 2 mm high (for an aspect ratio of 10), and the die was 50 mm long. The resulting maximum contact area between the die and the resin–fiber mass was 2200 mm^2. The die halves were secured by four bolts along each edge (Figure 2), but most of the tests were run with the entrance end bolts not tightened. This method allowed the die to open slightly so that, with the 0.8-mm radius on the cavity entrance edges, the transition of the tape from outside to inside the die was moderated to some extent. Indeed, some pultruder dies will have a slight taper at the entrance to facilitate this transition. The testing machine had specific cross-head speeds of which three, 12.5, 25, and 50 mm/min, were used. Thus, with a die length of 50 mm, the tape residence times were 4, 2, or 1 min, in the range of residence times used in full-scale pultrusion.

The test procedure was to load the required number of plies of prepreg tape (each ply being 20 mm wide) into the room temperature die, and to secure the upper end of the plies in the load cell grip. The die temperature was stabilized at

Table I. Properties of Epoxy Resin, Carbon Fiber Prepreg Tape

Property	*Tape A*	*Tape B*
Nominal ply thickness, mm	0.17	0.14
Fiber areal weight, kg/m^2	0.178	0.146
Resin content, weight percent	25	32
Volatiles, percent	1	1

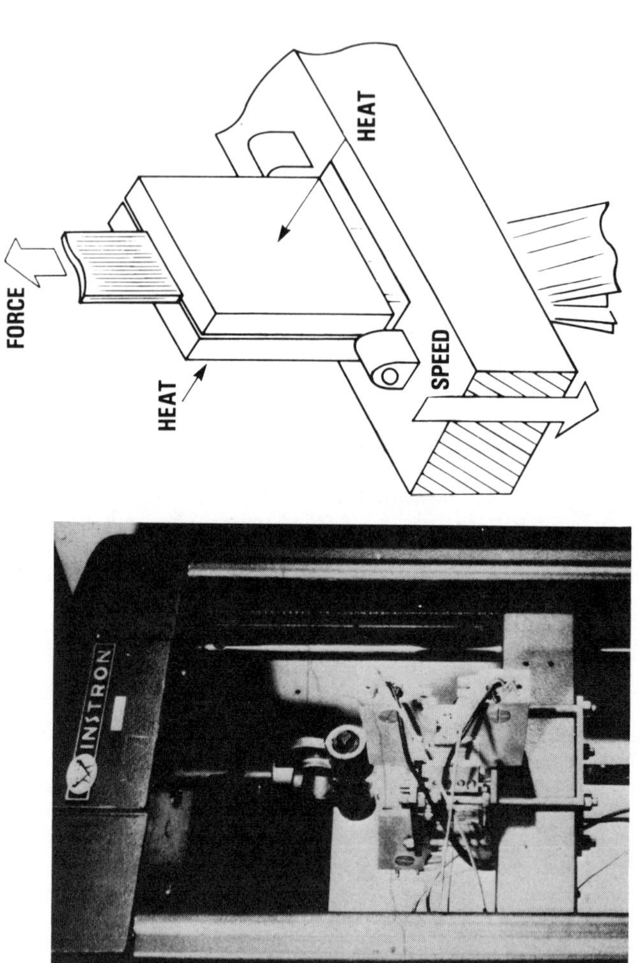

Figure 2. Laboratory-scale pultruder. (Reproduced with permission from Ref. 57. Copyright 1980, 1st Intl. Conf. on Reactive Processing of Polym.)

300 K, or in the 375–400 K range, and cross head and die were traversed downward at one selected speed. The die temperature was raised at a rate of 0.3–0.5 K/s to temperatures in the 400–525 K range. This procedure provided 0.25-m length pultruded stock with a nominal 2 × 20 mm cross section.

The test procedures show the differences between laboratory-scale and full-scale pultruders. In a full-scale pultruder, the die is fixed in space and the resin–fiber mass moves through it. To establish process parameters, the die temperature is fixed and the pulling speed is varied in a known way, but no measurements are made of the pulling force. By contrast, in the laboratory-scale pultruder, the die moves and the resin–fiber mass is fixed in space. The pulling speed is constant and the temperature is varied. In addition, the pulling force, a dependent variable, is measured. Although different, the two procedures are assumed to give equivalent results.

Results and Discussion

The pulling force in a pultruder must exceed the resisting forces resulting from collimation, bulk compaction, and temperature. Because strips of prepreg tape were used in this investigation, the collimation forces were effectively eliminated. The other resisting forces were then examined. First, the general procedure used was to hold the die temperature constant, to increase the volume of material being compacted into the die, and to measure the bulk compaction force. Then, with the material volume held constant, the die temperature was changed and the temperature-induced force was measured.

Bulk Compaction Force. The bulk compaction force arises from the volume change of the resin–fiber mass from outside to inside the die. A direct way of increasing the material volume (and hence the force) was to add plies to the prepreg tape as it entered the die. As additional plies were introduced into the die, the pulling force would be expected to increase. We increased the force by stacking the prepreg tape so that an additional ply was introduced in the center of the stack at 75-mm intervals that were equivalent to one and one-half die lengths. The thinnest end of the tape was fed into the die first. The die temperature was set at 375 K, a temperature high enough to melt the resin and low enough so the resin would not cure during the residence time (240 s at a speed of 12.5 mm/min) in the die. Measurements on resin samples with a cone-and-plate rheometer showed that, at a strain rate of 1 s^{-1}, viscosity began to increase only after 1700–2000 s at 375 K. Hence, any force increase in the compaction tests would be a result of the additional volume of material in the die.

The force measurements are listed in Table II and shown in Figure 3. Figure 3 also includes a representation of a longitudinal section of the tape entering the die. As the additional ply entered the die, the pulling force increased until the leading edge of that ply had passed through the die. Once the ply was through, the force remained constant (as listed in Table

Table II. Bulk Compaction Forces of Epoxy Resin, Carbon Fiber Prepreg Tape

	Tape A		*Tape B*	
Number of Plies	*Normalized Thickness*	*Pulling Force, N*	*Normalized Thickness*	*Pulling Force, N*
13	1.11	17.8, 19.6	—	—
14	1.19	57.8, 80.1	—	—
15	1.28	276, 316	1.04	11.6, 14.7
16	1.36	792, 947[a]	1.11	44.5, 48.9
17	—	—	1.18	129, 147
18	—	—	1.25	280, 343

NOTE: Measurements were made with die temperature of 375 K and pulling speed of 12.5 mm/min.

[a]Estimated

II) until the next additional ply entered the die. Each tape had a different thickness per ply (see Table I). The measurements were normalized by dividing the thickness of all the plies in the die by the nominal cavity height (2 mm) of the die. With the thickness normalized and the constant pulling force plotted on a logarithmic scale (Figure 3), the data fell along two straight lines. In other words, for a linear increase in prepreg thickness (or volume) the pulling force increased exponentially. Indeed, the pulling force for tape A at the highest relative thickness value approached or exceeded the 900-N limit of the pultruder. The data also showed that 14 plies

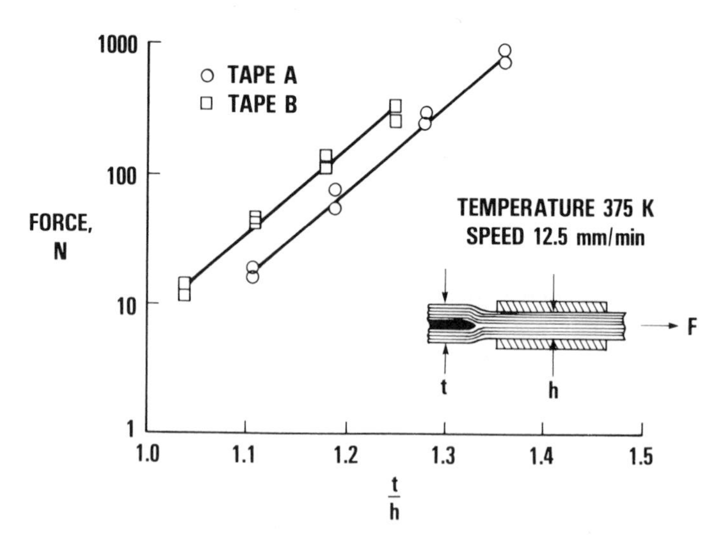

Figure 3. Pulling force as a function of normalized thickness of prepreg tape compacted into and pulled through the pultruder die.

of tape A required roughly the same pulling force (~70 N) as 16 plies of tape B (~59 N). These numbers of plies (except as noted below) were used in subsequent tests.

FORCE–MATERIAL VOLUME. The bulk compaction results (Table II and Figure 3) showed that the pulling force could be related to the prepreg thickness (or volume) by the expression

$$\frac{F}{F_1} = \exp\left[K\left(\frac{t}{h} - 1\right)\right] \tag{1}$$

In Equation 1, F is the pulling force for tape of any thickness t, and F_1 is the force for thickness $t = h$ (i.e., $t/h = 1$). When t equals h, the tape is not being compacted and the pulling force presumably would be a measure of the resin viscosity (*see* Appendix A). The empirically determined constant K is the slope of the straight lines in Figure 3. From the data in Table II, the values of the constants fell in the following range (with 95% confidence): Tape A, $F_1 = 2.5$–5.0 N, $K = 15.1$–15.8; and tape B, $F_1 = 5.3$–12.0 N, $K = 14.7$–15.6. Thus, although tape A had a lower resin content (Table I) and a lower reference force F_1, the force increased slightly more with increasing volume (higher K) than the force for tape B. Even so, for a value of $t/h = 1.5$ (tape entering the die was 50% thicker than the die cavity height), the estimated pulling force for tape B would still be nearly twice that for tape A (15.4 kN vs 8.1 kN). Both values are high and far in excess of the pulling capacity of the laboratory-scale pultruder used in this investigation. More important, however, is the wide range of pulling forces that we measured or estimated. The amount of material compacted into the die and, apparently, the resin content as well affect the pulling force. This material volume effect may help explain why previously reported values of pulling force have varied over such a wide range.

Other force–volume relationships, similar to Equation 1, have been used in connection with the compaction of polymer powders and granules (*61, 62*). These relationships have the form

$$\frac{F_a}{F_t} = \exp\left(C\frac{l}{D}\right) \tag{2}$$

where F_a is the force applied at one end, and $F_t < F_a$ is the force transmitted at the other end of powder confined in a cylinder of diameter D. The length l represents the displacement of the end of the powder when force is applied at that end, and C is an empirically determined constant. Thus, the applied force and the resulting displacement in the same direction have an exponential relationship. In the bulk compaction tests, with the tape displacement in the y direction and the force measurement in the x direction (*see* Figure 1), the force–displacement relationship also was exponential. Apparently, the pultruder die performs a y-force to x-force transformation that retains the exponential nature of the force–displacement relationship.

FORCE–DIE LENGTH. The data listed in Table II and plotted in Figure 3 are the constant force values obtained when a given number of plies (or a given amount of compressed material) was passing through the die. This situation is shown schematically in Figure 4 for two material thicknesses t_i and t_j. These two thicknesses extend over the entire length l of the die and have pulling forces of F_i and F_j. Now, consider an intermediate situation in which the thicker material (t_j) extends a distance x into the die. The pulling force in this case can be expressed by adapting Equation 1 to obtain

$$\frac{F}{F_1} = \exp\left\{K\left[\left(\frac{l-x}{l}\right)\frac{t_i}{h} + \frac{x}{l}\frac{t_j}{h} - 1\right]\right\} \tag{3}$$

The pulling force measurement would not indicate where the thicker material was located, although its location was known because the location along the prepreg tape and the pulling speed were known. The measured force was the sum of the pulling force for each thickness, weighted according to the relative length of each thickness in the die.

The importance of this pulling force–relative length effect can be illus-

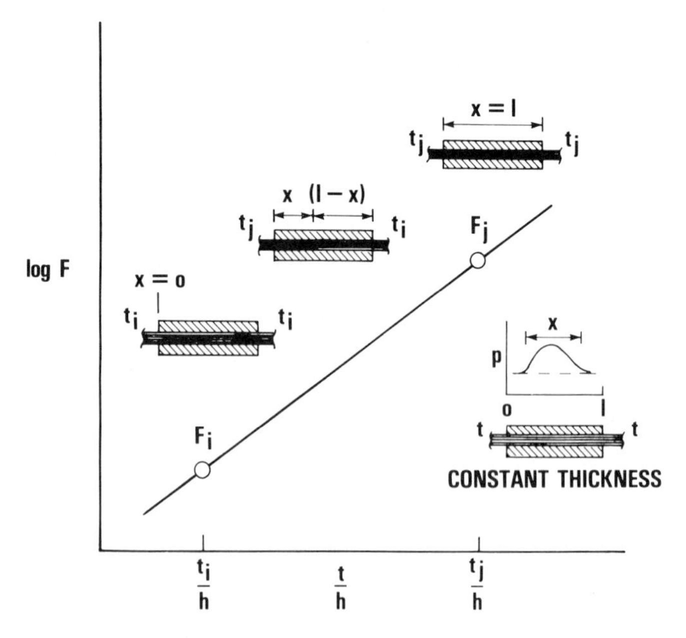

Figure 4. Representative pulling force–normalized thickness curve.

trated as follows: Consider the material to be initially the same thickness t throughout (same number of plies). Presumably the force on the die wall would be uniform and the pulling force constant. If that wall force changed (e.g., as a result of a chemical reaction) over part of the die length, then the pulling force also would change. This situation is shown schematically at the bottom of Figure 4 for a force increase over part of the die length. The greater the affected length, the greater will be the pulling force change.

As the pultruder die integrates rather than differentiates forces, the force distribution could have a variety of shapes that would result in the same pulling force. This situation is analogous to that found in a capillary rheometer that provides a force measurement but cannot, without additional information, indicate the velocity profile across the capillary. Even so, both the capillary rheometer and the pultruder die indicate, through force changes, that a change is taking place in the material.

The bulk compaction tests, then, showed the following results: (i) the pulling force increased exponentially with a linear increase in the volume of material fed into the die; (ii) the die transformed the forces normal to the die wall such that changes in these forces would be indicated by a change in the pulling force; and (iii) the die integrated the wall forces such that the greater the affected length along the die, the greater would be the pulling force change.

Temperature-Induced Force. Representative curves of the variation of pulling force with temperature are shown in Figure 5 for tape A and in Figure 6 for tape B, where a logarithmic axis is used for the force. These figures show the force variation for three pulling speeds of 12.5, 25, and 50

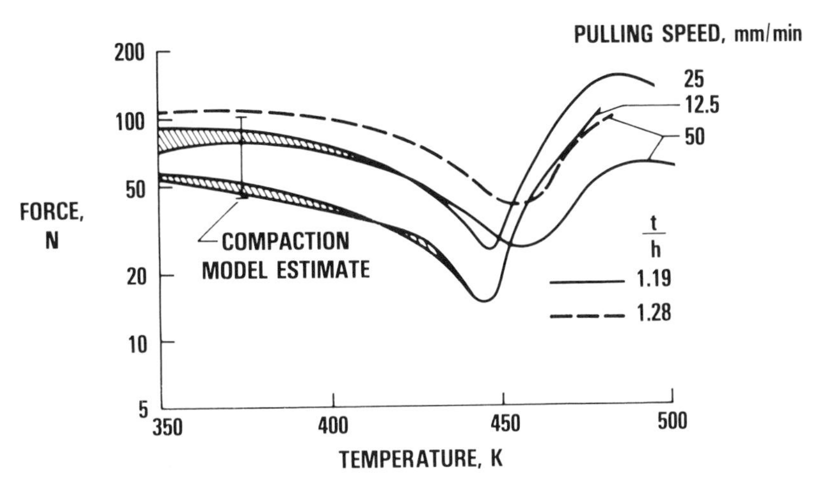

Figure 5. Pulling force as a function of die temperature for tape A.

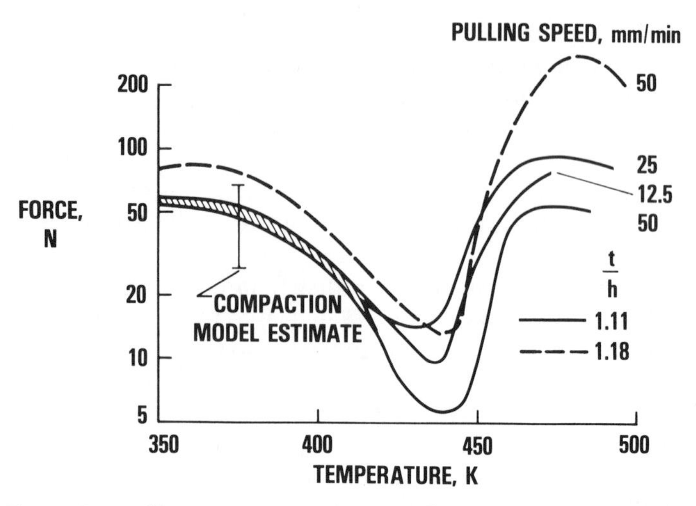

Figure 6. Pulling force as a function of die temperature for tape B.

mm/min. The dashed curves show the force variation at a pulling speed of 50 mm/min with the addition of one ply of prepreg tape to obtain a higher t/h ratio.

All of the force–temperature curves, whatever the differences in materials or testing, had a characteristic shape. At temperatures above the 350-K reference temperature, the pulling force decreased as the temperature increased. The force reduction continued and accelerated as the temperature approached the 430–450-K range. In this range, the force passed through a minimum value that was approximately 10–25% that of the 350-K pulling force. Once through the minimum, the pulling force increased sharply, approaching or exceeding the 350-K pulling force within the next 25°–50°.

An important operational parameter in pultrusion is the pulling speed. Some effects of pulling speed on pulling force are shown in Figures 5 and 6. At temperatures below the minimum force temperature, the most noticeable effect of speed was an increase in pulling force for tape A (Figure 5). Whereas the other curves (tapes A and B) started in the 50–60-N range, those for tape A, at speeds of 25 and 50 mm/min, began in the 80–90-N range. Part of the difference may be a result of the lower resin content (Table I) and, hence, the reduced lubrication of tape A. In the case of tape B (Figure 6), pulling speed had very little effect. The force curves essentially superimposed until the temperature approached the minimum force temperature. For temperatures above the minimum force temperature, however, the pulling speed had a noticeable effect on the pulling force for both tapes. For a given temperature, the 25-mm/min speed resulted in the highest force, the 50-mm/min speed resulted in the lowest force and the 12.5-mm/min speed provided an intermediate force.

The effect of adding an additional ply of prepreg tape is shown by the dashed curves in Figures 5 and 6. The results were obtained at a pulling speed of 50 mm/min. Additional plies would be of benefit in the use of the pultruded stock because a high fiber content should produce a strong part. However, the additional ply also increased the pulling force (by roughly 50%) over the entire temperature range. The increase in force was slightly larger at temperatures above the minimum force temperature than at lower temperatures. Consequently, a steady, even, pulling force for high fiber volume fraction stock would require an even die temperature and minimum variation from the set point.

The pulling force shown in Figures 5 and 6 can easily be changed to the average shear stress at the interface between the die wall and the resin–fiber mass. One size die with a cavity wall area of 2200 mm^2 was used throughout this investigation. With pulling forces ranging from 5 to 250 N, the calculated *average* shear stress ranged from 2 to 115 kPa. In another investigation in which the pulling force was measured (22), the average shear stress (not reported but estimated in this investigation) ranged from very low values such as those reported here up to 56 kPa. The pulling force was much larger (~1 kN) because the contact area between the die and resin–fiber mass was larger. Thus, the interface shear stresses appear to be fairly low so that fiber breakage or distortion ordinarily would not be a problem. The pulling stresses, i.e., pulling force divided by stock cross section, were higher than the shear stresses because of the difference in the respective areas. In this investigation, the shear area was 2200 mm^2 and the cross section area was 40 mm^2, a ratio of 55:1. Hence, the pulling stress ranged up to ~60 MPa, a value far below the 800 MPa or more that such a composite would be expected to sustain.

Characteristic Force–Temperature Curve. The curves of the variation in pulling force with temperature, shown in Figures 5 and 6, all had a characteristic shape (Figure 7). At temperatures above the 350-K reference temperature, the pulling force decreased as the temperature increased and passed through a minimum value that was approximately 10–25% that of the 350-K pulling force. The force then increased sharply and approached or exceeded the 350-K pulling force within the next 25°–50°. These large changes in the pulling force can be interpreted with the aid of data on the exothermic reaction that the resins undergo.

DIE TEMPERATURE EFFECTS. The rate and temperature range of the exothermic reaction of the epoxy resins was reported by Price et al. (57). At a temperature rise rate of 80 K/min, the reaction was initiated in the 430–440-K range, reached a maximum rate in the 470–480-K range, and was essentially completed in the 500–510-K range. The temperature range for reaction initiation corresponds very closely with that of the minimum in the force–temperature curves (Figure 7). Hence, the force data obtained at

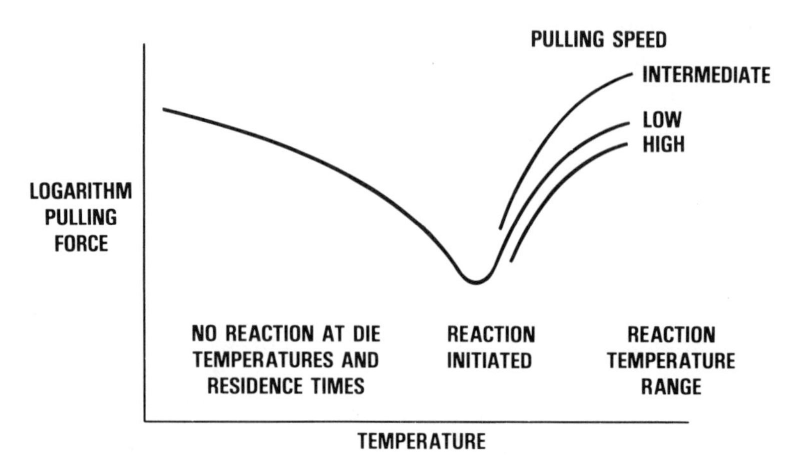

Figure 7. Characteristic pulling force–die temperature curve.

temperatures above that of the force minimum were affected by the resin reaction. The effect, of course, is an increase in resin viscosity and pulling force. At temperatures below the force minimum, the resin reaction and associated viscosity increase was much slower than the viscosity decrease resulting from the temperature increase. Consequently, the pulling force decreased with temperature increase.

The reduction in pulling force as the temperature increased toward the reaction initiation temperature range can be attributed to two causes. One cause, typical of all nonreacting polymers, is the reduction in viscosity as the temperature rises. The other cause, peculiar to the pultrusion process, is resin backflow at the die entrance. As the resin–fiber mass enters the die, the mechanical compaction and the attempted thermal expansion of the mass (mainly of the resin) generate pressure against the die wall. Under these pressures, the resin (the more mobile phase) will tend to flow from a higher to a lower pressure. The lowest pressure (atmospheric) is upstream, or outside the die, and results in the resin backflow. Although the resin backflow can be significant, as explained in the Appendix B, it was not in this investigation.

PULLING SPEED EFFECTS. The pulling speed had the most noticeable effect on pulling force in the reaction temperature range (Figure 7). In that range, the intermediate pulling speed (25 mm/min) gave the highest pulling force and the highest pulling speed (50 mm/min) gave the lowest pulling force at a given temperature. The pulling force obtained with the lowest speed (12.5 mm/min) fell between the two extremes. These results indicate that the highest pulling speed is the best one to use because it provides the highest throughput and the lowest force and associated stress. However, the pultruded stock made in this investigation was evaluated by specific gravity (an indication of material compaction) and short beam shear strength (an indication of resin cure). Good quality material would have high values of both properties. The room temperature measurements of these properties for stock made from both tapes are shown in Figure 8. The temperature axis refers to the pultruder die temperature. Each data point is the average of two or three tests except the filled symbol for tape A stock. This point is a 15-test average data point. Different symbols indicate the three different pulling speeds, and the trend for the 50-mm/min speed is shown by the dashed lines. The values in all cases for the high speed pultruded stock were less than or at most equal to those of the stock made at the lower speeds. Although there was little to indicate that either of the lower speeds provided especially better values, the high speed clearly provided lower quality material.

The pulling speed effect can be explained with the aid of Figure 9. This figure shows, for a given temperature, the general relationship between the logarithm of pulling force and pulling speed. This relationship is consistent with the results obtained in this investigation. Also shown in Figure 9 is a general pressure–time curve adapted from results reported by McGlone and Keller (*63*). The pressure is that exerted by a thermosetting molding compound on a constant volume die wall during compression molding as a function of time in the die. The pressure peak is associated with the heating and curing of the molding compound. A time θ' has been indicated at which a large part of the reaction seems to have been completed. Thus θ' may be thought of as a cure time.

In the case of a pultruder die, time in the die and distance along the die are directly related through the pulling speed by $s = x/\theta$. Then the resi-

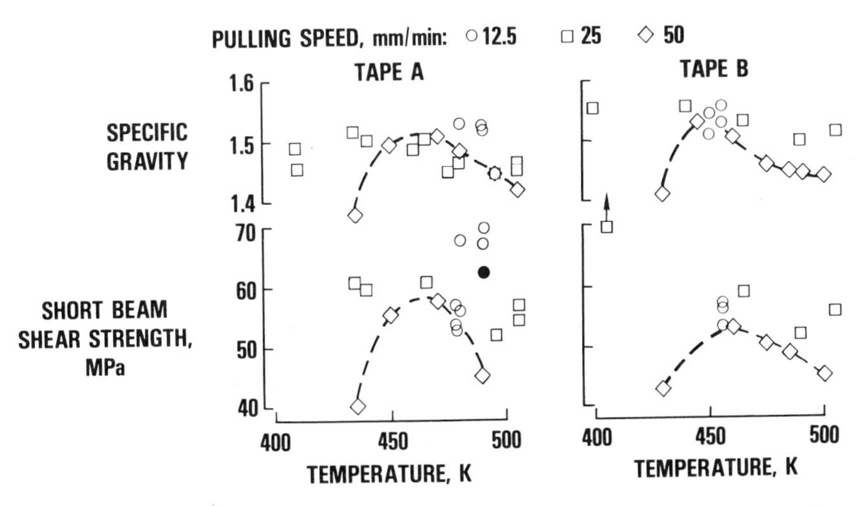

Figure 8. Room temperature specific gravity and short beam shear strength of pultruded stock made from tapes A and B at three different pulling speeds for a range of die temperatures.

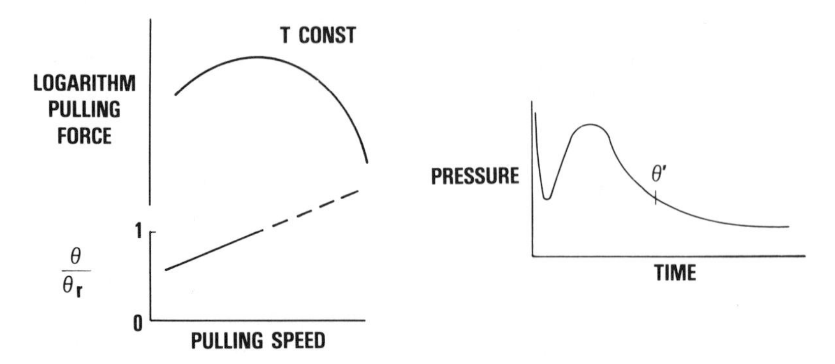

Figure 9. Representative force–speed curve obtained in this investigation, and representative pressure–time curve for constant volume compression molding as adapted from results reported by McGlone and Keller (63). The cure time–residence time ratio is shown as a function of pulling speed.

dence time in the die, the time available for compacting the resin–fiber mass and for reacting the thermosetting resin, is simply $\theta_r = l/s$. Each molding material at a given temperature will have a "cure time" θ' associated with the resin exothermic reaction, viscosity increase, and gellation. (This time is much shorter than the hours-long time often used for thermosetting resins.) If the pulling speed is high (Figure 9), the residence time θ_r is small relative to the cure time θ' and the ratio θ'/θ_r is greater than unity. The material passes through the die before the heating- and curing-associated pressure can develop and the pulling force is relatively low. If the pulling speed is very low, the residence time is larger than the cure time and the ratio θ'/θ_r is less than unity. Sufficient time is available for good compaction and curing, but throughput is reduced and die use is inefficient. In addition, making the time-to-distance transformation in Figure 9, the pressure peak occurs over a relatively short length of die. Thus the x-distance in Equation 3 and Figure 4 is small, and the pulling force is not as high as it might be. By contrast, if the pulling speed is adjusted so that the residence time is nearly the same as the cure time ($\theta'/\theta_r \approx 1$), then the highest throughput that produces good quality material is obtained. The pressure peak will then be spread out over most of the die length and, because of the force integrating action of a pultruder die discussed earlier, the pulling force will then be high. In other words, for a given material at a given temperature, the maximum quantity and the maximum quality are achieved at the maximum force.

Appendix A: Pultruder Die as a Band Viscometer

The pultruder die used in this investigation was similar to the band viscometer that has been used to measure the viscosity of printing inks (*64*, *65*). The band viscometer consists of a vertically mounted, thermostatted die with a rectangular cavity through which a strip of metal foil or plastic film is pulled. The foil/film substrate is the same width as the cavity, but has a known thickness t that is smaller than the cavity height h. Therefore, a clearance c exists between the die walls and substrate such that $h = t + 2c$. As the substrate enters the die, it passes through a reservoir of the test liquid, some of which adheres to the substrate surface. The adhering liquid is sheared in the clearance between the die walls and the substrate. A dead weight load supplies a constant substrate pulling force. The time required for the passage of the substrate through the die is a measure of the test liquid viscosity. This viscosity can be found from (65)

$$\eta = \frac{F_x \theta_r c}{Al}$$

where the expression has been cast in pultrusion die terms.

In a pultruder die containing resin-coated fiber bundles, the clearance c is not well defined. If $t/h = 1$ (which is associated with pulling force F_1), then the clearance should be zero. However, some virtual or equivalent clearance will exist in which the resin is sheared and produces a resisting force and a nonzero pulling force. An estimate of the equivalent clearance c was obtained using the viscosity value of a neat DGEBA resin obtained over a strain rate range of 100–1000 s^{-1} on a cone-and-plate rheometer. The measurements, made for another purpose, were taken at 350 and 400 K, and interpolated to the 375 K used in the bulk compaction tests. The interpolated value was found to be 0.03 Pa-s. Then, using the range of F_1 values given above, the range of estimated clearances and strain rates was as follows: Tape A, $c = 3$–$6\ \mu m$, $\dot{\gamma} = 38$–$74\ s^{-1}$; and tape B, $c = 1$–$3\ \mu m$, $\dot{\gamma} = 80$–$174\ s^{-1}$. Interestingly enough, tape A, which had the lower resin content, had a higher equivalent clearance.

Appendix B: Resin Backflow

Resin backflow at the pultruder die entrance is important for processing and composite properties. We observed very little backflow in this investigation because the prepreg tape was relatively dry. However, when a fiber bundle passes through a resin bath, as is typical of full-scale pultrusion, an excess of resin will cling to the fibers. Much of this excess will be mechanically stripped off as the fiber bundle enters the die. Resin backflow from within the die may take place as the resin attempts to expand on heating. When the thermally induced pressure (in the y direction) exceeds the resin flow pressure (in the x direction), the backflow will occur.

The conditions under which thermally induced backflow occurs can be found as follows: The pressure gradient in the y direction (Δp_y) (*see*

Figure 1), taken as the axis along which temperature moves into the resin–fiber mass, is

$$\Delta p_y = K\alpha\Delta T \tag{B1}$$

For a given material (specified K and α), the thermal pressure gradient is proportional to the temperature gradient (ΔT). In the negative x direction, the direction of backflow, the pressure gradient can be expressed (*66*)

$$\Delta p_x = \frac{k\eta xu}{m^2\epsilon} \tag{B2}$$

The resin flow speed u is taken relative to the fiber bundle that is moving with speed s in the positive x direction. Thus, backflow will not occur until u exceeds s thereby setting an upper limit on Δp_x when $u = s$. In addition, the distance x over which resin at a level y must flow back to reach the die entrance is

$$x = s\theta_y \tag{B3}$$

where θ_y is the time required for the wall temperature to diffuse in to level y. This time can be estimated from the Fourier number (NF)

$$\text{NF} = \frac{\theta_y}{y^2/D_T} = 1 \tag{B4}$$

Therefore, Equation B2 becomes

$$\Delta p_x = \frac{k\eta sy^2s}{m^2D_T\epsilon} = \left(\frac{k\eta}{m^2D_T\epsilon}\right)y^2s^2 \tag{B5}$$

The ratio of the two pressure gradients is

$$\frac{\Delta p_x}{\Delta p_y} = \left(\frac{k\eta u}{K\alpha m^2D_T\epsilon}\right)\frac{y^2s^2}{\Delta T} \tag{B6}$$

Resin backflow will take place when and where this ratio is less than unity, that is, the x-direction pressure gradient is less than the y-direction pressure gradient.

The pressure gradient ratio (Equation B6) is useful in assessing operating conditions and backflow. For example, the Δp_x establishes an upper limit for Δp_y and thus an upper limit on the resisting force generated at the die entrance. The values enclosed in parentheses are characteristics of a given material; those values following the parentheses are operational or geometric variables. Thus, for a given material, a higher die–material temperature difference at a fixed pulling speed will produce backflow. Similarly, at a fixed temperature a lower speed will also produce backflow. For a fixed speed and temperature, backflow will begin at the heated wall (y = 0). How far into the resin–fiber mass the backflow will extend depends on both material and operational values. Resin backflow does not necessarily take place over the entire cross section.

The results of backflow were observed in a commercially available pultruded rod that was large (higher temperature difference) and made at a low speed (provide time for die temperature to diffuse inward). The rod (Figure 10) was 38 mm in diameter and was made with unidirectional glass roving and a polyester resin that did not contain a filler. This rod had a nearly circular resin rich area in the center that extended over roughly two-thirds the diameter. The outer portion of the rod appeared to have a resin deficiency, presumably as a result of backflow. These observations were corroborated by specific gravity and fiber fraction measurements, listed in Table III, of the total cross section and the central core. The core was obtained by machining rod specimens to one-half the original diameter, which left the center one-fourth of the rod. The central core had a lower specific gravity and fiber content (and thus a higher resin content) than did the total cross section. From these measurements, the outer ring was estimated to have an even higher specific gravity and fiber content (and a lower resin content) probably a result of resin backflow.

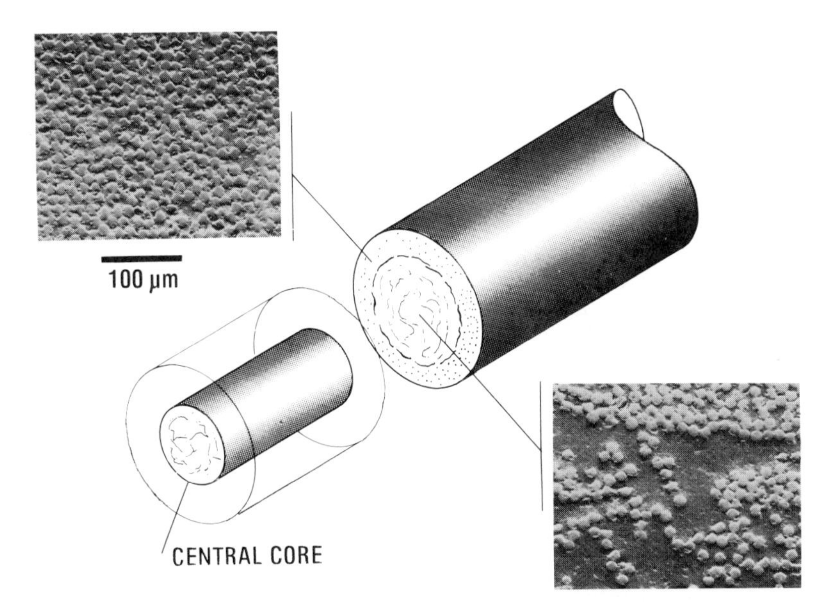

Figure 10. A cross section of the pultruded polyester unidirectional glass fiber rod showing the location of the central core. Also shown are the locations and micrographs of the resin rich center and the "drier" outer ring that were produced as a result of resin backflow.

Table III. Room Temperature Properties of Pultruded Circular Rod

Property	*Total Cross Section*	*Central Core*	*Outer Ring (est)*
Specific gravity	2.04	1.94	2.07
Fiber volume fraction	0.634	0.559	0.659
Diametral strength			
Number of tests	5	11	—
Mean, MPa	11.3	14.8	—
Coeff. of variation	0.24	0.06	—
Notched shear strength			
Number of tests	6	12	—
Mean, MPa	36.7	38.5	—
Coeff. of variation	0.05	0.12	—

As expected in a composite material in which the proportions of the resin and fiber change, the mechanical properties changed as well. Mechanical properties generally increase with an increase in fiber content. However, the composite must contain sufficient resin to bond the fibers together or else the properties will degrade. The rod specimens were evaluated by the diametral test, which generates a tensile load perpendicular to the fibers (*37, 44*), and by the notched shear test, which applies a shear load parallel to the fibers (*37, 38, 67*). The results, listed in Table III, showed that the diametral strength of the core was significantly higher than the diametral strength of the total cross section. Likewise, the notched shear strength of the core was marginally higher as well. If the strength of the total cross section represented some average over that section, the implication is that the strength of the outer portion of the rod was lower than that of the core. That is, the "drier," lower resin content outer portion was not as strong. The strength varied over the cross section as did the resin content, the latter variation was probably a result of resin backflow. Thus, resin backflow is important not only in the processing but also in the properties of the composite as well.

Nomenclature

A Area, m^2
c Clearance, m
C Constant
D Diameter, m; diffusivity, m^2/s
F Pulling force, N
h Cavity height, m
k Constant
K Bulk modulus, Pa; constant
l Die length, m; displacement distance, m

m Function of fiber radius and porosity, m
p Pressure, Pa
s Pulling speed, m/s
t Material thickness, m
T Die temperature, K
u Flow speed, m/s
x Distance, m
y Distance into resin–fiber bundle from heated wall, m
α Thermal expansion, 1/K
ϵ Porosity
η Resin viscosity, Pa-s
θ Time, s

Acknowledgment

The experimental part of this investigation was performed while H. L. Price was at NASA-Langley Research Center.

Literature Cited

1. Ewald, G. W. Pultrusion entry in "Modern Plastics Encyclopedia," Vol. 57 (10A), McGraw-Hill Inc.: New York, Oct. 1980; pp. 391–397.
2. Howald, A. M.; Meyer, L. S. U.S. Patent 2,572,717, October 16, 1951.
3. Meek, M. J. U.S. Patent 2,684,318, July 20, 1954.
4. Goldsworthy, W. B.; Landgraf, F. U.S. Patent 2,871,911, February 3, 1959.
5. Pancherz, H. J. J. U.S. Patent 2,948,649, August 9, 1960.
6. Boggs, L. R. U.S. Patent 3,244,784, April 5, 1966.
7. Goldsworthy, W. B. Soc. Plast. Ind. 23nd Annu. Tech. Conf. Proc., 1968, Paper 18-A.
8. Segawa, J.; Nakagowa, B. Ibid, Paper 18-B.
9. Meyer, L. S. Soc. Plast. Ind. 25th Annu. Tech. Conf. Proc., 1970, Paper 6-A.
10. Mackay, J. M.; McKee, R. B., Jr. U. S. Patent 3,533,870, October 13, 1970.
11. Goldsworthy, W. B. U.S. Patent 3,556,888, January 19, 1971.
12. Garnett, G.; Hancox, N. L.; Spencer, R. A. P. Plast. Inst. Conf. on Reinforced Plast. Research Proj. III, London, November 1971, Paper 10.
13. Comey, J. *Mach. Des.* **1971**, *43*, 45–49.
14. Kannebley, G. *Plast. Polym.* **1972**, 13–18.
15. Hardesty, E. E. Soc. Aerosp. Mater. Process. Eng., 4th National Tech. Conf. Proc., Vol. 4, October 1972, pp. 345–360.
16. Hill, J. E.; Goan, J. C.; Prescott, R. *SAMPE Quarterly* **1973**, *4*, 21–27.
17. Tickle, J. D. Soc. Automot. Eng., Int. Automot. Eng. Congr., January 1973, Detroit, MI, Paper 730172.
18. Meyer, L. S.; Eden, D. H. Soc. Plast. Ind. 28th Annu. Tech. Conf. Proc. 1973, Paper 6-F.
19. Hardesty, E. E. Ibid, Paper 15-D.
20. Dibb, J. J.; Wronski, A. S.; Watson-Adams, B. *Composites* **1973**, *4*, 227–228.
21. Morrison, R. S. *Mat. Engr.* **1973**, *77*, No. 4, 36–39.
22. Jones, B. H. Soc. Plast. Ind. 29th Annu. Tech. Conf. Proc. 1974, Paper 15-C.
23. Spencer, R. A. P. *Plas. Polym. Conf. Suppl.* **1974**, *6*, 140–147.
24. *Modern Plastics* **1975**, 52 No. 7, 30.
25. Jones, B. H. Soc. Plast. Ind. 31st Annu. Tech. Conf. 1976, Paper 17-D.

26. Halsey, N.; Marlow, D. E.; Mitchell, R. A.; Mordfin, L. AFML-TR-76-42, May 1976.
27. Comey, J. *Plast. Des. Process.* **1976**, *16*, 8–9.
28. Thompson, V.; Bradley, R. J. Soc. Manf. Eng. Conf. on Adv. Compos., Los Angeles, Calif., June 1976, Paper EM76–415.
29. Pepper, R. T.; Gigerenger, H.; Zack, T. A. Soc. Adv. Mater. Process. Eng. 8th Nat. Tech. Conf. Vol. 8, Seattle, Wash., October 1976, 338–342.
30. Jones, B. H.; Jakway, W. USAAMRDL-TR-76-5, December 1976.
31. Klein, T. H. U.S. Patent 4,012,267, March 15, 1977.
32. Fesko, D. G. *Polym. Eng. Sci.* **1977**, *17*, 242–245.
33. Tickle, J. D. *Mach. Des.* **1977**, *49*, 163–167.
34. Chung, H. H. NASA CR-145330, January 1978.
35. Post, C. T. *Iron Age* **1978**, *221*, 27–29.
36. Werner, R. I. Soc. Plast. Ind. 33rd Annu. Tech. Conf. Proc. 1978 , Paper 8-A.
37. Haarsma, J. C. *Ibid*, Paper 8-B.
38. Loveless, H. S. *Ibid*, Paper 8-C.
39. Sumerak, J. E.; Colangelo, F. V. *Ibid*, Paper 8-D.
40. McQuarrie, T. S. *Ibid*, Paper 8-E.
41. Tickle, J. D.; Halliday, G. A.; Lazaron, J.; Riseborough, B. *Ibid*, Paper 8-F.
42. Rolston, J. A. *Ibid*, Paper 8-G.
43. Martin, J. *Ibid*, Paper 8-H.
44. Haarsma, J. C. *Ibid*, Paper 22-E.
45. Bradley, R. NASA CR 158899, March 1978.
46. *Modern Plastics* **1978**, *55*, 50–52.
47. Dharan, C. K. H. *J. Mater. Sci.* **1978**, *13*, 1243–1248.
48. Van Werk, H. Aust. Inst. Eng. Nat. Conf. Publ. 79-8, 1979, 97–100.
49. Bonassar, J. J.; Lucas, J. J. USAAVRADCOM-TR-79-45, 1979.
50. Werner, R. I. Soc. Plast. Ind. 34th Annu. Tech. Conf. Proc. 1979, Paper 9-C.
51. Martin, J. D. *Plast. Eng.* **1979**, *35*, 53–57.
52. Shobert, S. M.; Fish, E. B. U.S. Patent 4,154,634, May 15, 1979.
53. Iddon, K. J.; Blundell, C. Proc. Symp. Fab. Tech. Adv. Reinf. Plast. Salford, Lancaster, England, April 1980, 58–68.
54. Morgan, P. E.; Trewin, E. M.; Watson, I. P. Ibid, the Manufacture and Use of Carbon Fibre Pultrusion. pp. 69–90.
55. O'Connor, J. E.; Roosz, S. A. 25th Natl SAMPE Symp. and Exh. Vol. 25, San Diego, CA, May 6–8, 1980, pp. 621–634.
56. Krutchkoff, L. *Plast. Des. Process.* **1980**, *20* July, pp. 34–38; August, pp. 37–41.
57. Price, H. L.; Cupschalk, S. G. 1st Intl. Conf. on Reactive Process. of Polym., Univ. Pittsburgh, October 1980.
58. LaRochelle, R. W.; Higgins, J. B. Soc. Plast. Eng. Natl. Tech. Conf., Cleveland, OH, 18–20 Nov. 1980, p. 213.
59. McQuarrie, T. S. U.S. Patent 4,252,696, 24 Feb. 1981.
60. Van Dreumel, W. H. M.; Kamp, J. L. M. *J. Compos. Mater.* **1977**, *11*, 461.
61. Beyer, C. E.; Spencer, R. S. In "Rheology–Theory and Applications"; Eirich, F. R., Ed.; Academic Press, 1960, Vol. 3, pp. 527–529.
62. Crawford, R. J. *Polym. Eng. Sci.* **1982**, *22*, pp. 300–306.
63. McGlone, W. R.; Keller, L. B. *Modern Plastics* **1957**, *March*, 173, *April*, 137.
64. Hull, H. H. *J. Colloid Sci.* **1952**, 7, pp. 316–322.
65. VanWazer, J. R.; Lyons, J. W.; Kim, K. Y.; Colwell, R. E. "Viscosity and Flow Measurement"; Interscience Pub. 1963, pp. 296–302.
66. Williams, J. B.; Morris, C. E. M.; Ennis, B. C. *Polym. Eng. Sci.* **1974**, *14*, pp. 413–419.
67. Loveless, H. S.; Ellis, J. H. *J. Test. Eval.* **1977**, 5, 369–374.

RECEIVED for review January 20, 1983. ACCEPTED June 29, 1983.

The Effect of Hygrothermal Fatigue on Physical–Mechanical Properties and Morphology of Graphite/Epoxy Laminates

JOVAN MIJOVIĆ and KING FU LIN
Department of Chemical Engineering, Polytechnic Institute of New York, Brooklyn, NY 11201

Changes in physical–mechanical properties of a unidirectional graphite/epoxy composite were evaluated during exposure to both static and cyclic hygrothermal fatigue. The moisture absorption process was adequately represented by Fick's law. Variations in glass transition and elastic modulus during absorption and desorption of moisture were accounted for by a diffusion concept on the morphological level. A permanent change in properties of the composite was recorded after one absorption–desorption cycle. The magnitude of the effect of thermal spikes on the composite properties was a function of the upper temperature of the spike and the number of spikes. Considerable microcracking was observed when the upper temperature of the spike was within the glass transition region. Microcracks appear to originate within the reinforcement–matrix boundary region.

GRAPHITE FIBER-REINFORCED EPOXY RESINS have undoubtedly become, within the past decade, the most widely used polymer–matrix structural composites in the aircraft and spacecraft industries. In actual service, these composites encounter various combinations of moist environments and temperatures. Hence, it is crucial to evaluate their performance, that is, changes in thermomechanical properties and dimensional stability under different hygrothermal conditions. This necessity has been recognized, and, consequently, studies on the influence of aggressive environments on properties of neat epoxy resins and their composites have been conducted.

The exposure of samples to an aggressive environment typically assumes the form of either static or cyclic hygrothermal fatigue. Static hygrothermal fatigue is defined by the conditions of preset constant temperature and relative humidity, whereas cyclic hygrothermal fatigue involves appli-

0065-2393/84/0206-0323$12.00/0

cation of various cyclic patterns as a function of time. However, because the exact duplication of actual service conditions requires impractically long times, accelerated testing methods are commonly used. They typically consist of shorter treatment times during which specimens are exposed to harsher environmental conditions than those encountered in service. Such hygrothermal fatigue treatments, either static and/or cyclic, are projected to be equivalent to a prolonged specimen exposure to milder conditions. During the hygrothermal fatigue, samples are periodically removed from the aggressive environment, and the ensuing changes in their physical/mechanical properties are evaluated as a function of time.

Most of the reported work, both experimental and theoretical, on the effect of hygrothermal fatigue on graphite/epoxy composites has been compiled in several recently published monographs (*1–4*). Nonetheless, a careful examination of published findings clearly reveals the lack of understanding of the effect of hygrothermal fatigue on composite properties on the morphological level. To overcome this lack one must develop a concept (model) of the morphology of highly cross-linked thermosets, and use experimental techniques capable of monitoring the hygrothermal fatigue-induced changes in physical–mechanical properties of composites in terms of the changes in composite morphology.

In general, a hygrothermal fatigue treatment involves diffusion of moisture in (absorption) and/or out (desorption) of the composite. Assuming that the moisture diffusion in the bulk graphite fibers is negligible, one is left with two regions in the composite material whose morphology must be elucidated: (1) the bulk resin matrix and (2) the reinforcement–matrix boundary region. Unlike graphite fibers, glass fiber surfaces are susceptible to degradation in the presence of moisture. The effect of moisture on the interactions in the reinforcement–matrix boundary region in glass-reinforced thermoset matrix composites has been studied (*5–13*), and two excellent reviews were published (*14, 15*).

The concept of a homogeneous thermosetting network has been used to describe the morphology of highly cross-linked neat thermosets. However, an apparent inadequacy of such a model in explaining various properties of thermosets and the ample experimental evidence collected within the last two decades have led to a formulation of the concept of inhomogeneous thermoset morphology. The model of coexisting regions of higher and lower cross-link density most adequately describes the morphology of highly cross-linked thermosets (*16*). The origin, size, and distribution of inhomogeneities depend on the type of the resin(s)/curing agent(s)/catalyst(s) formulation and the curing conditions. The likelihood of formation of inhomogeneous morphology is particularly pronounced in the commercially available high temperature epoxy formulations that are based on the tetraglycidyl molecules and contain various impurities (*17*). The most important consequence of an inhomogeneous morphology from the standpoint of hygrothermal fatigue would be the difference in the rate of moisture diffusion in the regions of different cross-link density.

An additional difficulty in the studies on composite materials arises from the fact that every treatment of diffusion in composites should consider the presence of an "interphase." The latter term, although clearly much more pragmatic than the two dimensional "interface," is nonetheless a misnomer, from the standpoint that the "interphase" is not a "phase" in the thermodynamic sense. Instead, this region includes and extends from the outer layer of the reinforcement to some distance into the thermoset matrix. The characteristics of this region depend on many factors: the morphology and surface energy of the fiber surface layer, the type of the fiber surface treatment and its interactions with the fiber and the matrix, the difference in the coefficients of thermal expansion of fibers and matrix, the state of residual stresses, and the morphological gradient in the matrix. The presence of fibers (loosely analogous to the presence of the mold surface during crystallization of semicrystalline polymers) during cure, is believed to play an important role in the formation of the morphological gradient in the resin and the subsequent response of the resin to various aggressive environments. The term *reinforcement–matrix boundary region* describes this unique and complex region more adequately than the term "interphase." The most prominent, but least understood, aspect of significance of this boundary region is its response to an aggressive environment and, hence, its role in determination of the long-term durability of composite materials. In the light of the complexity of the boundary region, one must be particularly careful when considering the theoretical predictions of mechanical behavior of composites. The main drawback of the great majority of such analytical models is the assumption of the presence of a two-dimensional interface. This interface is envisioned as an infinitely thin surface between the fibers and the matrix across which there exists a "perfect" adhesion between the phases.

Once the concept of the composite morphology is established, at least qualitatively, experimental work should be performed to detect and correlate the composite morphology to the hygrothermal fatigue-induced changes in the composite physical–mechanical properties. The majority of the reported investigations on the mechanical properties of hygrothermally fatigued composites have been conducted by comparing dry and wet static tensile and bending moduli, or dry and wet tensile, flexural, and creep strength. Although such measurements undoubtedly provide an indication of the overall effect of moisture, they reveal very little about the morphological aspects of absorption and desorption.

Therefore, our objectives were to investigate the effects of static and cyclic hygrothermal fatigue on the physical–mechanical properties of composites via optical and electron microscopy and to conduct thermomechanical measurements by using differential scanning calorimetry (DSC), thermogravimetric analysis (TGA), and dynamic mechanical analysis (DMA). DMA offers a distinct advantage over the other mechanical property tests. It provides the most sensitive response to various physical and chemical transitions and relaxations over a wide temperature range, to changes in the glass transition and dynamic moduli, and to morphological inhomogeneities.

Experimental

All specimens investigated were from the same batch of an eight-ply unidirectional graphite/epoxy (AS4/3502) composite. Curing of this material was typically done according to the following schedule: The minimum vacuum was set at 20 in. (508 mm), and the sample was placed in the autoclave. The temperature was raised to 275 °F (135 °C) at 3–5 °F/min (2–3 °C/min). The temperature was held for 15 min at 275 °F under vacuum pressure only, then the autoclave was pressurized to 85 psi (586 kPa). The temperature was held at 275 °F and 85 psi, and vacuum was maintained for 45 min. The temperature was then raised to 350 °F (177 °C) at 3–5 °F/min, and maintained for 2 h. The sample was then cooled to 150 °F (66 °C) in not less than 45 min, with pressure and vacuum maintained. Finally, the sample was removed from the autoclave and postcured at 400 °F (204 °C) for 4 h. This final step is recommended to develop optimum properties at 350 °F.

In addition to this curing schedule, two samples were given a second postcure at 240 °C for 30 and 60 min, but no further increase in glass transition was observed.

Samples for DMA were cut from the cured composite and had the following dimensions: 30 × 11 × 1.2 mm. Sample edges were lightly polished to remove occasional surface irregularities. All samples were maintained at room temperature for 60 days prior to the initiation of experiments. The specimens were then dried to a constant weight in a desiccator at 90 °C. For the absorption study, the dried specimens were placed into an aggressive environment at 90 °C and 100% relative humidity (RH). At chosen time intervals the samples were removed from the aggressive environment, and their properties were analyzed.

Desorption experiments were carried out by placing the moisture saturated samples in a desiccator at 90 °C. Again, the samples were removed from the desiccator and their properties evaluated at chosen time intervals.

For the cyclic hygrothermal fatigue (thermal spiking) study, the dried specimens were placed into an aggressive environment at 90 °C and 100% RH and maintained there for 2 weeks. Samples treated in this manner were assumed to be moisture saturated. Upon moisture saturation, the samples were exposed to a various number of thermal spikes. Each thermal spike cycle was composed of the following steps: The moisture-saturated samples were weighed and placed in preheated ovens at 180, 160, and 130 °C for the three sets of experiments for 10 min. The specimens were then placed in precooled tubes at − 80 °C (the tubes had been immersed in dry ice/acetone). After 10 min, the samples were removed and weighed. Then the samples were placed into the aggressive environment at 90 °C and 100% RH for 1 week to resaturate them with water. Finally, the samples were removed and weighed. An analytical balance (Mettler) was used.

Dynamic mechanical analysis (DMA) was performed in the Du Pont 981 model DMA connected to the 1090 thermal analyzer. Tests were run at a heating rate of 10 °C/min and at an oscillation amplitude of 0.2 mm peak-to-peak.

Thermogravimetric analysis (TGA) was performed in the Du Pont 951 model TGA connected to the 1090 thermal analyzer. Tests were run at a heating rate of 10 °C/min.

Differential scanning calorimetry (DSC) was performed in the Du Pont 910 model DSC connected to the 1090 thermal analyzer. Tests were run at a heating rate of 10 °C/min.

Samples for the microscopic investigations were first embedded in an amine-cured diglycidyl ether of bisphenol A (DGEBA) epoxy resin. The samples were then polished, etched with acetone for 90 min at 20 °C, and dried. Selective etching enhances the clarity of morphological features on the polymeric surfaces (*18*). A Bausch and Lomb bench metallographer Model 32 was used for light microscopy. Samples for scanning electron microscopy (SEM) were then gold shadowed and an AMR-1200 scanning electron microscope was used to investigate various composite surfaces.

Results and Discussion

Static Hygrothermal Fatigue. ABSORPTION. The dynamic mechanical spectrum of a dried (and not previously exposed to moisture) unidirectional (eight-ply) graphite/epoxy laminate is shown in Figure 1. The elastic

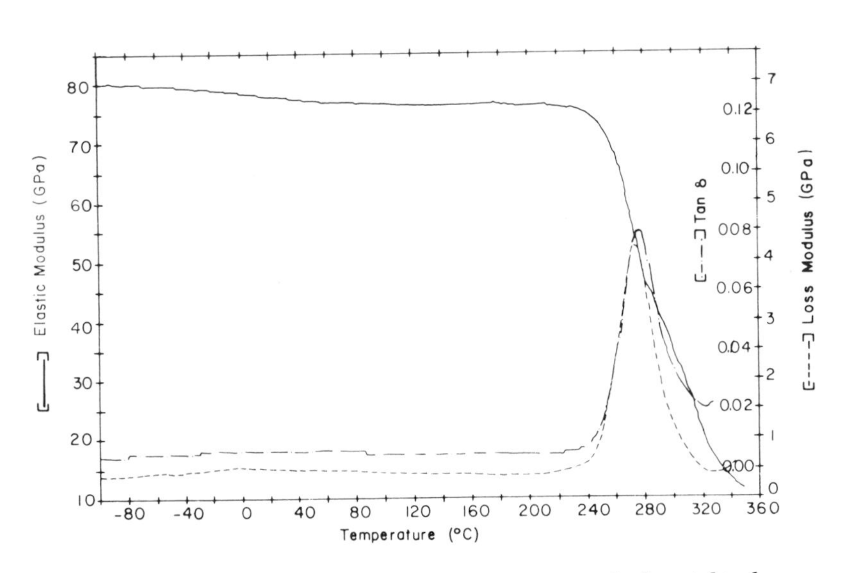

Figure 1. Dynamic mechanical spectrum of an untreated, dry eight-ply unidirectional AS/3502 graphite/epoxy composite.

modulus (E') of the dried composite decreases by only 2.6% in the temperature range from 20 to 200 °C. The first signs of an abrupt drop in elastic modulus, at approximately 230 °C, correspond to the onset of molecular motions in the resin, which eventually lead to the transformation from glassy into rubbery state. The glass transition (T_g) of this sample (and of all samples hereafter), defined by the location of the loss modulus (E'') peak in the dynamic mechanical spectrum, was 275 °C.

To immediately emphasize the effect of absorbed moisture, a dynamic mechanical spectrum of the moisture saturated unidirectional (eight-ply) graphite/epoxy laminate is shown in Figure 2. Over the entire temperature range, the E' of the moisture-saturated sample is noticeably lower than that of the dried specimen, as clearly seen by comparing Figures 1 and 2. The onset of molecular motions leading to the glass-to-rubber transformation, as determined by the beginning of an abrupt drop in E', was observed at approximately 140 °C. The T_g of the moisture-saturated sample was recorded at 210 °C. An additional peak was observed in the loss modulus curve at 280 °C, approximately 70 °C above the apparent T_g of the moisture-saturated sample.

Before beginning our discussion of results, we offer a brief qualitative description of the molecular mechanism of moisture penetration in cross-linked thermosets and their composites. A somewhat more detailed discourse is given elsewhere (*19*). To understand completely the process of absorption of water (or other small molecules) in cross-linked thermosets and their composites, one must correlate the ensuing changes in physical–

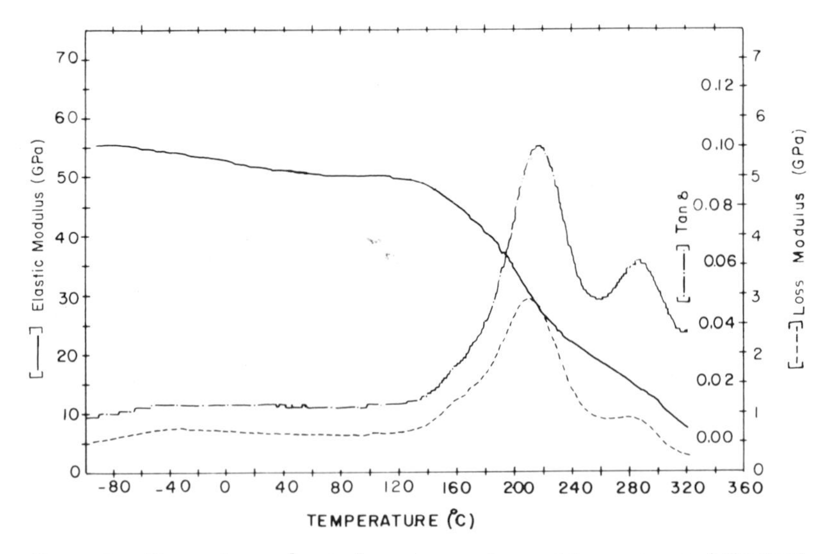

Figure 2. Dynamic mechanical spectrum of a moisture-saturated (90 °C/100% RH) eight-ply unidirectional AS/3502 graphite/epoxy composite.

mechanical properties to the composite morphology. In general, three possible sites of moisture absorption occur in thermoset–matrix composites:

First, regardless of the degree of sophistication of a particular processing technique, imperfections exist in every material and represent possible sites of moisture absorption. However, the identical thermal history (optimum curing conditions) of all samples is assumed to minimize the number of imperfections, as well as their variation from one sample to another.

Second, every study of diffusion in composites in terms of their morphology must consider the presence of the reinforcement–matrix boundary region. Although the thickness of that region is approximately several thousand Angstroms, its importance is paramount. Particularly important is its response to an aggressive environment and, hence, its role in the determination of the long-term durability of composite materials.

Third, the mechanism of moisture absorption in the thermosetting resin network should be clarified. If, in general, the phenomenon of diffusion in thermosets and thermoset–matrix composites is to be elucidated in terms of their morphology, the inhomogeneous character of thermosets should be considered.

Bearing in mind this qualitative concept of different sites of moisture penetration in cross-linked thermosets and their composites, we return to the discussion of our results. The increase in moisture content (weight gain) as a function of the square root of time of exposure to the aggressive environment (90 °C and 100% RH) is shown in Figure 3. The moisture content (M), a parameter of practical interest, was calculated from the following equation:

$$M = \frac{M_w - M_d}{M_d} \times 100\ (\%) \qquad (1)$$

where, M_w is the weight of moist sample, and M_d is the weight of dry sample. A thorough literature survey shows that, for the majority of graphite fiber-reinforced epoxy resins as well as for the neat resin, the mechanism of moisture diffusion is "Fickian"; that is, the mass transfer process is adequately described by the Fick's law (*3, 20–24, 28*). Nonetheless, the departure from Fickian behavior also has been reported (*24–26*). In one such case (*24*), good agreement was obtained between the experimental data and Fick's law for the neat 3501-5 resin and the unidirectional composite, whereas the bidirectional laminate displayed non-Fickian behavior. However, at long exposure times, a deviation from Fick's law has also been observed (*26*) in the neat resins and unidirectional composites. An apparent two-stage diffusion process has also been reported (*27*) in neat and reinforced DGEBA-type epoxies exposed to acetone. The majority of investigations, however, have reported Fickian behavior.

In the case of Fickian diffusion, the apparent composite diffusivity (D)

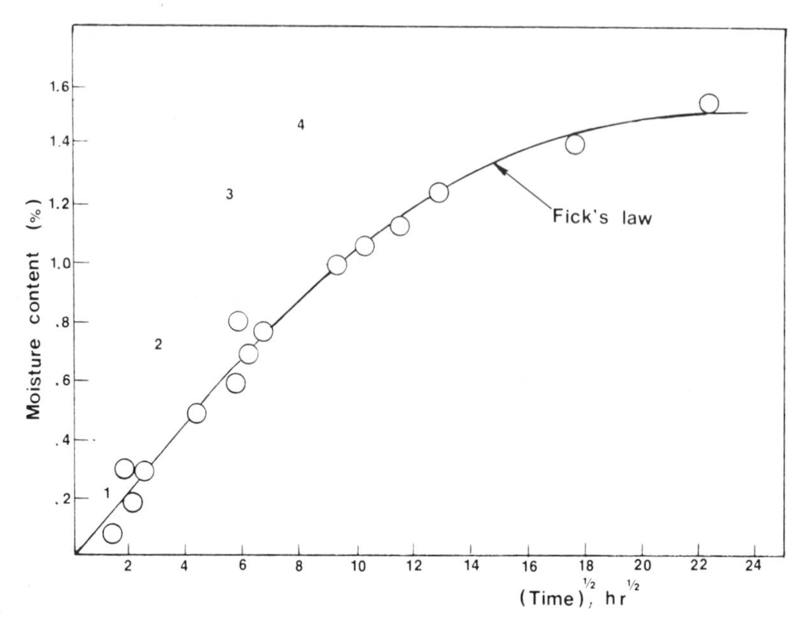

Figure 3. Moisture content as a function of square root of time of exposure to 90 °C/100% RH. Note four (1–4) stages of moisture absorption.

can be experimentally determined from the initial (linear) portion of the weight-gain vs. square-root-of-time curve. The D describes the overall diffusion into the bulk resin network, the reinforcement–matrix boundary region, and the imperfections. A value of $D = 3.78 \times 10^{-7}$ mm²/s was calculated from Figure 3. Furthermore, because unidirectional fiber composites are anisotropic materials, it is always important to calculate the values of diffusivities in the direction parallel (D_{11}) and normal (D_{22}) to the fiber in the composite. The solution to Fick's law using the corresponding set of boundary conditions for three-dimensional diffusion into a rectangular plate, has been given in the literature (*24*). The final form of that solution, used in our study, is given by Equation 2:

$$G = \frac{M_t - M_o}{M_\infty - M_o} = 1 - \frac{8}{\pi^2}\left[1 - 4\sqrt{\frac{R_1 t^*}{\pi}}\right]\left[1 - 4\sqrt{\frac{R_2 t^*}{\pi}}\right] \times \sum_{i=0}^{\infty} \frac{1}{(2i+1)^2} \exp[-(2i+1)^2 \pi t^*] \qquad (2)$$

where

$$R_1 = \frac{h^2 D_{11}}{l^2 D_{22}}, \quad R_2 = \frac{h^2}{n^2}, \quad t^* = \frac{D_{22} t}{h^2}$$

and l, n, and h are the specimen length, width, and thickness, respectively.

A good agreement between Fick's law (Equation 2) and the experimental data was obtained, as clearly seen in Figure 3. For the determination of D_{11} and D_{22}, however, the diffusivity of the resin (D_r) had to be determined first. The value of D_r was then used to calculate D_{11} and D_{22} (Equations 4 and 5) and to plot the theoretical curve in Figure 3. The diffusivity of the neat resin (D_n) and the diffusivity of that resin in the composite (D_r) are not the same even when the two systems are prepared under identical curing conditions. This discrepancy is caused by the difference in resin morphology (neat resin versus composite) that results from the existence of the reinforcement–matrix boundary region in the composite system. The apparent composite diffusivity follows the classical Arrhenius relationship from which the activation energy for diffusion can be determined (*20*, *22*). Interestingly, the activation energies for the neat 3501-5 resin and the corresponding AS 3501-5 composite differ significantly (*24*). Clearly, the mechanism of moisture diffusion differs on the morphological level. Considering the identical thermal history for the neat resin and the composite, the difference is explained by the existence of the reinforcement–matrix boundary region in the composite. Hence the correct value of the resin diffusivity (D_r), which accounts for the presence of the boundary region, must be obtained from the experimentally determined diffusivity of the composite (D). An equation for D from which D_r can be calculated for unidirectional composites is given as Equation 3:

$$D = D_r[(1 - vf)\cos^2\alpha + (1 - 2\sqrt{vf/\pi})\sin^2\alpha] \times \left[1 + \frac{h}{l}\sqrt{\frac{(1 - vf)\cos^2\beta + (1 - \sqrt{vf/\pi})\sin^2\beta}{(1 - vf)\cos^2\alpha + (1 - \sqrt{vf/\pi})\sin^2\alpha}} + \frac{h}{n}\sqrt{\frac{(1 - vf)\cos^2\gamma + (1 - \sqrt{vf/\pi})\sin^2\gamma}{(1 - vf)\cos^2\alpha + (1 - \sqrt{vf/\pi})\sin^2\alpha}}\right]^2 \quad (3)$$

where vf is the volume fraction of the fibers, and α, β, and γ are the angles between the fiber axis and the x, y and z axes (in our case $\alpha = \gamma = 90°$ and $\beta = 0°$). From the experimentally obtained value of D and Equation 3, the diffusivity of the resin in the composite was calculated to be $D_r = 2.86 \times 10^{-6}$ mm²/s. The diffusivity of a similar neat resin was studied by other researchers (*29*). From a plot of their values of D_n versus temperature for 100% RH we found, by interpolation, the diffusivity of the neat resin at 90 °C to be $D_n = 1.35 \times 10^{-6}$ mm²/s. This value, D_n, is significantly lower than our calculated value for the diffusivity of the resin in the composite (D_r). This difference is at least partially a result of the presence of the reinforcement–matrix boundary region in composite. The applicability of this morphological concept can be further checked against the findings reported. For instance, this concept can account for the lower value of the diffusivity of the neat resin and the discrepancy between calculated and experimentally obtained diffusivities of graphite/epoxy composites exposed to water (*30*). Using the value of D_r, one can calculate the values of D_{11} and D_{22} from the following equations:

$$D_{11} = (1 - vf)D_r \tag{4}$$

$$D_{22} = (1 - 2\sqrt{vf/\pi})D_r \tag{5}$$

where

$$D_x = D_{11}\cos^2\alpha + D_{22}\sin^2\alpha \tag{6}$$

$$D_y = D_{11}\cos^2\beta + D_{22}\sin^2\beta \tag{7}$$

and

$$D_z = D_{11}\cos^2\gamma + D_{22}\sin^2\gamma \tag{8}$$

The values calculated from Equations 4 and 5 ($D_{11} = 10.9 \times 10^{-7}$ mm^2/s and $D_{22} = 3.19 \times 10^{-7}$ mm^2/s) show that the diffusivity in the direction parallel to the fibers is higher than the diffusivity normal to the fibers. A similar ratio of D_{11}/D_{22} was also reported for a 1-mm thick AS 3501-5 composite after immersion in water at 75 °C (*31–33*). However, an accurate comparison of absolute diffusion rates in the directions parallel and perpendicular to the fibers must also specify the effect of tortuosity on the kinetics of moisture diffusion in the transverse direction. Only then can the preferential path of moisture absorption in the composite be defined.

Further examination of Figure 3 indicates that the entire weight-gain curve could be divided into four different stages (regions). A consideration of several stages in the weight-gain curve clarifies the molecular mechanism of moisture penetration into the composite. Simultaneously with the measurement of changes in moisture content, the changes in T_g were monitored. These changes in T_g are plotted in Figure 4 as a function of moisture content. The solid line represents the best fit to data, and the value of T_g at any given moisture content can be calculated from Equation 9, which is obtained by the least squares analysis:

$$T_g = 284.73 - 7.27(M) - 100.26(M)^2 + 48.22(M)^3 \tag{9}$$

In the first part of the curve, up to a moisture content of approximately 0.4% that corresponds to the first stage in the weight-gain curve in Figure 3, the T_g of the composite decreases only slightly. We believe that, during this stage, moisture penetrates into the reinforcement–matrix boundary region, the imperfections, and, to some extent, the less highly cross-linked regions in the resin. As a result of thermally induced longitudinal shrinkage during cooling, tensile stresses are formed in the reinforcement–matrix boundary region. The origin and nature of residual stresses in composites

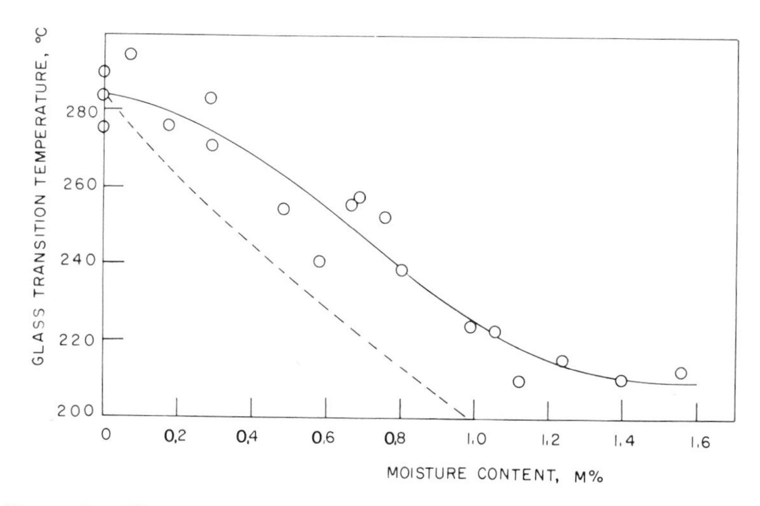

Figure 4. Glass transition temperature of the composite as a function of moisture content during absorption. Key: ---, Kelly–Bueche equation.

will be considered later. The tensile stresses are of interest because they contribute to an increase in free volume. This increase, in turn, enhances moisture absorption (*34*). A buildup of high osmotic pressure in voids and imperfections facilitates the absorption of moisture in these regions (*5*). Simultaneously, with initial absorption of moisture, the observed insignificant reduction in T_g (up to 0.4% moisture absorption) indicates the absence of considerable intramolecular plasticization of the resin network. Qualitatively, this observation is in excellent agreement with the reported results on the moisture-induced swelling strains in composites (*35*). The swelling strains are said to become detectable only above a certain threshold value of moisture concentration. More specifically, for an AS 3501-5 $[0]_{8T}$ graphite/epoxy composite, the threshold value of moisture gain above which the swelling strains start to increase linearly with moisture was approximately 0.4%. For several other graphite/epoxy composites (*23*), the threshold value of the moisture content below which no in-plane transverse swelling is observed was between 0.3 and 0.4%. In another instance (*30*), an abrupt initial moisture absorption was found to precede the classical Fickian diffusion mechanism. We therefore believe that the absence of detectable swelling strains in the early stages of absorption is indicative of moisture penetration into the reinforcement–matrix boundary region, the imperfections, and the parts of the less highly cross-linked resin characterized by relatively large free volume.

During the second stage of moisture absorption, the T_g decreases steadily. This decrease in T_g indicates that the moisture continues to penetrate into the resin network. The more highly cross-linked the parts of the resin, the slower the moisture penetration. Above the moisture content of approximately 1.0%, the T_g decreases very little and finally levels off. The dotted line in Figure 4 represents the glass transition of the composite as a function of moisture content, calculated from the Kelly–Bueche equation. In this equation we assume that all absorbed moisture acts as an intramolecular plasticizer (*36*). The following parameters were used in the Kelly–Bueche equation: T_g dry composite = 285 °C; T_g water = 4 °C; α_m = 3.78×10^{-4}/°C; and α_w = 4×10^{-3}/°C; where α_m and α_w are thermal expansion coefficients of the resin and water, respectively. The Kelly–Bueche equation has been derived for the saturation moisture content and thus should not be used for direct quantitative comparison with the data in this study. Nevertheless, the Kelly–Bueche line is qualitatively similar (parallel) to our experimental results in the region between 0.4 and 1.0% moisture content, where moisture is assumed to be absorbed within the resin network. Therefore, we maintain that the moisture absorbed in this stage (the second stage, Figure 3) directly determines the extent of intramolecular plasticization of the resin network and, hence, the T_g of the composite. A good agreement between the prediction of the Kelly–Bueche equation has been reported for the AS 3501-5 composite (*21*), although the small number of experimental points weakens the argument. A decrease in the T_g for carbon/epoxy composites was reported (*30*) after 11 years of exposure to water. An irreversible degradation of matrix was suggested but no explanation was given of the molecular mechanism by which such degradation causes a decrease in T_g. However, the observed good correlation between the experiment and the Kelly–Bueche equation for two neat resins (*21, 37, 38*) was questioned. Peyser and Bascom suggested (*39*) that the experimental techniques applied in the previous studies led to erroneous results. Of course, the results obtained with neat resins should not at all be used to directly characterize the behavior of composites, unless the presence of the reinforcement–matrix boundary region is accounted for. The presence of larger water clusters in our samples is unlikely because no physical transition due to water has been observed by thermal analysis.

Of primary concern in experimentally determining the effect of moisture on T_g of highly cross-linked thermosets and their composites is the loss (evaporation) of moisture at higher temperatures. Therefore, a question exists about whether the measured T_g is indeed representative of the glass transition of the wet specimen. To resolve this problem, two moisture-saturated samples were preheated at 180 °C for 20 and 40 min, and then tested in the DMA. Although the moisture content of the samples after the heat treatment dropped to 0.75% (20 min) and 0.55% (40 min), their glass transitions appeared at the same temperature. An explanation of this result is offered in terms of the diffusion mechanism on the morphological level. In fully saturated samples the moisture has penetrated the regions of high cross-link density in the resin network. The rate of moisture diffusion out of those regions is slow. The moisture that is rapidly lost during heating does not play the role of an intramolecular plasticizer and, therefore, should not affect the T_g. Because this heat treatment represents a more severe thermal environment than the one typically encountered during a dynamic mechanical test, we maintain that the moisture contained within the regions of higher cross-link density in the resin network will remain in the sample during the test and will exert a plasticizing effect on the composite. The initial rapid loss of moisture was also seen in an isothermal (180 °C) TGA thermogram of the weight loss of a moisture-saturated sample. The apparent abrupt moisture loss was followed by no detectable weight changes over a period of 1 h. Hence, once again, the rapidly lost moisture is not instrumental in the determination of T_g and will not invalidate the measurements of T_g of treated samples by DMA.

The elastic (storage) modulus at 20 °C (E'_{20}) is shown as a function of moisture content in Figure 5. A pronounced decrease in E'_{20} is observed during the first stage of moisture absorption (up to 0.4% moisture content). Simultaneously, an increase in weight gain (Figure 3) and insignificant changes in T_g (Figure 4) are observed. This moisture, initially absorbed in the reinforcement–matrix boundary region and less highly cross-linked regions in the resin, reduces the stiffness (hence the modulus). This moisture does not penetrate the bulk of the resin network and, hence, affects the T_g only slightly. Apparently, the observed reduction in stiffness at that moisture level is a consequence of some form of facilitated molecular motions between the fibers and the bulk resin (i.e., within the reinforce-

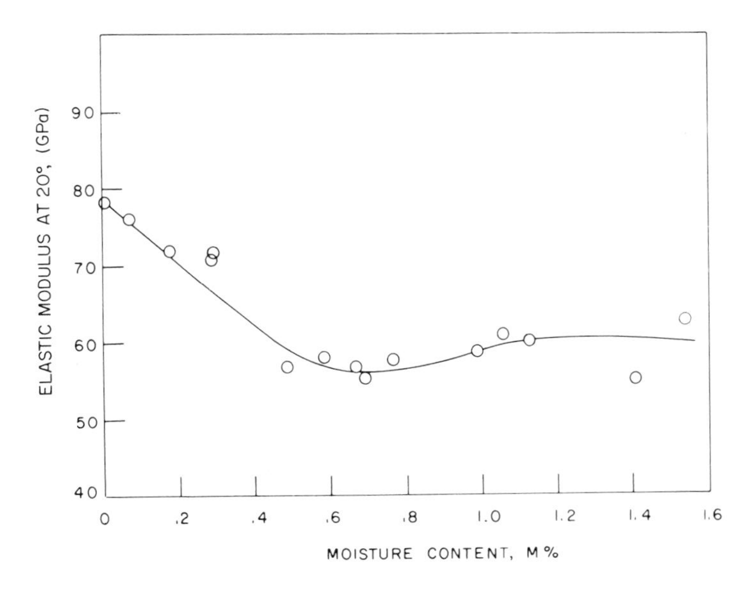

Figure 5. Dynamic elastic modulus at 20 °C as a function of moisture content during absorption.

ment–matrix boundary region). Most of the experimental data reported (*3*) show no dependence of elastic moduli on moisture content for 0° and, in some instances, for ±45° laminates. As expected, however, a strong dependence of elastic moduli on the moisture content was observed in various 90° laminates. The inability to detect moisture-induced changes in moduli is partly caused by the fact that the classical stress–strain measurements (e.g., in an Instron tensile tester) are not sensitive to subtle changes in polymeric materials on the morphological level. In that respect, DMA measurements are far superior. In another interesting report (*21*), the room temperature longitudinal elastic modulus (E_l) of a dry unidirectional AS/3501-5 composite was found to be higher than that of the sample with 1.6% moisture content but, surprisingly, lower than the modulus of the sample containing 1.05% moisture. It is difficult, however, to further explore this observation because only two data points were given and no mention was made of the time factor. In general, an unspecified time factor precludes the consideration of the effect of time-dependent rearrangements in composite morphology on changes in their physical–mechanical properties. We will consider this phenomenon later.

Figure 5 shows that beyond the moisture content of 0.4%, E'_{20} levels off and then slightly increases as the moisture content is further increased. A slight increase in E'_{20} could be possibly caused by the antiplasticization effect that the absorbed moisture exerts onto the room temperature E'. As moisture continues to penetrate into the resin network (above 0.4% moisture content), water molecules possibly flank the glycidyl ether groups thereby restricting their motion. This restriction, in turn, causes a decrease in the intensity of the low temperature relaxation (secondary transition – T_β) and, simultaneously, the observed increase in stiffness (E') at 20 °C. However, at higher temperatures where the chain mobility is enhanced, an increased amount of moisture will provide more intramolecular plasticization and lead to the observed decrease in T_g. Hence, in a certain range of moisture content, an increase in E'_{20} does not preclude a simultaneous decrease in T_g.

The third stage of moisture absorption (Figure 3) corresponds to the moisture content in the range between 1.0 and 1.4%. The glass transition of the composite decreases very slightly in this stage and then reaches an asymptotic value of approximately 210 °C. Similarly, the elastic modulus increases only slightly above the value of 60 GPa. On the other hand, the continuing increase in weight gain clearly indicates that moisture is still being absorbed, although apparently without a strong effect on either E'_{20} or T_g. We could give an explanation of this observation by considering the effect of static hygrothermal fatigue that the 90 °C/100% RH aggressive environment exerts on the composite. To elucidate that effect, however, we will first consider the existence of various nonmechanical stresses that could form in the composite materials in the absence of an externally applied mechanical stress field.

Residual stresses form in graphite/epoxy composites in the final stage of processing, i.e. during cooling from the postcure temperature to room temperature (*1*, *2*, *40*, *42*). The formation of residual stresses is a direct consequence of the difference in the coefficients of thermal expansion of the fibers (α_f) and the matrix (α_m). Even in a single ply (lamina), stresses are formed as a result of thermally induced longitudinal shrinkage (in the direction of fiber axis) during cooling: tensile stresses are created in the matrix; compressive stresses, in the fibers; and shear stress, within the reinforcement–matrix boundary region. Furthermore, in the laminates with the same fiber orientation, both radial tensile and compressive stresses are believed to exist perpendicular to the fiber direction (*41*). Additional residual stresses are encountered in laminates made up of plies with different fiber orientation, whereby deformation of one ply is constrained by the other plies (*2*, *42*). Finally, in unsymmetrical laminates, thermal stresses are also known to cause warping (*1*, *2*). The last two types of residual stresses, however, were not dealt with in this study, for only eight-ply unidirectional, $[0]_{8T}$ laminates were investigated.

Another type of nonmechanical stress, the swelling stress, is introduced into composite materials during their exposure to moisture. Theoretical analyses of the stresses in composites caused by hygrothermal fatigue have been reported (*2*, *42–44*). The swelling coefficient of the composite (β_c) is much larger in the transverse than in the longitudinal direction and consequently the latter is usually neglected (*2*). The magnitude and distribution of swelling stress are functions of the hygrothermal conditions of the environment and the duration of exposure. Because the swelling stresses are compressive in nature and the thermal stresses are primarily tensile, the two act in the opposite direction. Under certain conditions, a combination of these stresses would result in a zero net residual stress in the composite. Mathematically, this condition is expressed as Equation 10.

$$\epsilon_y^T + \epsilon_y^H = 0 \tag{10}$$

where ϵ_y^T and ϵ_y^H are the transverse strains associated with thermal and swelling stresses, respectively (*2*). A temperature–relative humidity correlation for the residual stress-free state in AS3501-5 composites has been shown graphically (*40*). According to that correlation, the samples used at the conditions of our study (similar graphite/epoxy formulation) would apparently be far removed from the state of zero residual stress. A simple schematic representation of the coupling between residual curing stresses and swelling stresses has also been shown (*23*). Although the residual stress-free state could be achieved in an aggressive environment under certain conditions, such a situation would not represent a permanent status quo, because the changes in the composite on the morphological and then macroscopic level (physical–mechanical properties) would continue to occur. The reason for this condition is twofold: First, during the static hygrothermal fatigue chemical reactions such as gradual hydrolytic degradation could occur. Second, time-dependent molecular relaxations occur in reinforced thermosets even at temperatures well below the glass transition of the resin (*45–47*). The long sample conditioning period in this study was designed to allow most of the time-dependent molecular rearrangements to occur prior to the exposure of samples to moisture. Hence, all samples were kept at 90 °C under dry conditions until no further changes in their dynamic mechanical spectra were observed.

Bearing in mind the presence of nonmechanical stresses, we believe that, as the resin network absorbs moisture and swells, the existing residual stresses are locally magnified to the point at which the microcracking occurs. Microcracking, of course, is preceded by an increase in free volume during the continuation of moisture absorption. The formation of microcracks during the static hygrothermal fatigue of neat epoxies and graphite/epoxy composites has been reported (*24*). For instance, surface cracks have been observed in the neat 3501-5 resin upon long-term exposure to 71 °C and 95% RH (*24*). The initial moisture absorption in virgin specimens of AS/3501-5 composite was reported to induce microcracking (*35*). Nonetheless, the extent of microcracking was assumed to be small, because only a small additional weight gain was observed and the microcracks were not visible under 300 × magnification. Ashbee et al. (*5*) suggested that the diffusion of moisture in voids can cause locally high osmotic pressure that, in turn, could lead to the development of internal stresses and cracks. In other instances (*48*, *49*), microcracking has been invoked to explain the observed increase in sorption rate (upon desorption and subsequent absorption) and the corresponding increase in the diffusion coefficient. Microstructural damage has also been postulated (*23*) for several graphite/epoxy composites subjected to absorption–desorption cycles although no supporting evidence was obtained from the microscopic investigation in that study. Severe swelling and delamination of plies are said to be responsible for the observed increase in the equilibrium moisture content after an absorption–desorption cycle. The predicted value of the equilibrium moisture content in the composite was calculated on the basis of the equilibrium moisture content of the neat resin, which does not consider the presence of the reinforcement–matrix boundary region. Also, the concentration gradients during moisture absorption could cause unequal swelling stresses within the composite, thus leading to the formation of microcracks (*20*). If microcracking indeed occurs in our study as a result of static hygrothermal fatigue, it would account for the observed weight gain above 1.0% moisture content, simply because of the moisture penetration into additionally formed voids in the composite. The glass transition and the E' at 20 °C, at the conditions of this study, are apparently not influenced by the formation of microcracks during moisture absorption. Our absorption results suggest that, at 1.0% moisture content, the resin network becomes saturated with moisture. The higher observed "equilibrium" moisture content (1.55%), achieved during the third and fourth stage of the weight-gain curve (Figure 3), takes into account the additional moisture absorbed within the subsequently formed microcracks. The term equilibrium is somewhat of a misnomer here; the absorption of moisture has been reported to continue as a function of time, albeit at a very slow rate. The phenomenon of microcracking is very important and will be further considered in the discussion of results of the cyclic hygrothermal fatigue. However, later we will advance a much more thorough description of the concept of initiation and propagation of microcracks on the morphological level.

DESORPTION. In addition to absorption, desorption measurements are needed to understand completely the effect of moisture on the physical–mechanical properties of composites. In actual service, the absorbed moisture is known to desorb at high temperature and in the space environment. Desorption studies reveal the rate and extent of reversibility and permanence of changes induced by the absorbed moisture. The absorption measurements in this study have indicated that the moisture absorbed during stages 1, 3, and 4 contributes to the total weight gain but influences the T_g very slightly. On the other hand, during the second stage of absorption, moisture gradually penetrates into the resin network, and acts as an intramolecular plasticizer and thus decreases the T_g. With this concept of moisture absorption in mind, the desorption studies were undertaken.

In Figure 6, the moisture content during desorption is shown to decrease exponentially as a function of the square root of desorption time. The initial rate of desorption was high. The following relationship was found between the desorption time (t) and the weight percentage of moisture in the composite (M_t) over the entire range of the experiment:

$$M_t = -0.21 \ln \frac{t}{176} \tag{11}$$

for $t < 176$ h.

Desorption studies by others have yielded interesting results. For example, in a $[0]_{8T}$ AS/3501-5 composite, analogously to our observation, the rate of desorption was much faster than the rate of absorption (*35*). The faster rate of desorption was concluded to be a direct consequence of the existence of tensile residual stresses in the first (in particular) and the second ply of the laminate. These stresses, in turn, facilitate and accelerate the moisture desorption. In other instances, however, the rates of absorption and desorption were roughly equivalent for several graphite/epoxy composites (*3, 23*). We shall further address the question of relative rates of absorption and desorption, together with the effect of absorption–desorption cycles on the properties of composites, later in the chapter.

A simultaneous change in T_g as a function of moisture content during

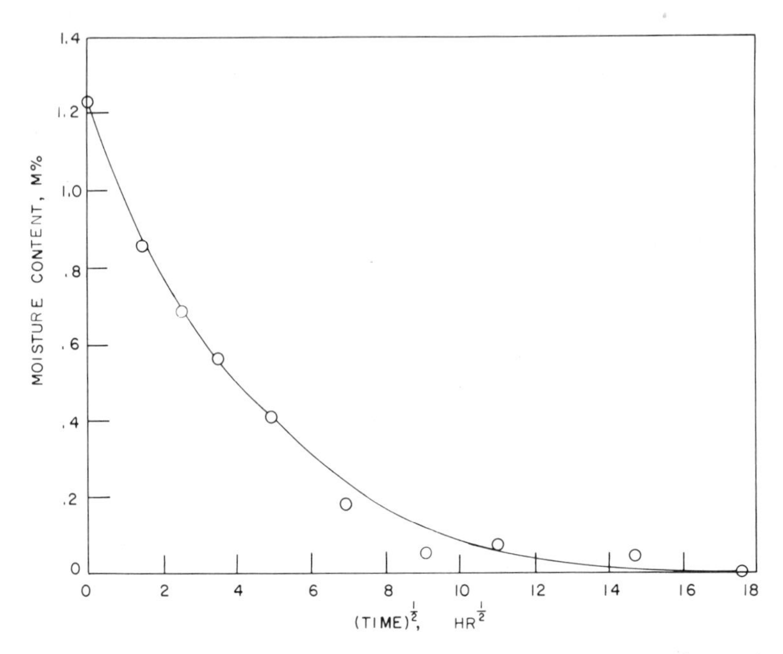

Figure 6. Moisture content as a function of square root of time during desorption in the dry environment at 90 °C.

desorption is shown in Figure 7. The solid line connects the experimentally obtained data points. Apparently, T_g increases only slightly as desorption proceeds until the moisture content is reduced to approximately 0.4%. Desorption beyond that point, however, significantly increases the glass transition. Figure 7 also shows that the experimental values of T_g agree very well with those calculated from the Kelly–Bueche equation (dotted line) for the moisture content below 0.4%, whereas considerable deviation occurs above that value. The portion of the curve below 0.4% moisture content describes a slow desorption from the regions of higher cross-link density in the resin network. As discussed previously, the amount of moisture absorbed within those regions directly determines the T_g. Consequently, the removal of this moisture has a pronounced effect on the composite glass transition. Moisture absorbed during stages 1, 3, and 4 is relatively quickly desorbed and accounts for the observed high initial rate of desorption.

Changes of E' at 20 °C (E'_{20}) as a function of moisture content during desorption are shown in Figure 8. The value of E'_{20} increases initially until the moisture content is reduced to approximately 0.55%. The loss of moisture in this range probably reduces the extent of moisture-induced lubrication within the reinforcement–matrix boundary region and hence contributes to an increase in stiffness. A slight decrease in E'_{20} is noted as the moisture content is reduced below 0.55%. Possibly, in this stage (when the moisture is essentially located within the resin network), the value of E' is

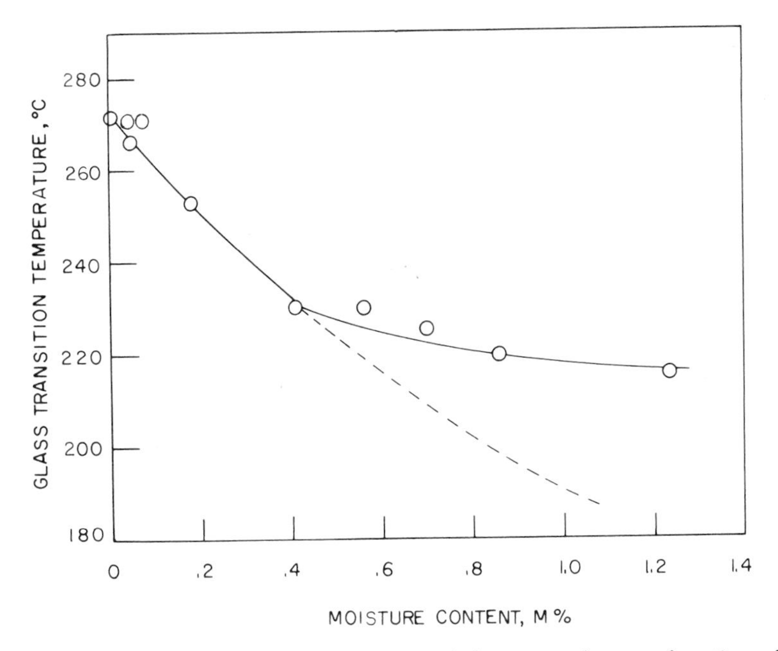

Figure 7. Glass transition temperature of the composite as a function of moisture content during desorption. Key: ---, Kelly–Bueche equation.

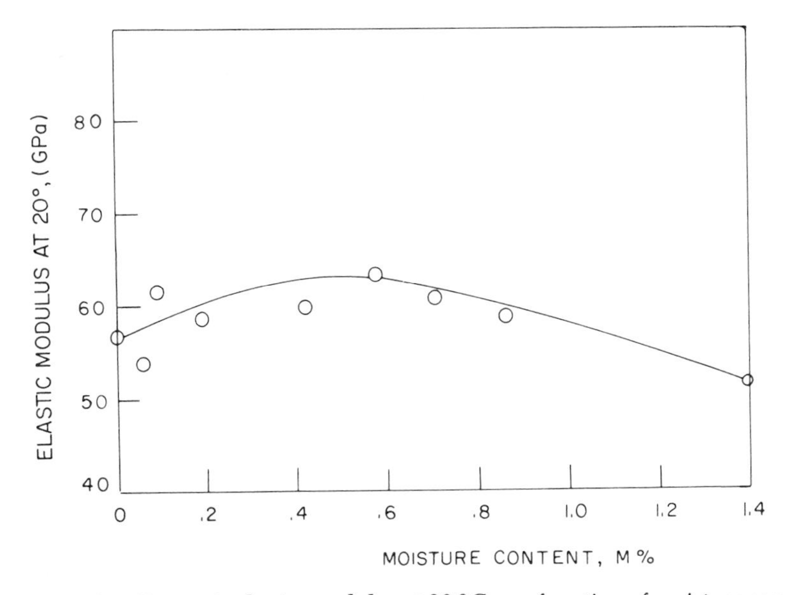

Figure 8. Dynamic elastic modulus at 20 °C as a function of moisture content during desorption.

affected by the extent of antiplasticization. The latter decreases as the moisture content is decreased, hence reducing the stiffness of the sample. The net effect on E' of coupling between curing and swelling stresses within the reinforcement–matrix boundary region is difficult to assess. The preceding explanation of the mechanism of moisture diffusion on the morphological level is, at least qualitatively, consistent with the one used to explain the absorption process.

Interestingly, after the completion of desorption, the glass transition and the elastic modulus at 20 °C of the dried sample were lower than those of the untreated dry sample (Figure 1). Hence the absorption–desorption process (one cycle) was not completely reversible because it produced a permanent change in the composite properties. More specifically, the T_g of the moisture-saturated and then dried sample was recorded at 271 °C. This value represents a 5% decrease in comparison with the untreated dry sample, or an 81.4% recovery when compared with the moisture saturated sample. The change in T_g as a function of moisture content during absorption and desorption is characterized by a pronounced hysteresis, as clearly seen in Figure 9.

The observed permanent changes in the composite occur mostly during the absorption of moisture. On the morphological level, they are the result of irreversible damage, which occurs primarily above 1.0% moisture content, at the conditions of the study. This damage causes an increase in moisture content but yields no appreciable change in E'_{20} and T_g. The extent of irreversible damage, of course, depends on the nature and number

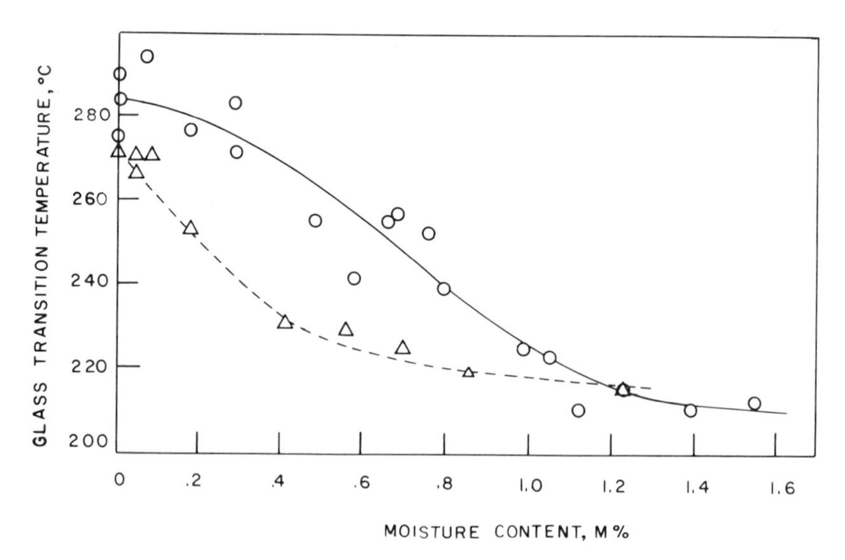

Figure 9. Changes in the composite glass transition as a function of moisture content during absorption (—) and desorption (---).

of fatigue cycles. This concept is further supported by other results with the AS/3501-5 composite (*31*, *32*). After each of the three absorption–desorption cycles, a large effect on the rate of moisture absorption (D_a) and no effect on the rate of moisture desorption (D_d) from the equilibrated state were found. Comparable results have been reported by others. For instance, when a $[0_4]_T$ AS/3501-5 composite was exposed to an absorption–desorption cycle at 71 °C and 75% RH, an increase in the diffusion coefficient during reabsorption led to the conclusion that some permanent internal damage had taken place (*24*). Hence, considering our findings and the reported results, we conclude that the mechanism responsible for the moisture uptake (during which permanent changes occur) differs from the one that governs the release of moisture from saturated samples.

As the next step in the evaluation of the effect of one absorption–desorption cycle on the composites' properties, TGA was performed to compare the total weight loss upon high temperature degradation of treated and untreated samples. We speculated that an irreversible damage accrued during the absorption–desorption cycle and could have affected the thermal degradation characteristics of the composite. The measurements of the weight loss as a function of temperature for three different samples are shown in TGA thermograms in Figure 10. Curve 1 represents

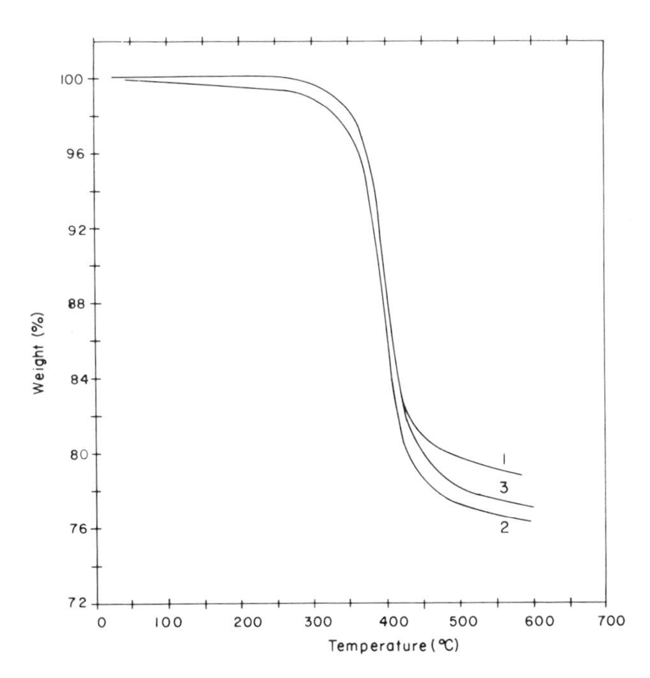

Figure 10. TGA thermogram of weight loss as a function of temperature for (1) untreated, dry composite, (2) saturated composite, and (3) saturated and dried composite.

the untreated dry sample; curve 2 is the moisture-saturated sample; and curve 3 is the moisture-saturated and then dried sample. All three samples have the same degradation temperature (T_d = 370 °C), which is defined as the temperature at which 5% of the initial weight has been lost. The value of T_d obtained from the TGA thermogram (Figure 10) agrees quite well with the results of DSC studies, shown in Figure 11. In the untreated dry sample (curve 1), the first signs of weight loss have been observed at approximately 250 °C. The moisture-saturated sample (curve 2) starts to lose moisture quickly and, not surprisingly, exhibits a greater total weight loss at 600 °C. The moisture-saturated and then dried sample (curve 3) is characterized by a greater weight loss than the untreated dry sample. Hence, one absorption–desorption cycle appears to have enhanced the total weight loss at high temperature (approximately 600 °C), as clearly seen in Figure 10.

An explanation for this observation is again offered on the morphological level. Some chain segments (particularly those characterized by the highest residual stresses) act as precursors for the formation of microcracks. The latter are formed during absorption, and this process is likely to be enhanced by hydrolysis in the presence of aggressive environment. The permanent changes in properties of graphite/epoxy composites may be caused by hydrolytic degradation (*30*). Also, commercially available tetraglycidyl

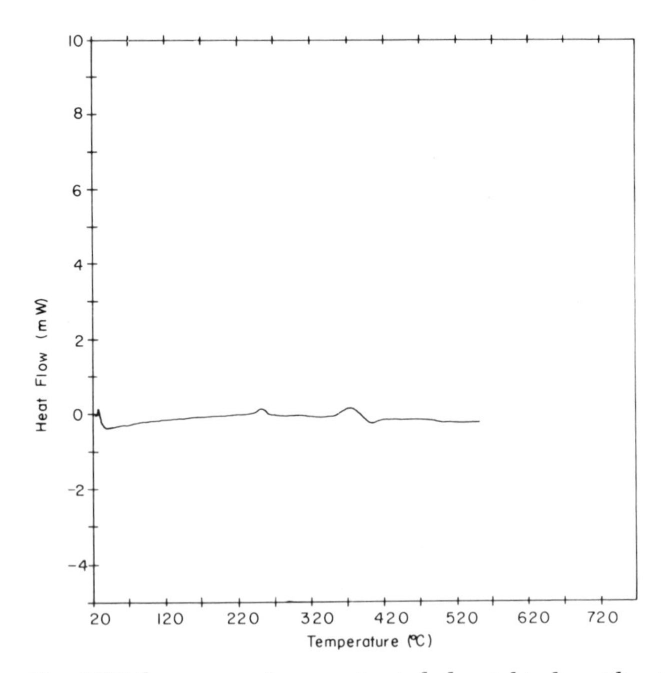

Figure 11. DSC thermogram for an untreated, dry eight-ply unidirectional AS/3502 graphite/epoxy composite.

methylene dianiline (TGMDA) resins have been shown to undergo considerable hydrolysis in a matter of days during exposure to 60 °C and 96% RH (*17*). Therefore, partial hydrolysis of chain segments characterized by the highest residual stress could facilitate subsequent chain scissions during degradation at higher temperature. Thus, in the moisture-saturated and then dried sample (Figure 10, curve 3), some hydrolysis took place, and consequently that sample showed a larger weight loss than the untreated dry sample. If, prior to hydrolysis, the hydrolyzed chain segments had been in a highly strained state, a decrease in T_g would be expected as a result of hydrolysis. This decrease in T_g is envisioned as a result of the increase in ease (due to decrease in strain) that begins the molecular (chain) motion leading to the glass-to-rubber transition. The plasticizing effect of degradation products is another possible cause for the drop in T_g. Hence, hydrolysis most likely contributes to the observed decrease in T_g of the moisture-saturated and then dried sample in comparison to the untreated, dry sample. The magnitude of the additional weight loss recorded at 600 °C in TGA thermograms, becomes time-independent above a certain limit of exposure to the aggressive environment. In our case the time limit was approximately 2 weeks, which corresponds to the "equilibrium" of moisture saturation. Therefore, a particular aggressive environment appears to be characterized by an upper limit of effect on the physical–mechanical properties of composites after one absorption–desorption cycle. A possibility of deviations at long exposure times should not be excluded, however.

Cyclic Hygrothermal Fatigue. THERMAL SPIKING. Prior to exposure to thermal spikes, all samples were saturated with moisture. Hence, we begin this section by offering a brief morphological description of a moisture-saturated sample.

Considerable interest has been generated in understanding the origin and nature of the reinforcement–matrix boundary region in composite materials. An important factor that influences the character of the boundary region in multiphase systems is the difference between thermal expansion coefficients of the two phases. The effect of induced thermal stresses on the thermal expansion coefficient of filled polymers has been investigated (*50*). The thermal expansion coefficient of a glass fiber-reinforced styrene–acrylonitrile copolymer was studied as a function of the volume fraction of glass (*51*). A deviation from the rule of mixture relationship was found. This result may have been caused by the reduced frequency of conformational changes of polymeric chains in the vicinity of the interface. In a study of polymer adhesion to various substrates (*52*), the emphasis was on the restrictions of the vibrational motion of the polymeric segments on and near the surface of the substrate. In graphite/epoxy composites, because of the difference in thermal expansion coefficients between the fiber and the matrix, residual thermal stresses will build up preferentially within the boundary region during cooling from the postcure temperature. If the adhesion or cohesion forces at any point within the reinforcement–matrix boundary region are exceeded by the local contractive force of the epoxy resin, microcracks will form. The latter could be initiated along the fiber–matrix interface, or at some other location within the boundary region. Actually, the weakest part of the boundary region probably lies within the resin in the close proximity of the surface to which it is bonded (*53*). Microscopic investigations have also revealed that a visible amount of matrix remains on the reinforcement upon fracture. According to our concept of the character of composite morphology, the restrictions to the mobility of chain segments within the resin network decrease with the distance from the reinforcement–matrix boundary region.

We have shown that the T_g of the composite decreases only when the absorbed moisture begins to penetrate the resin network and act as an intramolecular plasticizer. Simultaneously with the observed increase in the moisture content, the slope of the E' curve in the glass transition region becomes less steep. This change in E' slope indicates that the moisture absorption takes place to a different degree in different parts of the resin network. Because of the restricted molecular mobility within the reinforcement–matrix boundary region, the extent of intramolecular plasticization in moisture saturated samples is smaller in the boundary region than in the bulk matrix. Consequently, in the moisture-saturated epoxy composites, the glass transition of the plasticized epoxy network outside the boundary region is lower than that of the epoxy network within the boundary region. This phenomenon has been detected by DMA (Figure 2). The peak at 280 °C represents the T_g of the epoxy network in the reinforcement–matrix boundary region. Apparently, by preferential penetration, the absorbed moisture has enlarged the difference in the local mobility of chain segments in the different parts of the resin network, to a degree detectable by DMA.

The E'' peak observed at 280 °C in the dynamic mechanical spectrum is not caused by further curing or degradation. Some additional curing of the resin takes place prior to the T_g, as evidenced by the DSC exotherm and the observed increase in elastic modulus in DMA spectrum. However, the curing reactions occur at a temperature considerably lower than 280 °C. Any significant composite degradation, on the other hand, does not take place until a much higher temperature is reached as recorded by both DSC (Figure 11) and TGA (Figure 10) studies.

The thermal spiking study was performed next. The application of thermal spikes was intended to simulate the repeated exposure to temperature extremes encountered during the service of commercial and supersonic aircraft or the space shuttle. Three different thermal spikes were applied. Each spike differed from the other two by the upper temperature limit. The upper temperature limit was carefully chosen as a function of its relation to the molecular mobility of the polymeric network. The highest temperature (180 °C) was well within the T_g region, 160 °C was approximately at the onset of the T_g region, and 130 °C was below the onset of the T_g region. Upon exposure of moisture saturated samples to thermal spikes, numerous interesting observations were made. Figure 12 shows the change in moisture content as a function of the number of thermal spikes for 180°, 160°, and 130 °C cycles. The moisture content increases initially for all cycles but the subsequent behavior is different. For instance, in the 180 °C cycle the moisture content reaches an apparent maximum after six cycles and then drops slightly before increasing again. Qualitatively, a similar trend was observed in the 160 °C cycle with an apparent maximum after eight spikes. In the 130 °C cycle, however, the increase in weight gain is slow and after 10 spikes it reaches approximately 1.5%. Furthermore, when the previously spiked sample in the 180 °C cycle was resaturated with moisture, its water content increased by 18.5% in comparison to the nonspiked water-saturated sample. Meanwhile, the same treatment in the 130 °C cycle produced an increase of only 6%.

Changes in T_g as a function of the number of thermal spikes are plotted in Figure 13. A 20 °C drop in glass transition was recorded after three thermal spikes in 180 and 160 °C cycles. Because the glass transition is determined by the extent of intramolecular plasticization, the drop in T_g indicates that, as a consequence of thermal spiking, additional sites within the resin network have become accessible to moisture. The moisture could then

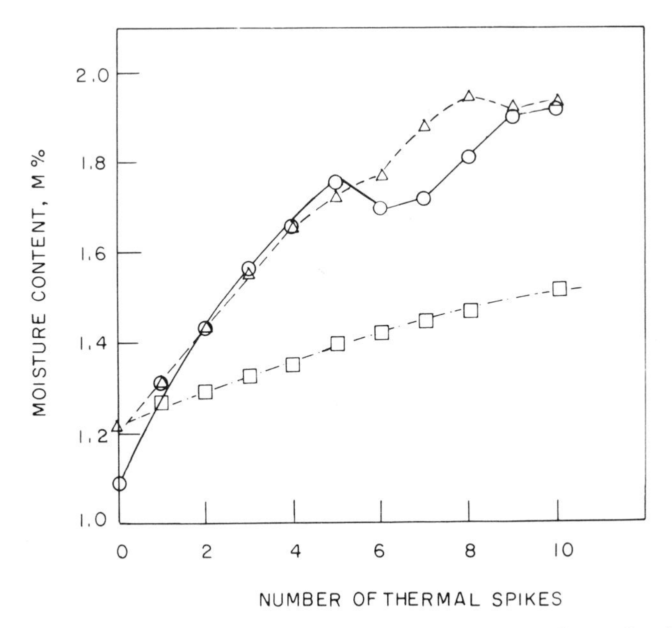

Figure 12. Moisture content as a function of number of thermal spikes at 180 °C (—), 160 °C (---), and 130 °C (·-·).

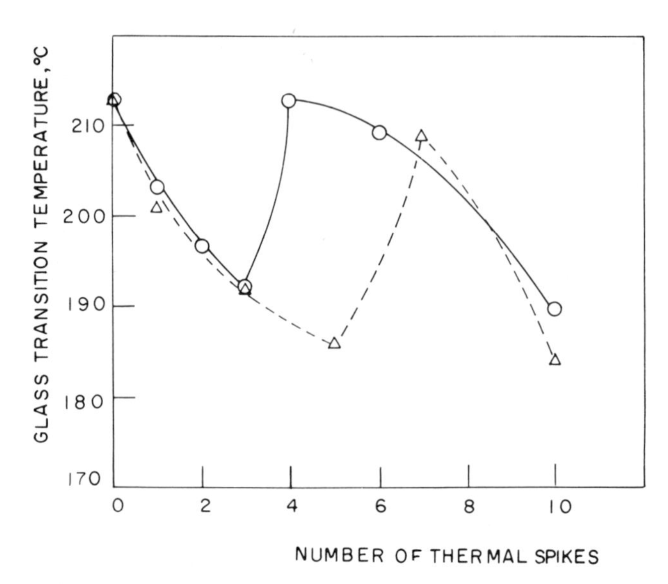

Figure 13. Glass transition temperature of the composite as a function of number of thermal spikes at 180 °C (—) and 160 °C (---).

act as an intramolecular plasticizer. Interestingly, with further spiking, the T_g increases and then drops again. We will attempt to explain the observed changes in moisture content in conjunction with changes in the T_g and E', with microscopic evidence.

It was difficult to compare our results with the findings of other researchers. Only a very limited amount of information on the effect of thermal spikes on physical–mechanical properties of graphite/epoxy composites has been reported, and the nature of thermal spikes varies from one study to another. A tabulation of studies on thermal spiking of composites (*3*) clearly reveals that no fundamental investigation has been undertaken. Consequently, often contradictory conclusions have been made. An increase in the rate of moisture absorption and the equilibrium moisture content as a function of the number of thermal spikes for a series of T300/5208 composites has been reported (*54*). Furthermore, although the additionally absorbed moisture was removed during desorption, permanent changes occurred in the composite. Nonetheless, an explanation of such changes on the molecular level has not been advanced. Several other studies (*55*) have also reported an increase in the rate of moisture absorption as a function of number of thermal spikes. On the other hand, the moisture absorption characteristics and tensile properties of T300/1034 composites were reported to have changed insignificantly as a result of thermal spiking (*3*). Although the composite diffusivity did not change during spiking, it increased during the next remoisturization of the dried sample. Interestingly, no further changes in the composite diffusivity were observed upon subsequent remoisturizations. Different results were reported (*31*) when a neat 3501-5 resin and a series of AS/3501-5 composite samples were exposed to an increasing number of absorption–desorption cycles. Diffusion coefficients of absorption, measured in various directions, increased upon each subsequent absorption–desorption cycle. On the other hand, diffusion coefficients for desorption had absolute values higher than those of absorption coefficients, but remained almost unchanged over the three absorption–desorption cycles. In partially saturated composites, the highest moisture concentration may exist in the form of a ridge or layer at some distance beneath the surface of the sample. As a consequence of such moisture distribution, different states of stress are encountered on either side of the high moisture content ridge. Furthermore, the nonuniformity of stress could cause microcracking during the application of thermal spikes. In another study (22), boron–graphite/epoxy hybrid composites were immersed in water at 23 and 77 °C and periodically exposed to thermal spikes. Only the samples at 23 °C showed an increase in moisture content over that of the control sample.

In general, when moisture-saturated samples are exposed to thermal spikes, two phenomena could occur: an increase in free volume and a formation of microcracks. Both phenomena contribute to the experimentally observed additional moisture absorption. In that respect, however, the effect of the change in free volume is believed to be less pronounced than the formation of microcracks, primarily for the following reasons. First, the highest upper temperature limit of a thermal spike is 180 °C, which is still below the T_g (measured by the E'' peak) of all specimens. Hence, during the application of thermal spikes, although thermal equilibrium is quickly reached, the resin is never completely transferred into the rubbery state in which the thermal expansion coefficient is higher and the increase in free volume would be more pronounced. Second, upon quenching, the thermal spike-caused increase in free volume will be gradually offset by the ensuing time-dependent decrease in free volume. This phenomenon, as already mentioned, has been observed in many epoxy resins. Therefore, the thermal spikes primarily cause a relief in the restrictions to local molecular motions in the resin network. These motions are particularly pronounced within the reinforcement–matrix boundary region. As a result, microcracking is induced within the boundary region. Because of the mismatch in the thermal expansion coefficients between the epoxy matrix and graphite fibers, the contractive force of the epoxy matrix during quenching may exceed either the local cohesive force within the boundary region (some distance from the fiber surface) or the adhesion force along the fiber–matrix interface. Either case leads to the formation of microcracks. The latter case facilitates additional absorption of moisture into previously largely inaccessible regions of the network and, as a consequence, the T_g drops. Several observations can be easily made from the dynamic mechanical spectrum of the sample exposed to one thermal spike at 180 °. The slope of the E' curve in the T_g region is steeper than in the nonspiked sample. This difference in slope indicates that the onset of molecular motions occurs over a narrower range. Moreover, the glass transition is lower, the T_g peak is sharper, and the 280 °C transition (the T_g of the epoxy network in the reinforcement–matrix boundary region) is reduced and less pronounced than in the nonspiked sample. All these observations point toward a decrease in the molecular mobility restrictions and further intramolecular plasticization of the resin network.

In the next step in the course of our investigation, both light and scanning electron microscopy were employed to elucidate the origin and character of microcracks induced by thermal spiking. Photomicrographs of the cross section of an untreated composite are shown in Figure 14. Although some interpenetration of fibers between the adjacent plies and the presence of resin rich regions were clearly seen, no microcracks were detected. Upon exposure of samples to various thermal spikes, however, the following observations have been made. The mildest thermal spike (130 °C), had apparently induced very few microcracks in the samples after five cycles, as shown in Figures 15 and 16. In these figures, the maximum crack width was estimated at approximately 2.5 μm. Also, the cracks initiate somewhere in the region between two plies and then continue to propagate across one of the plies. The observation that the crack growth does not continue through the resin-rich region indicates that the latter is less brittle and hence characterized by higher fracture energy. On the other hand, in the fiber-rich areas, a significant proportion of the resin is contained within the reinforcement–matrix boundary regions. These regions are characterized by higher restrictions to molecular mobility and higher stiffness, but lower fracture energy (i.e., toughness). The temperature of 130 °C is well below the onset of the molecular motions in the glass transition region. This condition explains the observation that only a small number of cracks were produced and that no apparent effect of the number of thermal spikes on T_g was seen.

However, when the upper temperature of thermal spike was within the T_g region, much more pronounced microcracking was observed. As seen in Figure 17, a sample exposed to 10 cycles at 180 °C contains cracks about 10 times wider than those found upon exposure to 10 cycles at 130 °C. Furthermore, after 10 cycles at 180 °C, the cracks were seen to traverse the resin rich regions and to enter the adjacent ply (Figure 18). Again, the highest temperature of the cycle (180 °C), although within the glass transition region, is still below the wet T_g as defined by the E'' peak in dynamic mechanical spectra. In that respect, our findings are different from the results reported with a neat resin (*56*). In that study, microcracks were observed only at the thermal spike temperatures above the wet T_g of the resin. There is, however, no contradiction in the noted difference. Ac-

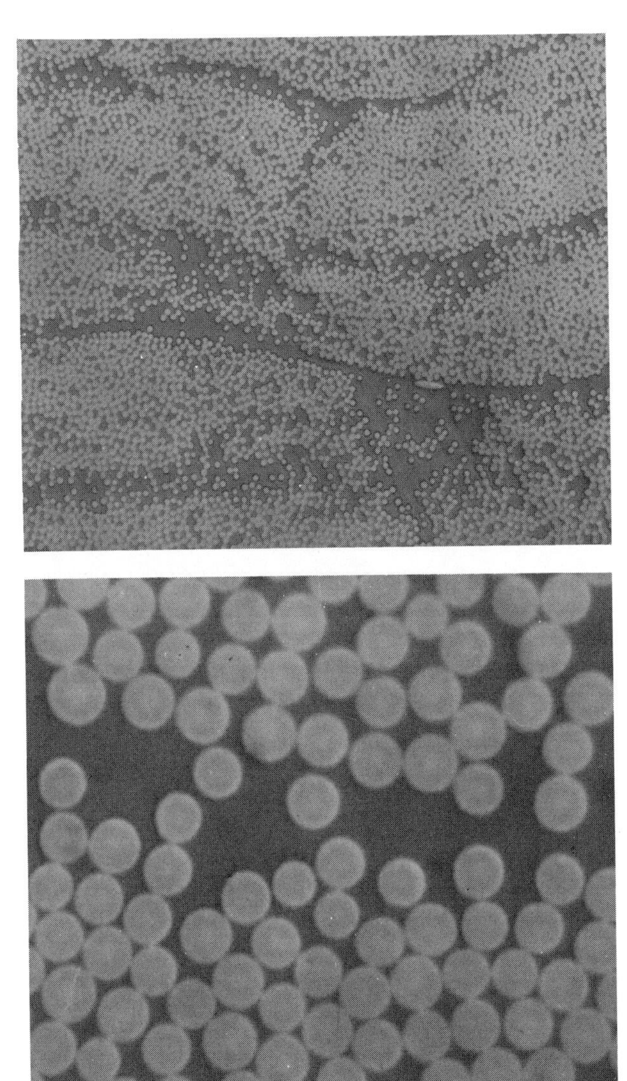

Figure 14. Photomicrograph of a cross-sectional area of an untreated, dry specimen: (top) 100× magnification; (bottom) 800× magnification.

cording to our morphological concept, the cracking in composites (vs. the neat resin) is expected to occur at lower thermal spike temperatures because of the presence of additional restrictions to molecular mobility within the reinforcement–matrix boundary regions. These regions, of course, are absent in the neat resin.

The size of a crack produced during thermal spiking is a function of the highest temperature of the spike and the number of spikes. Figure 19 shows a small microcrack formed within the reinforcement–matrix bound-

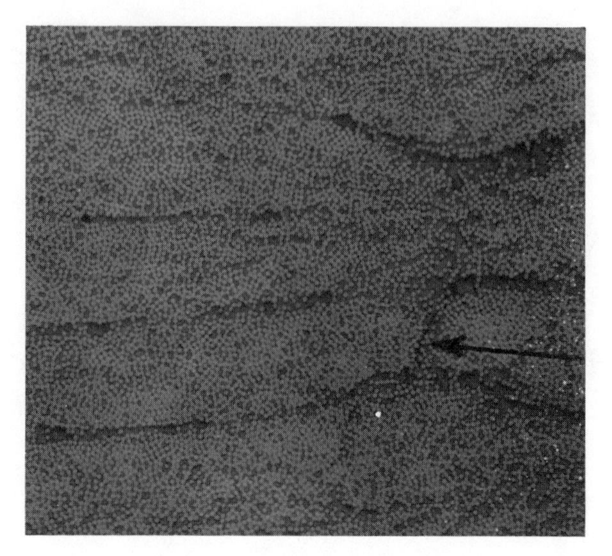

Figure 15. Photomicrograph of a cross-sectional area of a specimen after five cycles at 130 °C. Magnification 80×. Note a small number of microcracks.

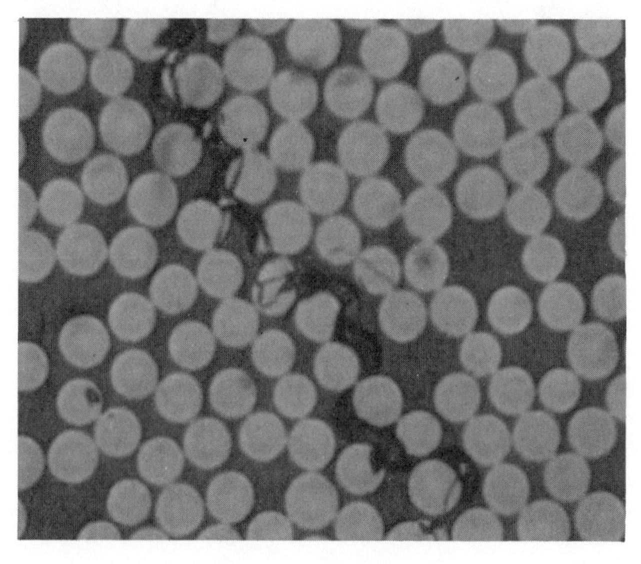

Figure 16. Photomicrograph of a cross-sectional area of a specimen after five cycles at 130 °C. Magnification 800×.

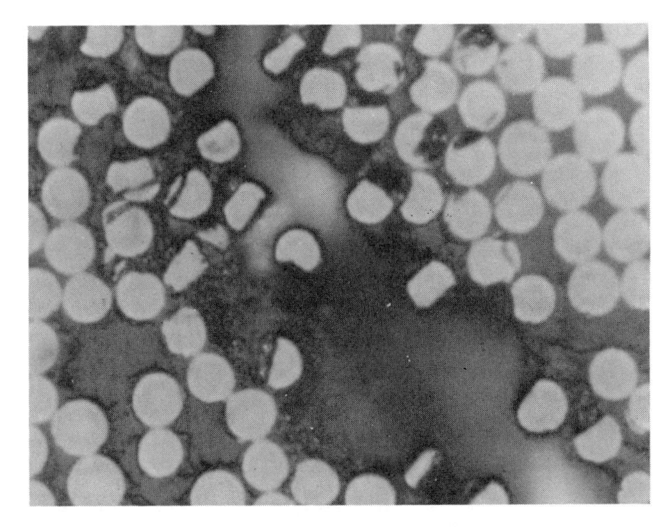

Figure 17. Photomicrograph of a cross-sectional area of a specimen after 10 cycles at 180 °C. Magnification 800×. Note the difference with Figure 16.

ary region as a result of thermal spiking at 180 °C. The width of that microcrack is approximately 50 nm. The vast majority of microcracks initiate in the reinforcement–matrix boundary region and propagate through the resin toward an adjacent fiber, as shown in Figures 20 and 21.

As the number of thermal spikes at 180 °C is increased, the T_g con-

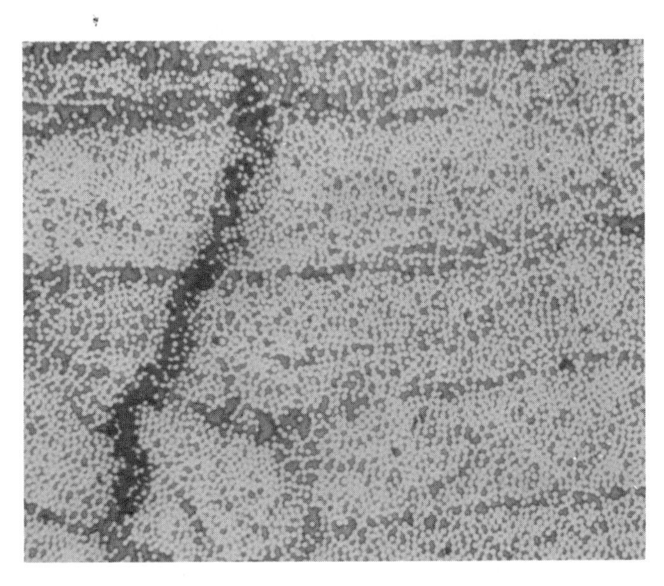

Figure 18. Photomicrograph of a cross-sectional area of a specimen after 10 cycles at 180 °C. Magnification 100×. Note how cracks traverse several plies.

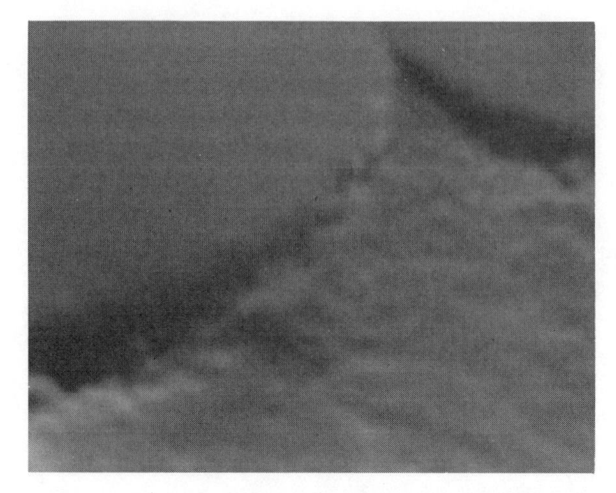

Figure 19. Scanning electron micrograph of a specimen after seven cycles at 180 °C. Magnification 18,000 ×. Note a small microcrack in the reinforcement–matrix boundary region.

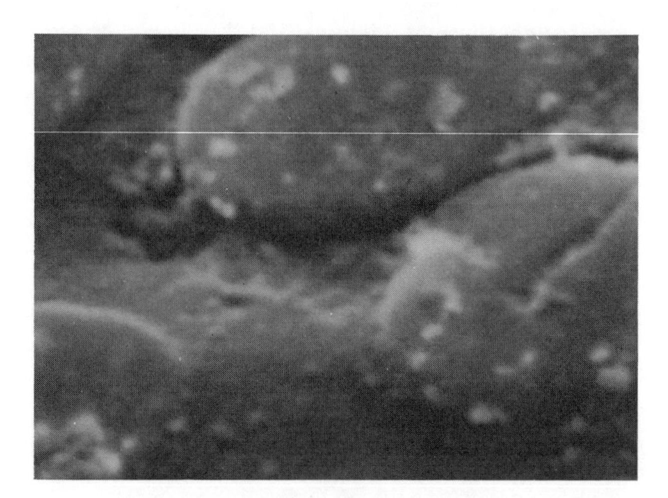

Figure 20. Scanning electron micrograph of a specimen after 10 cycles at 180 °C. Magnification 7500 ×. Note the crack propagation path.

tinues to drop and eventually (after three or four cycles) approaches the highest temperature of the thermal spike, as clearly seen in Figure 13. Actually, at this point the matrix is well above the temperature of the onset of glass-to-rubber transition and the likelihood of composite failure in service becomes high. Beyond what we refer to as "the first critical number of spikes" (three or four cycles) the T_g starts to increase again as also seen in Figure 13. During the first three or four thermal spikes, various microcracks were generated in the sample. After three or four cycles, the temper-

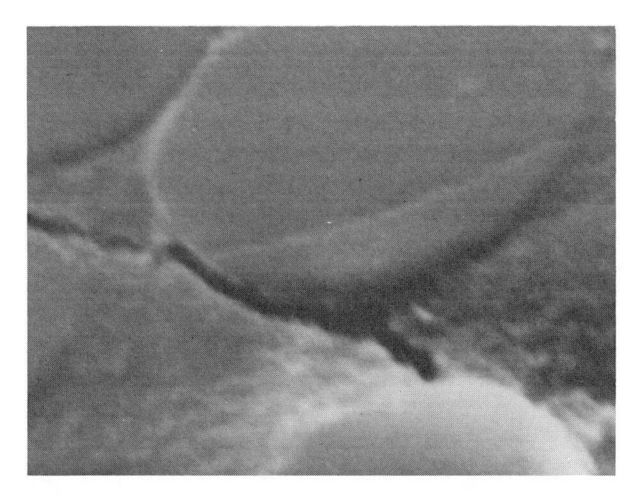

Figure 21. Scanning electron micrograph of a specimen after seven cycles at 180 °C. Magnification 11,200×. Note crack propagation path.

ature of the onset of molecular motions in the epoxy network is well below the upper temperature limit of the spike (180 °C) and the resin, at that temperature, can expand significantly. The latter phenomenon is a result of the well-known abrupt increase in the coefficient of thermal expansion in the vicinity of the glass transition. Hence, as it expands, the resin can close some smaller microcracks within the reinforcement–matrix boundary region. This closure does not proceed by the crack healing mechanism that has been observed in various thermoplastics (57). The resin first expands during the high temperature step of the thermal spike and closes some small microcracks. It then rapidly contracts during quenching and causes the formation of curing stresses, predominantly within the boundary region. These nonmechanical stresses, in turn, contribute to an increase in the restrictions to molecular mobility in the resin and, hence, to the observed increase in T_g (Figure 13). Several small microcracks that appear to have been closed during thermal spiking, are shown in Figures 22 and 23. Simultaneously, some moisture is driven out of the sample, indicated by the sharp increase in the amount of moisture released during the duration of thermal spike (Figure 24) and by the small change in the total moisture content (Figure 12). Larger cracks, however, are not closed. Instead, they continue to propagate; the total moisture content of all spiked samples upon saturation is always higher than that of the nonspiked sample.

Eventually, the T_g increases to almost the same level as in the nonspiked specimens, that is, well above the highest temperature of thermal spike. The corresponding moisture content, however, remains above the original level of the nonspiked sample, because some large microcracks form and remain in the composite as a result of repeated thermal spiking and are filled with moisture during saturation. Simultaneously, some of the microcracks grow into large cracks during the continuation of thermal spiking, as clearly seen in Figure 25. Once the T_g is again above the highest temperature of the thermal spike, microcracking resumes. The moisture content increases and the T_g starts to drop in accordance with the molecular mechanism analogous to the one that governs the behavior of composites during the first three or four thermal spikes. Also, as a result of the static hygrothermal fatigue (samples are maintained at 90 °C and 100% RH between the spikes), moisture continues to penetrate the newly formed

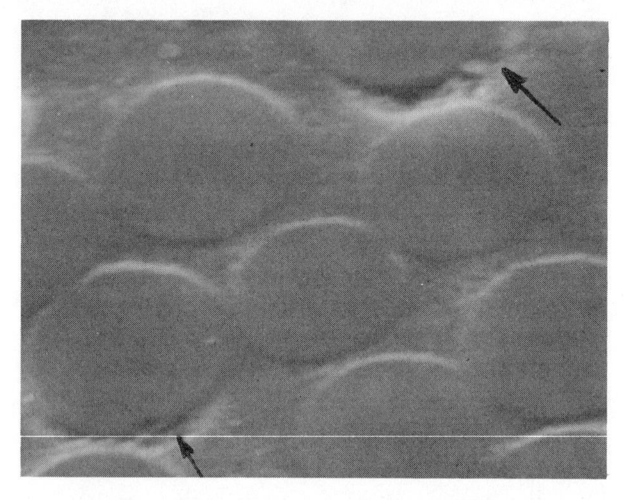

Figure 22. Scanning electron micrograph of a specimen after seven cycles at 180 °C. Magnification 5000×. Arrows indicate microcrack closure.

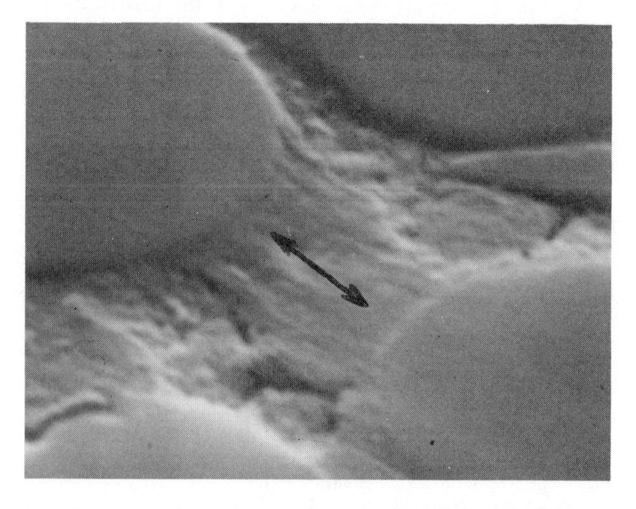

Figure 23. Scanning electron micrograph of a specimen after seven cycles at 180 °C. Magnification 11,300×. Arrows indicate microcrack closure.

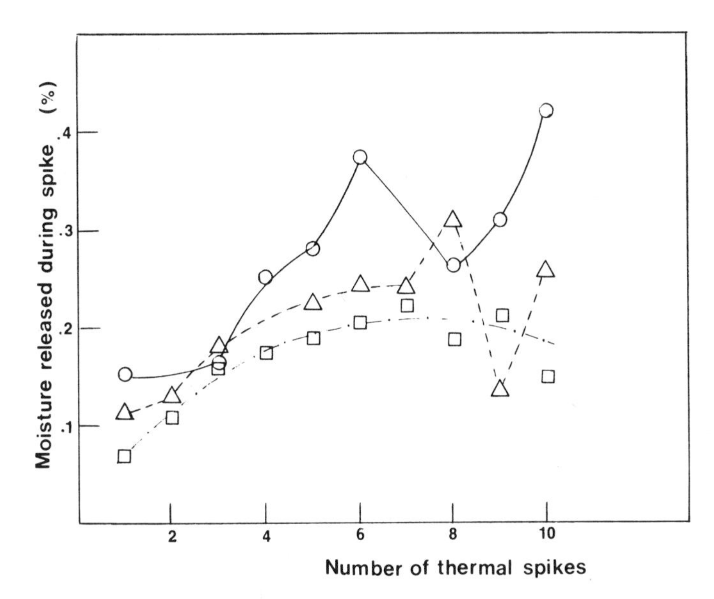

Figure 24. Amount of moisture released during each spike as a function of number of thermal spikes at 180 °C (—), 160 °C (---), and 130 °C (·--·).

and enlarged cracks, and the total weight gain increases (Figure 12). Some very large cracks have been observed after 10 cycles at 180 °C as shown in Figure 26.

With the upper temperature limit of the thermal spike at 160 °C, the lowest glass transition was observed after six thermal spikes. Nonetheless,

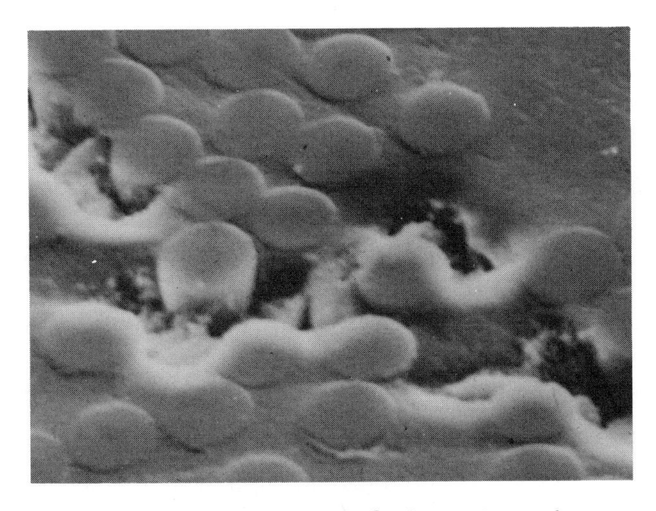

Figure 25. Scanning electron micrograph of a specimen after seven cycles at 180 °C. Magnification 1400×.

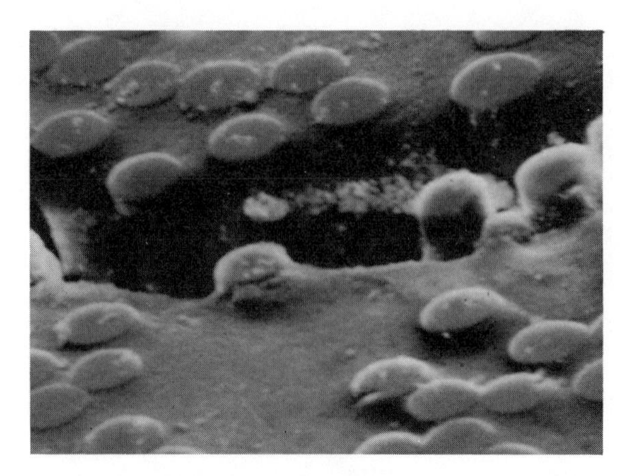

Figure 26. Scanning electron micrograph of a specimen after 10 cycles at 180 °C. Magnification 1420×.

both the T_g and the moisture content follow, at least qualitatively, the same patterns as those observed in the 180 °C spike. Experimental results obtained in the 160 °C cycle are represented by the dashed lines in Figures 12, 13, and 24. The fact that six spikes are needed before the T_g starts to rise again suggests that the effect of 160 °C cycle is milder and that it takes more thermal spikes to bring about the closure of some microcracks. Finally, the 130 °C thermal spike was chosen because that temperature is well below the onset of the glass transition region for all specimens. Not surprisingly, the experimental results show a considerably different pattern from those of 160 and 180 °C cycles as clearly seen in Figures 12 and 24. Moreover, no appreciable difference in the equilibrium value of moisture content was observed between the samples spiked at 130 °C and those exposed to static hygrothermal fatigue only. The amount of moisture released during the 130 °C thermal spiking is small, as seen in Figure 24. All these observations suggest that thermal spiking at temperatures below the onset of the T_g range does not produce severe microcracking. This statement has been corroborated by microscopic evidence.

Another interesting observation was made in the dynamic mechanical spectra of samples thermally spiked at 180 °C. As the specimens passed through the glass transition region of the resin network in the reinforcement–matrix boundary region, an abrupt drop in E' was observed (Figures 27–29). The magnitude of that drop appears to increase as the number of thermal spikes is increased from 4 (Figure 27) to 6 (Figure 28) and then to 10 (Figure 29). In the glass-to-rubber transition in moisture-saturated samples, according to our concept of the composite morphology, the onset of molecular motion starts within the bulk resin and gradually encompasses the reinforcement–matrix boundary region. The latter is characterized by the more restricted molecular mobility. At the point when the resin net-

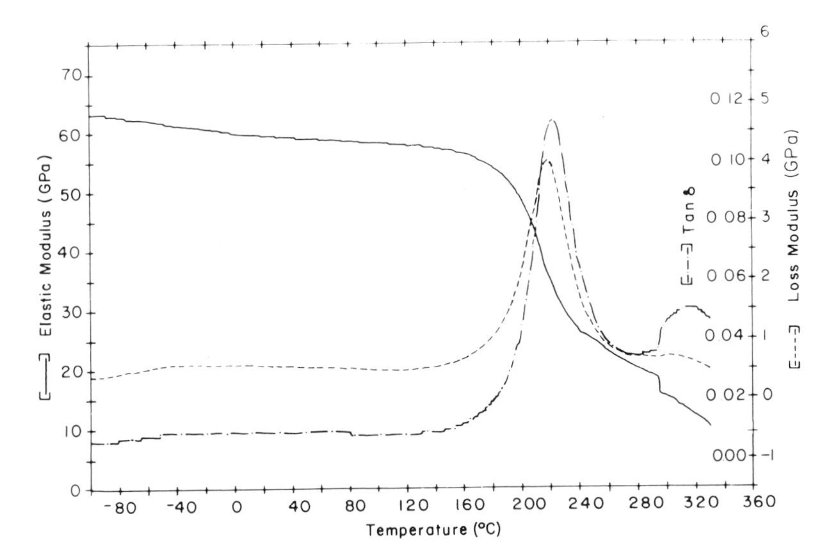

Figure 27. Dynamic mechanical spectrum of the moisture-saturated specimen after four cycles at 180 °C. Note the characteristic drop in dynamic elastic modulus at the end of glass transition region

work in the vicinity of larger microcracks within the boundary region becomes rubbery, the microcracks actually become gaps across which the stress cannot be transferred from the matrix to the fibers. Simultaneously, the resonant frequency of the sample, which is directly proportional to the elastic modulus, drops abruptly to maintain a constant amplitude of oscil-

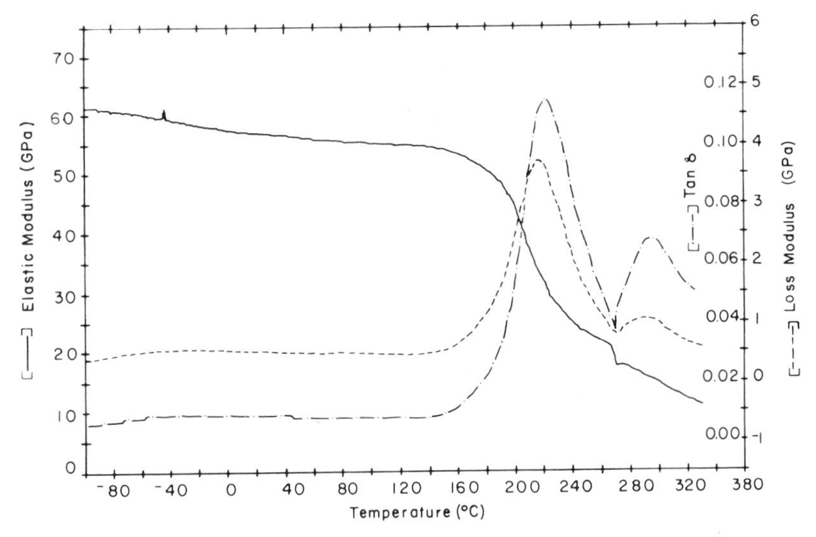

Figure 28. Same as Figure 27; six cycles at 180 °C.

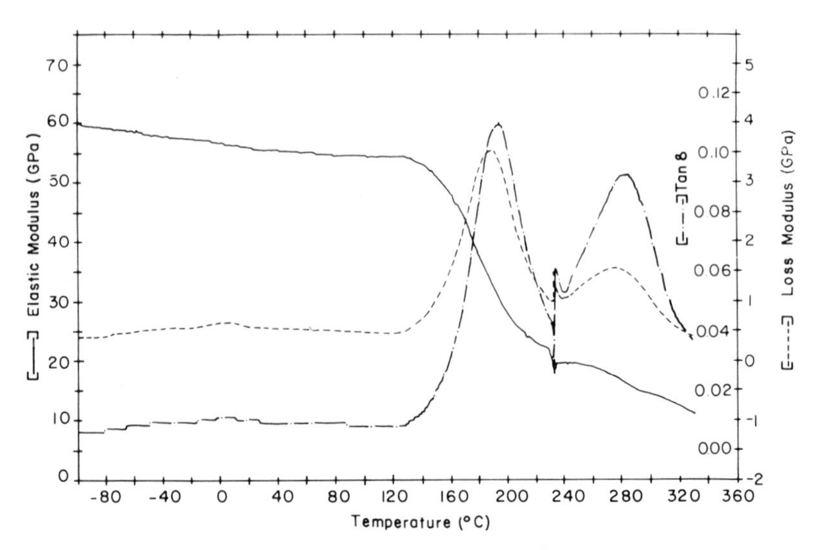

Figure 29. Same as Figure 27; 10 cycles at 180 °C.

lation. As the number of thermal spikes increases, the size of larger microcracks also increases (as clearly seen in our micrographs), and, consequently, the magnitude of the observed drop in E' increases, as shown in Figures 27–29. The temperature at which this abrupt drop in elastic modulus occurs is plotted in Figure 30 as a function of the number of thermal

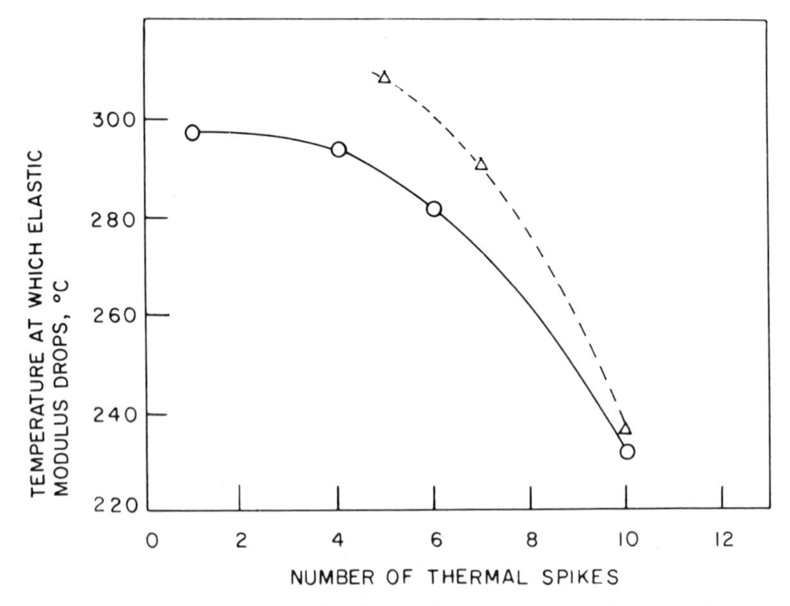

Figure 30. Temperature of the abrupt drop in dynamic elastic modulus as a function of the number of thermal spikes at 180 °C (—) and 160 °C (---).

spikes for 160 and 180 °C cycles. After approximately seven cycles in the 180 ° C spike, the temperature at which this abrupt drop in E' takes place starts to decrease rapidly. Simultaneously, the amount of moisture released during the duration of the thermal spike shows an abrupt upward trend (Figure 24). Both of these observations indicate that, beyond approximately seven thermal spikes, larger microcracks form rapidly and continue to grow during further spiking. No evidence of the drop in E' was observed in the 130 °C cycle.

Changes in the 20 and 177 °C value of E' as a function of the number of thermal spikes in the 180 °C cycle are shown in Figure 31. The temperature of 177 °C was of interest as the highest thermal environment at which the resin can be used in service (58). Interestingly, this curve follows the same trend as the T_g curve of Figure 13. The initial decrease in the 20 °C elastic modulus, during the first three or four thermal spikes, is caused by the moisture absorbed in microcracks in the reinforcement–matrix boundary region. The absorbed moisture is believed to provide some form of "lubrication" in the boundary region, thus decreasing the stiffness. After the first critical number of thermal spikes (three or four), some smaller microcracks had closed and some moisture had desorbed. Hence, the lubricating effects of the moisture were diminished. Consequently, a small increase in E' was noted. At a higher number of thermal spikes, the modulus decreases

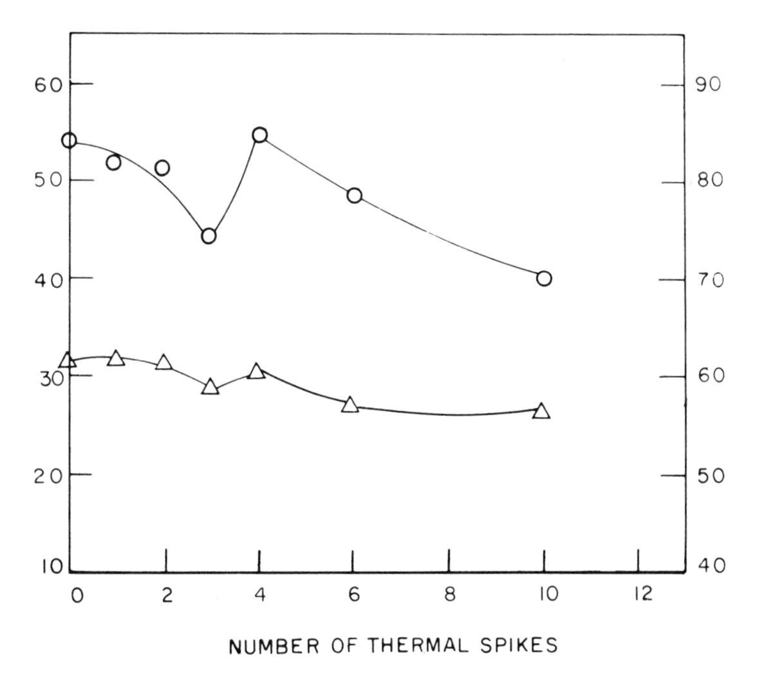

Figure 31. Dynamic elastic moduli at 20° (Δ) and 180 °C (○) as a function of the number of thermal spikes in the 180 °C cycle.

again according to the same molecular mechanism responsible for its initial drop. However, the total change in the value of E'_{20} between the nonspiked sample and the sample spiked 10 times at 180 °C is only 8%. The values of elastic moduli at 20 and 180 °C showed insignificant changes as a function of the number of thermal spikes at 130 °C, as clearly seen from Figure 32. Other researchers, using standard tensile testers, have not observed changes in the room temperature value of E' as a function of the number of thermal spikes.

In the final step, we dried a series of thermally spiked (180 °C specimens to a constant weight and then measured their T_g values. The results of this experiment are shown in Figure 33. Clearly, upon drying the T_g of thermally spiked samples decreases linearly as a function of the number of thermal spikes. The observed decrease in T_g is believed to be a consequence of degradation (hydrolysis) of some chain segments during the application of thermal spikes. Most likely, the hydrolysis preferentially attacks the more highly strained chain segments, reduces the overall restrictions to molecular mobility, and shifts the onset of glass-to-rubber transition to a lower temperature.

Summary

A thorough investigation was conducted on the effect of hygrothermal fatigue on physical–mechanical properties of an eight-ply unidirectional

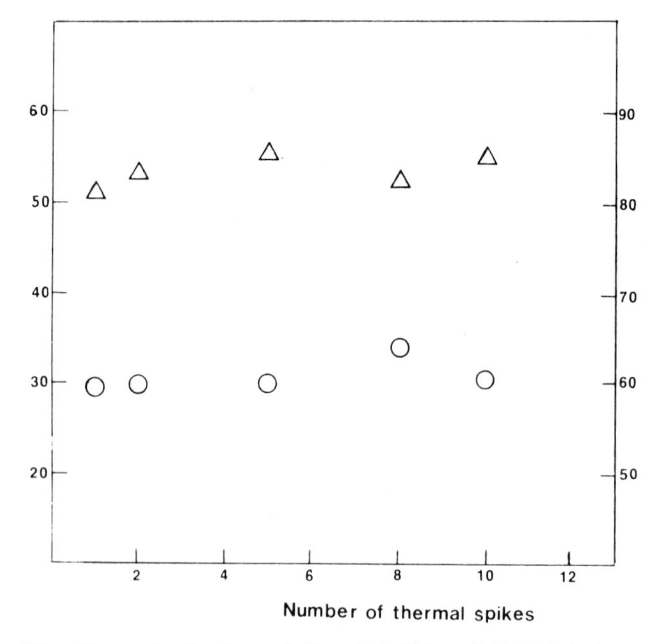

Figure 32. Dynamic elastic moduli at 20° (○) and 180 °C (△) as a function of the number of thermal spikes in the 130 °C cycle.

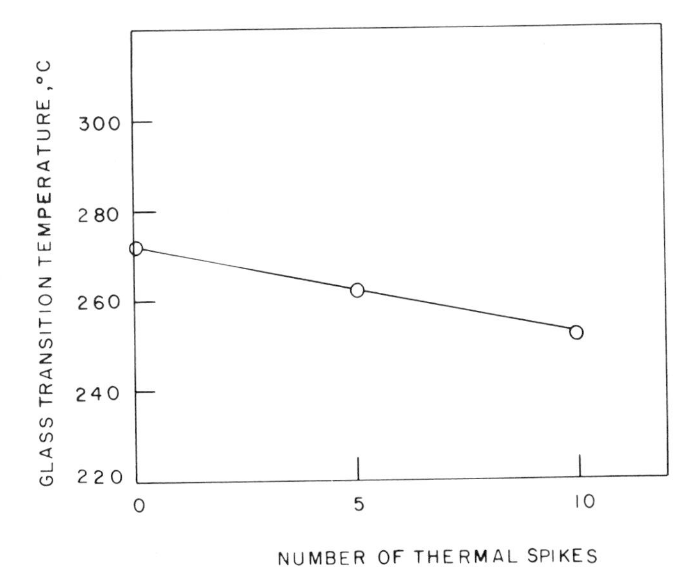

Figure 33. Glass transition of the dried composite as a function of the number of thermal spikes in the 180 °C cycle.

graphite/epoxy composite. During exposure to static hygrothermal fatigue (90 °C and 100% RH) the absorption of moisture was found to proceed according to Fick's law. Four stages were defined on the moisture absorption curve, in terms of the diffusion process on the morphological level. Moisture desorption occurred at a faster rate than absorption, and a permanent change in the composite was detected after one absorption–desorption cycle. The effect of various thermal spikes on physical–mechanical properties of the composite was also evaluated. An increase in moisture content was noted after the application of thermal spikes. Also, during thermal spiking, numerous microcracks were introduced in the composite. The initiation of microcracks occurred within the reinforcement–matrix boundary region that is characterized by highest restrictions to molecular mobility. The extent of the subsequent propagation of microcracks was a function of the type and number of thermal spikes. No significant microcracking was observed when the upper temperature limit of the thermal spike was below the onset of the glass transition region, as determined by the loss modulus trace in dynamic mechanical spectra.

Nomenclature

D	Apparent composite diffusivity, mm^2/s
D_n	Diffusivity of the neat resin, mm^2/s
D_r	Diffusivity of the resin in the composite, mm^2/s
D_{11}	Composite diffusivity in the direction of fibers, mm^2/s

D_{22}	Composite diffusivity in the direction normal to fibers, mm^2/s
E'	Elastic modulus, Pa
E''	Loss modulus, Pa
M	Moisture content, %
M_w	Weight of the moist sample, g
M_d	Weight of the dry sample, g
M_o	Initial sample weight, g
M_t	Sample weight at time t, g
M_∞	Weight of the saturated sample, g
T_d	Degradation temperature, °C
T_g	Glass transition temperature, °C
h	Specimen thickness, mm
l	Specimen length, mm
n	Specimen width, mm
t	Time, s
V_f	Volume fraction of fibers in composite
α, β, γ	Angles between fiber axis and the X, Y, and Z axes, respectively
α_f	Coefficient of thermal expansion of the fiber, cm^3/cm^3 °C
α_m	Coefficient of thermal expansion of the matrix, cm^3/cm^3 °C
α_w	Coefficient of thermal expansion of water, cm^3/cm^3 °C
β_c	Coefficient of swelling of composite, dimensionless
ϵ_y^T	Transverse strain associated with thermal stress, dimensionless
ϵ_y^H	Transverse strain associated with swelling stress, dimensionless

Acknowledgment

This material is based upon work supported by the National Science Foundation under Grant No. MEA-8120211.

Literature Cited

1. Agarwal, B. D.; Broutman, L. J. "Analysis and Performance of Fiber Composites"; J. Wiley and Sons: New York, 1980.
2. Tsai, S. W.; Hahn, H. T. "Introduction to Composite Materials"; Technomic Publishing Co.: Westport, Conn., 1980.
3. "Environmental Effects on Composite Materials"; Springer, G. S., Ed.; Technomic Publishing Co.: Westport, Conn., 1981.
4. "Handbook of Composites"; Lubin, G., Ed.; Van Nostrand Reinhold Company: New York, 1982; pp. 238–243.
5. Ashbee, K. H. G.; Frank, F. C.; Wyatt, R. C. *Proc. R. Soc. London Ser. A* **1967**, *300*, 415.
6. Ashbee, K. H. G.; Wyatt, R. C. *Proc. R. Soc. London Ser. A* **1969**, *312*, 553.
7. Field, S. Y.; Ashbee, K. H. G. *Polym. Eng. Sci.* **1972**, *12*, 30.
8. Ishai, O.; Mazor, A. *J. Compos. Mater.* **1975**, *9*, 370.
9. Ishai, O.; Arnon, U. *J. Test. Eval.* **1977**, *5*, 320.
10. Ishai, O.; Arnon, U. *J. Test. Eval.* **1978**, *6*, 364.
11. Hogg, P. J.; Hall, D. *Met. Sci.* **1980**, *Aug.–Sept.*, 441.
12. Carswell, W. S.; Roberts, R. C. *Composites* **1980**, *April*, 95.
13. Gitschner, H. W.; Menges, G. In "Recent Advances in the Properties and Ap-

plications of Thermosetting Materials"; The Plastics and Rubber Inst.: London, England, 1979.
14. Ishida, H.; Koenig, J. L. *Polym. Eng. Sci.* **1978**, *18(2)*, 128.
15. Ishida, H. *Polym. Compos.*, in press.
16. Mijović, J.; Koutsky, J. A. *Polymer* **1979**, *20*, 1095.
17. Hagnauer, G. L.; Pearce, P. J. *ACS Org. Coat. Appl. Polym. Sci. Proceedings* **1982**, *46*, 580.
18. Mijović, J.; Koutsky, J. A. *Polym.-Plast. Technol. Eng.*, **1977**, *9(2)*, 139.
19. Mijović, J., *Polym. Compos.* **1983**, *4*, 73.
20. Shirrell, C. D.; Halpin, J. In "Composite Materials: Testing and Design"; (Fourth Conference), ASTM STP 617, Am. Soc. Test. Mater.; 1977, pp. 514–528.
21. Browning, C. E.; Husman, G. E.; Whitney, J. M. In "Composite Materials: Testing and Design," (Fourth Conference), ASTM STP 617, 1977; pp. 481–496.
22. Hedrick, I. G.; Whiteside, J. B., presented at the AIAA Conference on Aircraft Composites, San Diego, Calif., March, 1977.
23. Mauri, R. E.; Crossman, F. W.; Warren, W. J. "Assessment of Moisture Altered Dimensional Stability of Structural Composites," *SAMPE Symposium*, **1978**, *23*, 1202.
24. Whitney, J. M.; Browning, C. E. In "Advanced Composite Materials—Environmental Effects," ASTM STP 658, Vinson, J. R., Ed., Am. Soc. Test. Mater.; 1978; pp. 43–60.
25. Gillat, O.; Broutman, L. J. In "Advanced Composite Materials—Environmental Effects." ASTM STP 658, Vinson, J. R., Ed., Am. Soc. Test. Mater.; 1978; pp. 61–83.
26. Browning, C. E. *Polym. Eng. Sci.* **1978**, *13(1)*, 16.
27. Mijović, J. *Ind. Eng. Chem. Prod. Res. Dev.* **1982**, *21*, 290.
28. Crank, J. "The Mathematics of Diffusion," Oxford University Press: London, England, 1956.
29. DeIasi, R.; Whiteside, J. B. In "Advanced Composite Materials—Environmental Effects, ASTM STP 658, Vinson, J. R.; Ed.; ASTM; 1978; pp. 2–20.
30. Mazor, A.; Broutman, L. J.; Eckstein, B. H., *Polym. Eng. Sci.* **1978**, *18*, 341.
31. Leung, C. L.; Dynes, P. J.; Kaelble, D. H. In "Nondestructive Evaluation and Flaw Criticality for Composite Materials," ASTM STP 696, Pipes, R. B., Ed.; Am. Soc. Test. Mater.; 1979; pp. 298–315.
32. Leung, C. L.; Kaelble, D. H. In "Advanced Composites and Applications," U.S. NBS Spec. Publication, 1979, No. 503, pp. 32–39.
33. Leung, C. L.; Kaelble, D. H. In "Resins for Aerospace," May, C.A., Ed.; ACS SYMPOSIUM SERIES, No. 132, ACS: Washington, D.C., 1980, pp. 419–434.
34. Fahmy, A. A.; Hunt, J. C. *Polym. Compos.* **1980**, *1*, 77.
35. Hahn, H. T.; Kim, R. Y. In "Advanced Composite Materials—Environmental Effects," ASTM STP 658, Vinson, J. R., Ed., Am. Soc. Test. Mater.; 1978; pp. 98–120.
36. Bueche, F., "Physical Properties of Polymers"; Interscience: New York, 1962.
37. Browning, C. E., Ph.D. Thesis, Univ. of Dayton, 1976.
38. McKague, E. L., Jr.; Reynolds, J. D.; Halkias, J. E. *J. Appl. Polym. Sci.* **1978**, *22*, 1643.
39. Peyser, P.; Bascom, W. D. *J. Mater. Sci.* **1981**, *16*, 75.
40. Hahn, H. T. *J. Compos. Mater.* **1976**, *10*, 266.
41. Scola, D. A. In "Interfaces in Polymer Matrix Composites"; Plueddeman, E. P., Ed.; Academic Press: New York, 1974.
42. Novak, R. C.; DeCrescente, M. A. *J. Eng. Power* **1970**, *92*, 377.
43. Pipes, R. B.; Vinson, J. R.; Chou, T. W. *J. Compos. Mater.* **1976**, *10*, 129.
44. Crossman, F. W.; Wang, A. S. D. *J. Compos. Mater.* **1978**, *12*, 2.
45. Mijović, J. *J. Appl. Polym. Sci.* **1982**, *27*, 1149.
46. Mijović, J. *J. Appl. Polym. Sci.* **1982**, *27*, 2919.
47. Kong, E. S. W.; Lee, S. M.; Nelson, H. G. *Polym. Compos.* **1982**, *3*, 29.

48. Crossman, F. W.; Mauri, R. E.; Warren, W. J. "Hygrothermal Damage Mechanisms in Graphite–Epoxy Composites," NASA Contractor Report 3189, December 1979.
49. Delmonte, J.; McCrory, B., 36th Annual Conf. RP/C Inst. SPI Inc. Section 12-C, p. 1, Washington, D.C., February, 1981.
50. Wang, T. T.; Kwei, T. K. *J. Polym. Sci.* **1969**, A-2, 7, 889.
51. Kajiyama, T.; Yoshinaga, T.; Takayanagi, M. *J. Polym. Sci. Polym. Phys. Ed.* **1977**, *15*, 1557.
52. Kumins, C. A. In "Advances in Organic Coatings Science and Technology", Vol. 4; Parfitt, G. D.; Patsis, A. V., Eds.; Technomic Publishing Co.: Westport, Connecticut, 1982; p. 108.
53. Racich, J. L.; Koutsky, J. A. *J. Appl. Polym. Sci.* **1976**, *20*, 2111.
54. McKague, E. L., Jr.; Halkias, J. E.; Reynolds, J. D. *J. Compos. Mater.* **1975**, *9*, 2.
55. Augl, J. M.; Berger, A. E. "Moisture Effects on Carbon Fiber Epoxy Composites," SAMPE Technical Conference; 1976, *8*, 383.
56. Browning, C. E. In "Diversity–Technology Explosion," Proceedings of the 22nd National Symp. Exhibition, SAMPE, San Diego, CA, 1977, pp. 365–394.
57. Wool, R. *Org. Coat. Plast. Chem.* **1979**, *40*, 271.
58. Hercules Inc., Product Data Sheet No. 855, 1978.

RECEIVED for review January 20, 1983. ACCEPTED August 3, 1983.

INDEXES

AUTHOR INDEX

SUBJECT INDEX

A

B

D

E

Q

R

S

T

U

V

W

Y

Editing and indexing by Susan Robinson
Production by Anne Riesberg
Jacket design by Anne G. Bigler
Managing Editor: Janet S. Dodd

Typeset by Action Comp Co., Baltimore, Md.
and Hot Type Ltd., Washington, D.C.
Printed and bound by Maple Press Co., York, Pa.